# CONVERSION FACTORS FOR MASS

| multiply number of → <br><br> to obtain ↓     by ↘ | (lb-sec$^2$/in) | Grams (gr) | Kilograms (kg) | Pounds mass (lbm) | Slugs (lb-sec$^2$/ft) |
|---|---|---|---|---|---|
| (lb-sec$^2$/in) | 1 | $5.711(10^{-6})$ | $5.711(10^{-3})$ | $2.591(10^{-3})$ | $8.333(10^{-2})$ |
| Grams (gr) | $1.751(10^5)$ | 1 | $1(10^3)$ | 453.6 | $1.459(10^4)$ |
| Kilograms (kg) | 175.1 | $1(10^{-3})$ | 1 | 0.4536 | 14.59 |
| Pounds mass (lbm) | 386.0 | $2.205(10^{-3})$ | 2.205 | 1 | 32.17 |
| Slugs (lb-sec$^2$/ft) | 12 | $6.853(10^{-5})$ | $6.853(10^{-2})$ | $3.108(10^{-2})$ | 1 |

# CONVERSION FACTORS FOR FORCE

| multiply number of → <br><br> to obtain ↓     by ↘ | Dynes (gr-cm/sec$^2$) | Newtons (kg-m/sec$^2$) | Pounds force (lb) | Poundals (lbm-ft/sec$^2$) |
|---|---|---|---|---|
| Dynes (gr-cm/sec$^2$) | 1 | $1(10^5)$ | $4.448(10^5)$ | $1.383(10^4)$ |
| Newtons (kg-m/sec$^2$) | $1(10^{-5})$ | 1 | 4.448 | 0.1383 |
| Pounds force (lb) | $2.248(10^{-6})$ | 0.2248 | 1 | $3.108(10^{-2})$ |
| Poundals (lbm-ft/sec$^2$) | $7.233(10^{-5})$ | 7.233 | 32.17 | 1 |

# MECHANICS OF MACHINES

# MECHANICS OF MACHINES

W. L. Cleghorn
*University of Toronto*

New York ■ Oxford
OXFORD UNIVERSITY PRESS
2005

Oxford University Press

Oxford   New York
Auckland   Bangkok   Buenos Aires   Cape Town   Chennai
Dar es Salaam   Delhi   Hong Kong   Istanbul   Karachi   Kolkata
Kuala Lumpur   Madrid   Melbourne   Mexico City   Mumbai   Nairobi
São Paulo   Shanghai   Taipei   Tokyo   Toronto

Published by Oxford University Press, Inc.
198 Madison Avenue, New York, New York 10016
www.oup.com

Oxford is a registered trademark of Oxford University Press

**Library of Congress Cataloging-in-Publication Data**

Cleghorn, W. L. (William L.)
    Mechanics of machines / W. L. Cleghorn.
       p.  cm.
    Includes bibliographical references and index.
    ISBN-13: 978-0-19-515452-8
    ISBN 0-19-515452-5
    1. Mechanical engineering. I. Title.

TJ170.C58 2005
621.8—dc22                          2004057660

Printing number: 9  8  7  6  5  4  3  2  1

Printed in the United States of America
on acid-free paper

*To My Parents*

# Contents

# Preface

This book is intended mainly for use as an undergraduate text. It contains topics and material sufficient for a one-half-year course. It may also be useful to practicing engineers.

Conventional notation is employed throughout, primarily using the International System of Units. Approximately one-quarter of the examples employ the U.S. Customary System of Units.

Many examples included in each chapter relate to situations of practical significance. Problems of varying length and difficulty are included at the end of each chapter. In addition, twenty-six computer design projects are supplied in Appendix A. It is suggested that these be completed using the software Working Model 2D, Version 5.0.

A CD-ROM accompanies this textbook. It includes numerous files for animations of mechanisms based on the software Working Model 2D, Version 5.0. These require a file player of the software. Other files are in a movie format based on the software Visual Nastran, Version 7.0, and they have been stored in VRML format. They may be viewed using Flux Player. Within the text, each of these files is identified either in boldface in square brackets (e.g., **[Model 1.1]**) or within a text box located in the margin as shown here.

**Model 1.1**
Single-Cylinder
Piston Engine

Also included on the CD-ROM are video clips of real mechanical systems in operation. Within the text, these are identified in boldface in square brackets (e.g., **[Video 1.29]**) or within a text box located in the margin. They may be viewed using QuickTime Movie Player. An index of all animations and video segments are listed in Appendix E.

**Video 1.29**
Prosthetic Hand

Graphical and analytical methods of analysis are included for the kinematic and kinetic analyses of mechanisms. Graphical methods may be applied in situations where analysis for only one position of a mechanism is required. They often require a scaled drawing in the configuration for which an analysis is required. For several of the end-of-chapter problems, an associated scaled drawing may be obtained from the CD-ROM. Such problems are identified by an icon of a CD-ROM as shown here in the margin. Analytical solutions are useful in instances when solutions are required for a series of mechanism configurations.

Chapter 1 covers basic concepts, including linkage classification by motion characteristics, as well as degrees of freedom of planar joints of mechanisms. Some common examples of mechanisms along with their associated animations are provided.

Chapter 2 provides the background material required to carry out static and dynamic analyses of planar mechanical systems. Expressions for relative velocity and acceleration in the radial-transverse coordinate system are covered. The instantaneous center of velocity of a body is presented. Equations for the kinetics of a rigid body and associated commonly employed sets of units are also provided.

Chapter 3 covers a traditional graphical analysis of planar mechanisms. Both velocity and acceleration analyses are presented using vector polygons for one position of the mechanism. Velocity analyses implementing the method of instantaneous centers are also covered.

Chapter 4 presents an analytical method based on complex numbers for the kinematic analysis of a planar mechanism. The equations generated using this technique may be programmed on a computer for completing an analysis in a series of positions.

Chapter 5 gives a comprehensive synopsis of gears. This includes many common types of gears and related animations of the gears in meshing action. This chapter also includes some of the common methods of gear manufacture.

Chapter 6 presents an analysis of gear trains. For analysis of planetary gear trains, an algorithm suited for computer implementation is provided.

Chapter 7 presents design procedures of cam mechanisms. Both graphical and analytical methods are covered. In addition, the computer program Cam Design is included on the CD-ROM. This program can analyze a wide variety of disc cams. For a given set of input parameters and prescribed motion of the follower, all pertinent kinematic parameters are provided as a function of input motion.

Chapter 8 covers graphical force analyses of planar mechanisms. Each graphical analysis may be applied for one configuration of a mechanism—for either static or dynamic conditions.

Chapter 9 covers analytical force analyses of planar mechanisms. The governing equations of motion are derived for an arbitrary configuration of a mechanism and may be programmed on a computer to determine results for multiple configurations. Means to balance a four-bar mechanism and a slider crank mechanism are also provided.

Chapter 10 covers the analysis and design of flywheels. This includes determining the size of a flywheel required to keep speed fluctuations within a desired tolerance.

Chapter 11 presents some common methods for the synthesis of mechanisms. This includes graphical and analytical techniques for function synthesis and rigid-body guidance synthesis of four-bar and slider crank mechanisms.

Appendix A includes a set of design projects. They can be ideally solved using the Working Model 2D software. Appendix B covers an extensive set of commonly employed mechanisms and machines. Appendix C provides background reference material related to scalars and vectors. Appendix D gives a brief review of mechanics, which is the basis of much of the material presented in this textbook. Appendix E lists the files included on the CD-ROM that accompanies this textbook.

## Acknowledgments

I express my sincere appreciation to the many colleagues and faculty from the University of Toronto and the University of Manitoba who played a significant part in the preparation of this book. I am most grateful to the following individuals for reviewing the text, improving figures, developing software, and offering their comments and suggestions: Dr. Nikolai Dechev, Mr. Sean Voskamp, Ms. Mina Hoorfar, Dr. Homayoun Najjaran, Professor Kenneth C. Smith, Ms. Laura Fujino, Dr. Leif E. Becker, Professor James K. Mills, Mr. Hoi Sum (Sam) Iu, Mr. Peter Bahoudian, Mr. Martin Côté, Mr. Andy K. L. Sun, Professor Robert G.

Fenton, Professor John Van de Vegte, Professor Ron P. Podhorodeski, Dr. George Tyc, Mr. Masoud Alimardani, Mr. Daniel Ohlsen, and Mr. K. K. (John) Mak. I acknowledge the assistance of numerous undergraduate students who reviewed portions of drafts and provided useful feedback.

Valued help and contributions were received from the following individuals in industry: Mr. Richard Houghton, General Gear Limited; the late Mr. Hagop Artinian, Swissway Machining Limited; Mr. John Augerman, Mr. Will J. Bachewich, Mr. Michael W. Borowitz, Mr. Steven J. Grave, and Mr. Bruce M. Kretz, General Motors of Canada Limited; and Mr. William Frey, HD Systems, Inc.

I would like to thank Media Machines for allowing Oxford University Press to distribute Flux Player software with this book. I would also like to thank the MSC Software Corporation for allowing Oxford University Press to distribute Working Model® Textbook Edition with this book.

The Department of Mechanical and Industrial Engineering and the Information Technology Courseware Development Fund at the University of Toronto are gratefully acknowledged for providing funding for the development of much of the software.

A special expression of my appreciation is extended to Dr. Clarice Chalmers and the late Mr. Wallace G. Chalmers. Completion of this book would not have been possible without their supportive sponsorship and encouragement.

*W. L. Cleghorn*

# MECHANICS OF MACHINES

# 1 Introduction

## 1.1 PRELIMINARY REMARKS

A *machine* is defined as an apparatus that transmits energy through its parts to perform desired tasks. This definition does not restrict the form of energy or the size and rigidity of the parts. It therefore encompasses a vast array of mechanical and electrical devices, many of which have a profound effect on our lives. An automobile is a machine that transfers energy from the fuel into the driving of the wheels to provide motion. Alternatively, the "parts" of a machine may be electrons. Therefore, a machine can be a nuclear station for transferring energy from atoms for use in generating electrical energy.

The wheel and the lever were among the earliest inventions of machines. Engineers have since designed numerous ingenious machines for our aid and entertainment. Machines have been designed to enhance the field of medicine. Doctors employ machines to reattach accidentally severed limbs. Man-made machines such as prosthetic hands, arms, and legs are used to replace natural limbs. In recent years, robotic machines have had a significant influence in manufacturing methods. Some are able to rapidly perform repetitive tasks in a reliable and accurate manner and can be used to work in hostile and dangerous environments.

Some machines are a combination of mechanical and electrical components. An automobile is composed of mechanical parts of the drive train, suspension, and steering systems, along with electrical parts used to ignite the air-fuel mixture in the engine and to provide lighting and sensory feedbacks for the driver.

Machines having electrical components usually incorporate at least some mechanical parts. A computer has electric motors with rotating shafts for driving discs, and there is a mechanical arm to enable storing and accessing data from the hard drive. The keyboard may be considered a machine with mechanical parts. Printers and scanners have machine components such as gears and lever arms to handle the paper. Many electronic watches

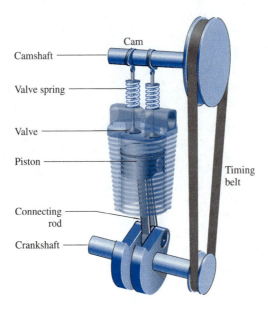

**Figure 1.1** Single-cylinder piston engine [Model 1.1].

Camshaft

Cam

Valve spring

Valve

Piston

Connecting rod

Crankshaft

Timing belt

have hands operated by miniature ratchet wheels that are pushed around one tooth at a time.

This textbook will consider only the mechanical parts of machines. They may be either moving or stationary. Even with this restriction, there are countless types of machines.

*Mechanics* is a science that predicts the conditions of a body either at rest or in motion, when under the action of forces and moments. This textbook, entitled *Mechanics of Machines*, deals with the determination of the forces and motions of machines. To conduct such analyses, it is usually convenient to divide a machine into subsystems, referred to as *mechanisms* (or *linkages*), rather than attempting to analyze all parts of a machine simultaneously.

Figure 1.1 illustrates a machine, a single-cylinder internal combustion piston engine. An animation of this machine is provided by **[Model 1.1: Single-Cylinder Piston Engine],** included on the CD-ROM accompanying this textbook. A description of the operation of this machine is given in Appendix B, Section B.8.1. Figures 1.2(b), 1.2(c), and 1.2(d) highlight three mechanisms contained in the machine shown in Figure 1.2(a): a *timing belt drive*, a *cam mechanism*, and a *slider crank mechanism*.

A simple machine may also be considered as a single mechanism. For instance, the tongs shown in Figure 1.3(a) can be considered either as a machine or as a mechanism. Figure 1.3(b) shows a free body diagram of the system used to analyze the manual force required to generate sufficient gripping force.

Mechanisms are widely used in applications where precise relative movement and transmission of force are required. Motions may be continuous or intermittent, linear and/or angular. Examples of several types of machines and mechanisms are presented in Section 1.2 and Appendix B. Engineers, and those particularly concerned with design, should ensure that they are aware of the wide range of machines and mechanisms and be familiar with their behavior. There is a good reason for this; it is often possible to apply or slightly modify an existing machine design to meet the needs of new problems.

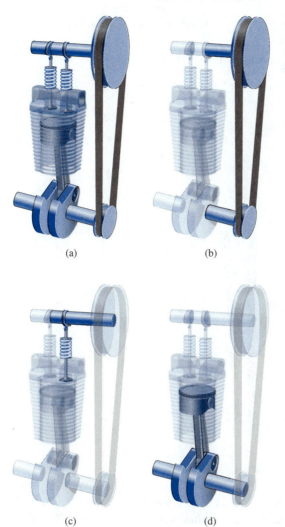

(a)

(b)

(c)

(d)

**Figure 1.2** Mechanisms in a single-cylinder piston engine: (a) engine, (b) timing belt drive, (c) cam mechanism, (d) slider crank mechanism.

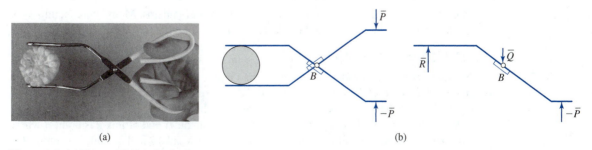

(a)

(b)

**Figure 1.3** (a) Tongs. (b) Free body diagram.

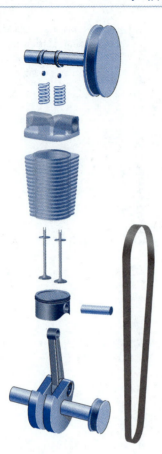

**Figure 1.4** Engine assembly [Model 1.4].

Individual parts of a machine or mechanism are also referred to as *links*. They may be nonrigid, such as cables or belts. Alternatively, they may be rigid bodies such as cranks, levers, wheels, bars, or gears. Figure 1.4 shows an exploded view of an engine. **[Model 1.4]** provides an animation of assembling this engine, link by link. All links of this mechanism may be considered rigid, except for the flexible belt that is wrapped around both pulleys.

| Model 1.4 |
| Engine |
| Assembly |

## 1.2 COMMONLY EMPLOYED MECHANISMS

This section provides a small sampling of common mechanisms. Many more examples of machines and mechanisms and their descriptions are contained in Appendix B.

### 1.2.1 Slider Crank Mechanism

| Model 1.5 |
| Slider Crank |
| Mechanism |

A *slider crank mechanism* is illustrated in Figure 1.5. This mechanism incorporates a stationary *base link*, designated as link 1. All portions of the base link are depicted with hatched lines. The other links can move relative to the base link. The *crank*, designated as link 2, rotates about a *base pivot*, denoted as $O_2$. At the other end of link 2 is point $B$ where a pivot point or bearing allows relative rotation between links 2 and 3. Link 3 is referred to as the *coupler* or *connecting rod*. The coupler is connected to link 4, through a bearing,

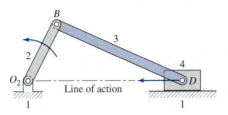

**Figure 1.5** Slider crank mechanism [Model 1.5].

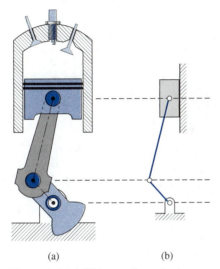

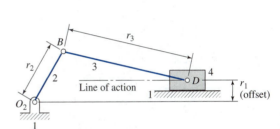

(a) (b)

**Figure 1.6** (a) Slider crank mechanism. (b) Skeleton representation.

**Figure 1.7** Slider crank mechanism with offset [Model 1.7].

denoted as point $D$. Link 4 is called a *slider* or *piston*. The slider moves with respect to the *slide*, which is part of the base link. The straight-line path of the center of the slider is referred to as the *line of action*.

When analyzing the motions of mechanisms, it is often convenient to employ highly simplified drawings, called *skeleton diagrams*. The dimensions of a skeleton diagram are the ones critical for determining motions. Figure 1.6(a) shows a slider crank mechanism as it would appear in an engine. The skeleton form of this mechanism is given in Figure 1.6(b). Bearings in skeleton diagrams are represented by either small circles or dots. The crank and connecting rod are each drawn simply as a straight line. The piston is shown as a rectangle. Hatched lines indicate the base link.

Another slider crank mechanism is shown in Figure 1.7. In this instance, base pivot $O_2$ is not on the line of action of point $D$. Dimension $r_1$ is referred to as the *offset*. Dimensions $r_2$ and $r_3$ are lengths of the crank and coupler, respectively.

### 1.2.2 Four-Bar Mechanism

*Four-bar mechanisms* are among the most common and useful mechanisms. A typical four-bar mechanism is illustrated in Figure 1.8(a). It has four *bars* or links: the stationary base link and three moving links. Links 2 and 4 are connected to the base link through base

**Model 1.7**
Slider Crank
Mechanism with
Offset

**Model 1.8**
Four-Bar
Mechanism

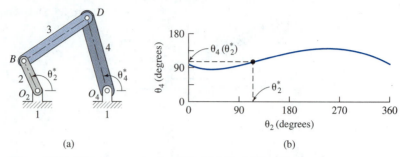

(a)                                                    (b)

**Figure 1.8** (a) Four-bar mechanism [Model 1.8]. (b) Function graph.

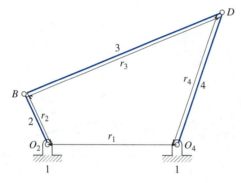

**Figure 1.9** Dimensions of a four-bar mechanism.

pivots $O_2$ and $O_4$, respectively. The *coupler,* link 3, is attached to links 2 and 4 through moving pivots, designated as points $B$ and $D$. For the mechanism shown, if link 2 is driven to rotate about its base pivot, then link 4 will in turn be forced to also rotate about its base pivot.

For the mechanism shown in Figure 1.8(a), angular displacements of links 2 and 4 are designated as $\theta_2^*$ and $\theta_4^*$. Figure 1.8(b) shows the angular displacement of link 4 as a function of the angular displacement of link 2. This mechanism is also known as an *angular function-generating mechanism.*

Figure 1.9 shows a skeleton diagram of a four-bar mechanism. Also shown are the lengths of the links. These lengths dictate the type and extent of motion that may be achieved. The types of motion are presented in Section 1.7.

Many mechanisms can be developed from a single skeleton form. Examine the skeleton diagram of the four-bar mechanism shown in Figure 1.10(a). All other mechanisms illustrated in Figure 1.10 are essentially the same four-bar mechanism and are therefore considered to be *equivalent mechanisms.* As shown through the animation provided using [**Video 1.10**], all of these mechanisms transmit the same motions. When analyzing a motion, without consideration of the associated forces, it is not necessary to know the widths of links and sizes of bearings between links. The critical information in determining motions are the *center-to-center distances.* Figures 1.10(a) and 1.10(c) illustrate a typical distance $r$ that is the same for two of the equivalent mechanisms.

Mechanisms can be built up by starting with a single mechanism, then adding links to create more complicated mechanisms. Figure 1.11 shows a four-bar mechanism used in a

**Video 1.10**
Equivalent Four-Bar Mechanisms

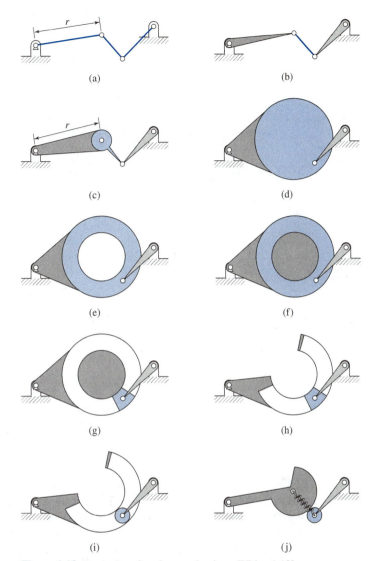

**Figure 1.10** Equivalent four-bar mechanisms [Video 1.10].

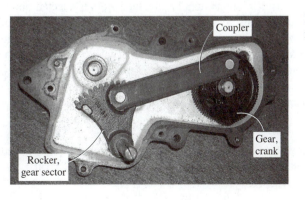

**Figure 1.11** Washing machine mechanism.

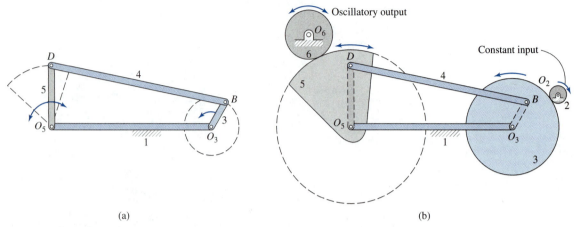

(a)                                                     (b)

**Figure 1.12** Washing machine mechanism [Model 1.12].

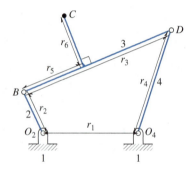

**Figure 1.13** Dimensions of a four-bar mechanism with a coupler point.

**Model 1.12**
Washing
Machine
Mechanism

domestic washing machine. Figure 1.12 illustrates the same system. Figure 1.12(a) shows a four-bar mechanism that is part of the mechanism shown in Figure 1.12(b). Constant input motion is supplied through a gear, link 2, which drives a larger gear, link 3. Both gears are represented as circles. Links 1, 3, 4, and 5 constitute the four links of the four-bar mechanism. Link 5 is a gear sector that meshes with link 6. The proportions of the links are such that when link 2 is driven at a constant rotational speed, link 6 will have oscillatory rotational motion. In operation, link 6 drives an agitator to provide the washing action.

Various paths of motions may be obtained from a selected point on the coupler of a four-bar mechanism, also called the *coupler point*. These mechanisms are referred to as *path-generating mechanisms*. The shape of the path traced by the coupler point is a function of the dimensions shown in Figure 1.13. These consist of distances $r_1$, $r_2$, $r_3$, and $r_4$ between pivots. In addition, dimensions $r_5$ and $r_6$ define the location of the coupler point, $C$, with respect to the pivot between links 2 and 3. For instance, Figure 1.14 shows a four-bar mechanism for which

**Model 1.14A,**
**Model 1.14B**

Four-Bar with
Coupler Point

$$r_1 = 7.2 \text{ cm}; \qquad r_2 = 2.5 \text{ cm}; \qquad r_3 = 6.0 \text{ cm}; \qquad r_4 = 7.2 \text{ cm}$$

Also shown are three different *coupler curves* that were obtained by employing distinct values of $r_5$ and $r_6$. Using **[Model 1.14A]** and **[Model 1.14B]**, it is possible to specify

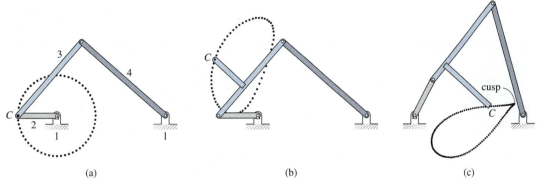

**Figure 1.14** Four-bar mechanism with coupler point [Model 1.14A, 1.14B].

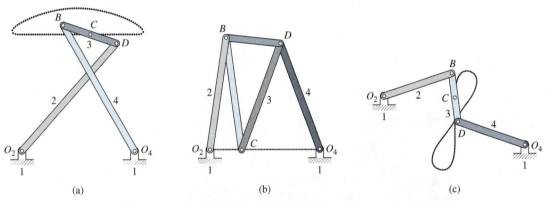

**Figure 1.15** Straight-line mechanisms: (a) Chebyshev [Model 1.15A], (b) Robert [Model 1.15B], (c) Watt [Model 1.15C].

> **Model 1.15A**
> Chebyshev
> Straight-Line
> Mechanism

> **Model 1.15B**
> Robert Straight-
> Line Mechanism

> **Model 1.15C**
> Watt Straight-
> Line Mechanism

> **Model 1.16**
> Film Transport
> Mechanism

the position of the coupler point, $C$, with respect to the coupler, and view the corresponding coupler curve.

Some four-bar mechanisms with a properly selected coupler point can generate a path that is very close to a straight line, for a portion of the cycle. Such mechanisms are referred to as *straight-line mechanisms*. Figure 1.15 shows three types of such mechanisms, called the *Chebyshev straight-line mechanism*, the *Robert straight-line mechanism*, and the *Watt straight-line mechanism*.

Figure 1.16 illustrates an application of a four-bar mechanism incorporating a coupler point. This mechanism is employed to intermittently advance film in a movie projector. During each cycle, the coupler point enters a perforation (Figure 1.16(a)) in the film and advances it to the next picture frame. When the end of the arm is not engaged in a perforation (Figure 1.16(b)), the film remains stationary, and a shutter (not shown) is then opened to allow light to pass through to project an image momentarily.

## 1.2.3 Belt Drive and Gearing

A common requirement in the design of machinery is to smoothly transmit rotational motion from one shaft to another. When the shafts are parallel, this may be accomplished

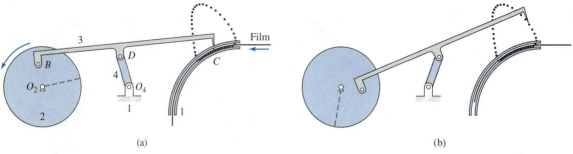

(a)                                    (b)

**Figure 1.16**  Film transport mechanism [Model 1.16].

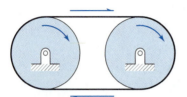

**Figure 1.17**  Open loop friction drive.

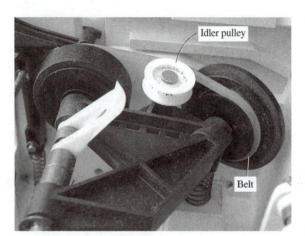

**Figure 1.18**  Open loop friction drive.

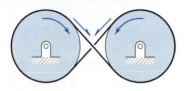

**Figure 1.19**  Cross belt friction drive.

by mounting pulleys on the shafts and wrapping a continuous belt around both pulleys under tension, as shown in Figure 1.17. For this system, known as an *open loop friction drive*, both pulleys rotate in the same direction. Figure 1.18 shows another arrangement of an open loop friction drive. The *idler pulley* enables the input and output pulleys to turn in different planes. Still another arrangement is achieved by crossing the belt between the pulleys, as shown in Figure 1.19. Now the pulleys will rotate in opposite directions. This arrangement is called a *cross belt friction drive*.

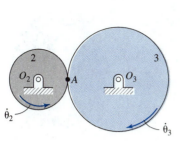

Figure 1.20 Friction gears.

Figure 1.21 Toothed gears [Model 1.21].

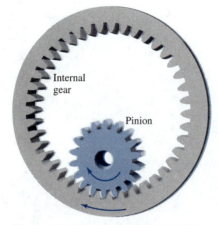

Figure 1.22 Internal gear [Model 1.22].

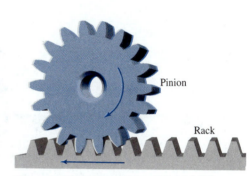

Figure 1.23 Rack and pinion [Model 1.23].

Figure 1.20 shows an alternative means of transferring rotational motion through a pair of rolling cylinders in physical contact. Transfer of motion in this instance relies on friction between the cylinders. This mechanism is known as a pair of *friction gears.* Smooth transmission is achieved as long as the friction capacity between the friction gears is not exceeded. If the friction capacity is exceeded, slippage occurs, and the motion of the follower will not have a constant ratio relative to that of the driver.

Slippage is avoided if the gears have interlocking teeth. These are called *toothed gears.* Figure 1.21 shows three meshing toothed gears that form a *gear train.* The gear in the middle, being the smallest of the set, is referred to as the *pinion.* It is also referred to as an *idler gear.* In the gear set shown in Figure 1.22, one of the gears, known as an *internal gear,* has its teeth on the inside of a ring. When an internal gear meshes with a gear having teeth on its periphery (referred to as an *external gear*), both gears turn in the same direction. Internal gears are commonly employed in planetary gear trains, which are discussed in Chapter 6.

Figure 1.23 shows a *rack and pinion* gear set. The *rack* provides for straight-line motion and is equivalent to a portion of a gear of infinite radius. Figure 1.24(a) illustrates a rack and pinion gear set employed in the steering mechanisms of automobiles. Figure 1.24(b) shows a close-up view of this gear set.

Gearing and gear trains are discussed in more detail in Chapters 5 and 6.

**Model 1.21**
Toothed Gears

**Model 1.22**
Internal Gear

**Model 1.23**
Rack and Pinion

**Model 1.24**
Rack and Pinion
Steering

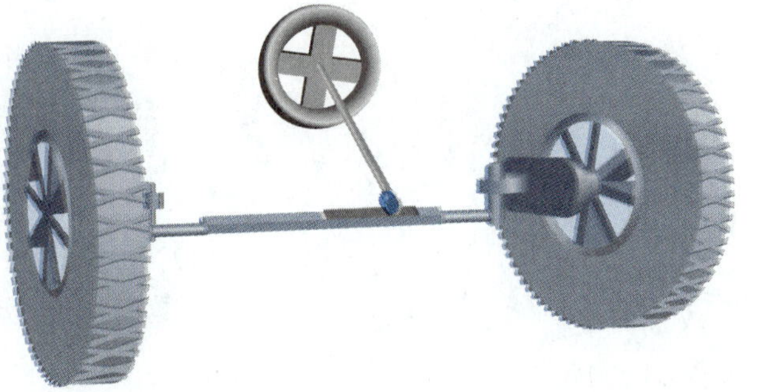

(a)

(b)

**Figure 1.24** Rack and pinion steering [Model 1.24].

## 1.2.4 Cam Mechanism

A *cam* is a link of a mechanism used to transmit motion to another link, known as a *follower*, by direct contact. One type of cam mechanism, known as a *disc cam mechanism*, is illustrated in Figure 1.25. Starting from the position shown in Figure 1.25(a), rotation of the disc cam, link 2, causes the follower, link 4, to translate. Figure 1.25(b) illustrates a configuration of the mechanism after the follower has undergone a translation. If the roller remains in contact with the cam, and the cam is driven at constant speed, then motion of the follower is periodic. An application of a disc cam mechanism is in the valve train of a piston engine. A typical valve train, including the intake and exhaust valves, is given in Figure 1.26. This illustration is a close-up view of a portion of Figure 1.1.

**Model 1.25** Disc Cam Mechanism

**Model 1.26** Disc Cams

**Model 1.27A** Wedge Cam Mechanism

Cam mechanisms are simple and inexpensive, have fewer moving parts, and can be designed to occupy a smaller space compared to other mechanisms that provide similar output motions. Furthermore, the shape of a cam may be determined in a straightforward manner to produce a wide variety of follower motions. For these reasons, cam mechanisms

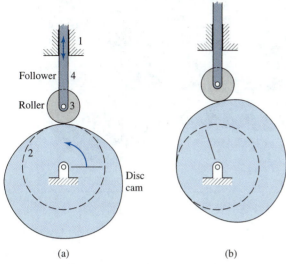

Figure 1.25 Disc cam mechanism [Model 1.25].

Figure 1.26 Disc cams [Model 1.26].

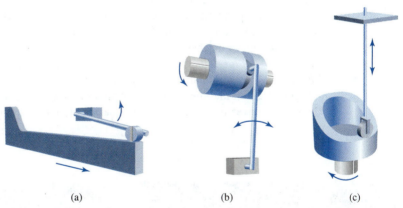

Figure 1.27 Types of cams: (a) wedge cam [Model 1.27A], (b) cylindrical cam [Model 1.27B], (c) end cam [Model 1.27C].

**Model 1.27B**
Cylindrical Cam
Mechanism

**Model 1.27C**
End Cam
Mechanism

**Model 1.28**
Fishing Reel

are used extensively in modern machinery. The disadvantages of cam mechanisms are poor wear resistance and the noise generated by impacts between the cam and follower when operated at high speed.

Figure 1.27 illustrates additional types of cam mechanisms: the *wedge cam mechanism*, the *cylindrical cam mechanism*, and the *end cam mechanism*. An application of a cylindrical cam is in a fishing reel, as illustrated in Figure 1.28. As the cylindrical cam rotates, the follower translates back and forth, causing the fishing line to evenly wind onto the *spool*.

The analysis and design of disc cam mechanisms are presented in Chapter 7.

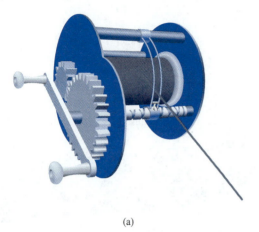
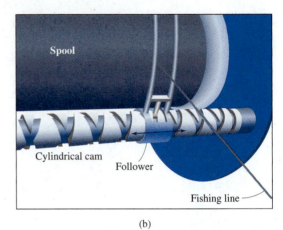

(a)            (b)

**Figure 1.28** Fishing reel [Model 1.28].

## 1.3 PLANAR AND SPATIAL MECHANISMS

Video 1.29
Prosthetic Hand

*Planar motion* is restricted to a plane. For a *planar mechanism*, the motions of all of its links must take place either in the same plane or in planes that are parallel to one another. The slider crank mechanism and four-bar mechanism illustrated in Figures 1.5 and 1.8 are examples of planar mechanisms.

In a *spatial mechanism*, links move in three dimensions. For example, Figure 1.29 illustrates a prosthetic hand [1]. Fingers of the hand start from the configuration shown in Figure 1.29(a), and then they can wrap around an object as shown in Figure 1.29(b). The thumb moves in a plane that is not parallel to the planes of motion of the other four fingers.

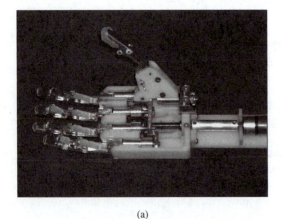
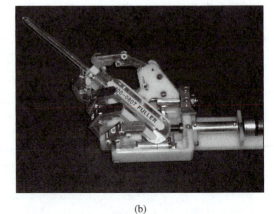

(a)            (b)

**Figure 1.29** Prosthetic hand [Video 1.29].

## 1.4 KINEMATIC CHAINS AND KINEMATIC PAIRS

A *kinematic chain* is an assembly of links connected together without specifying the base link. Figure 1.30 illustrates a four-bar kinematic chain. All links are connected by pivots. Figure 1.30(b) shows the skeleton diagram representation of this kinematic chain.

A series of alternative mechanisms may be produced by, in turn, holding one of the links of the kinematic chain in a fixed position to become the base link. Usually, the type and amplitude of absolute motions (i.e., with respect to the base link) depend on the choice of the base link. Section 1.6 presents examples of various types of motion that may be produced from a single kinematic chain.

The links of a mechanism are connected together by *kinematic pairs* or *joints*. Three common types of kinematic pairs in planar mechanisms are as follows:

- *Turning pairs* allow relative turning motion between two links. Such pairs are also called *bearings*, *pivots*, or *pin joints*. A four-bar mechanism (Figure 1.8) has a total of four turning pairs. Figure 1.31 illustrates other examples of links connected by turning pairs, along with their skeleton representations.
- *Sliding pairs* allow relative sliding motion between two links. For example, in Figure 1.5, link 4 is permitted to undergo sliding motion with respect to link 1. Figure 1.32 illustrates other examples of links connected by sliding pairs.
- *Rolling pairs* allow relative rolling motion between two links, such as employed in a pair of friction gears (Figure 1.20). Figure 1.33 illustrates other examples of links connected by rolling pairs. For a rolling pair, it is assumed that there is no slippage between the links.

Each turning pair, sliding pair, and rolling pair permits one relative motion between adjacent links. All of these kinematic pairs are referred to as *one degree of freedom pairs*.

Another type of kinematic pair is referred to as a *two degree of freedom pair*. It allows *two* relative motions between the adjacent links. Examples are shown in Figure 1.34. Figure 1.34(a) demonstrates two degrees of freedom between links 2 and 3. One of the degrees of freedom is the relative turning between the links. The other degree of freedom is the translational motion of the pin at one end of link 3 in the slot of link 2. As shown in Figure 1.34(a), this two degree of freedom pair may be represented in skeleton form using one sliding pair and one turning pair.

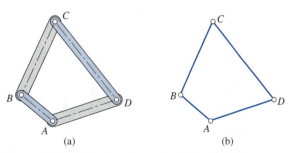

**Figure 1.30** (a) Four-bar kinematic chain. (b) Skeleton representation.

Mechanism Links

Skeleton Representations

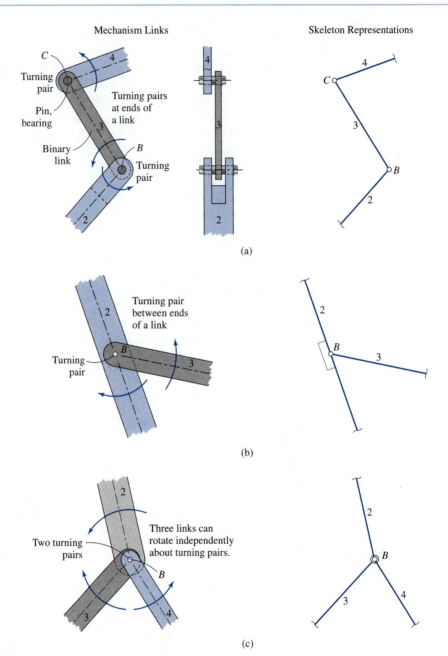

Figure 1.31 Examples of turning pairs.

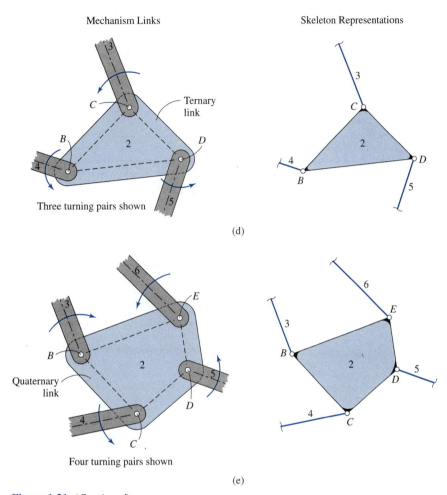

Mechanism Links

Skeleton Representations

Ternary link

Three turning pairs shown

(d)

Quaternary link

Four turning pairs shown

(e)

**Figure 1.31**  (*Continued*)

## 1.5 MECHANISM MOBILITY

The *mobility* of a mechanism is defined as the number of independent parameters required to specify the position of all links of the mechanism. This section is restricted to considering the mobility of planar mechanisms.

The location of a rigid body executing planar motion can be described by specifying three independent parameters. A planar mechanism with $n$ links has $n - 1$ links that can move. Thus, $3(n - 1)$ parameters would need to be specified if the links moved independently. Figure 1.35(a) shows a four-bar mechanism. In Figure 1.35(b), one of the links has been isolated from the rest of the mechanism. The location of this link may be described by specifying the $x$ coordinate of point $B$, the $y$ coordinate of point $B$, and angle $\theta$. Similar specifications may be applied for the other movable links.

Links of a mechanism do not move independently but instead are coupled by kinematic pairs. Each kinematic pair between links provides one or more constraints that reduce the

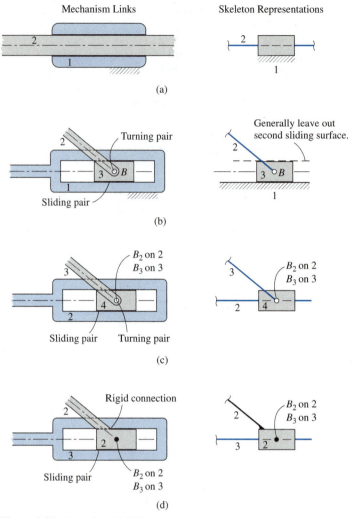

**Figure 1.32** Examples of sliding pairs.

number of independent motions of the mechanism. Each one degree of freedom pair (i.e., turning pair, sliding pair, or rolling pair) allows just one relative motion between links, while two relative motions are constrained. For a turning pair (Figure 1.31), only relative rotation is permitted. The constrained relative motions are the linear translations at the turning pair in the horizontal and vertical directions. For a sliding pair (Figure 1.32), the only relative motion permitted is translation of the slider, tangent to the direction of the slide. The constrained motions consist of translation perpendicular to the direction of the slide, and relative rotation between the links. For a rolling pair (Figure 1.33), the links remain in contact. Therefore, one constraint is that the center of rotation of one link must remain a fixed distance from the surface of the other link forming the pair. Also, for a rolling pair, no slippage is allowed between the links. Therefore, the second constraint arises due to restricted relative rotations that occur between the links. Each two degree of freedom pair (Figure 1.34) allows

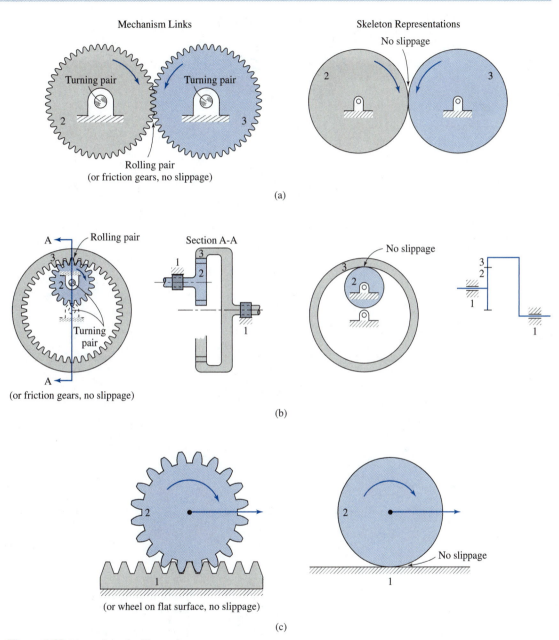

Figure 1.33 Examples of rolling pairs.

two relative motions between links, and just one relative motion is constrained. For the two degree of freedom kinematic pair shown in Figure 1.34(d), there can be both relative turning and sliding between links. Relative sliding takes place along the *common tangent* to the two surfaces at the point of contact. The only constraint when considering these two links is that there cannot be any relative motion along the *common normal*.

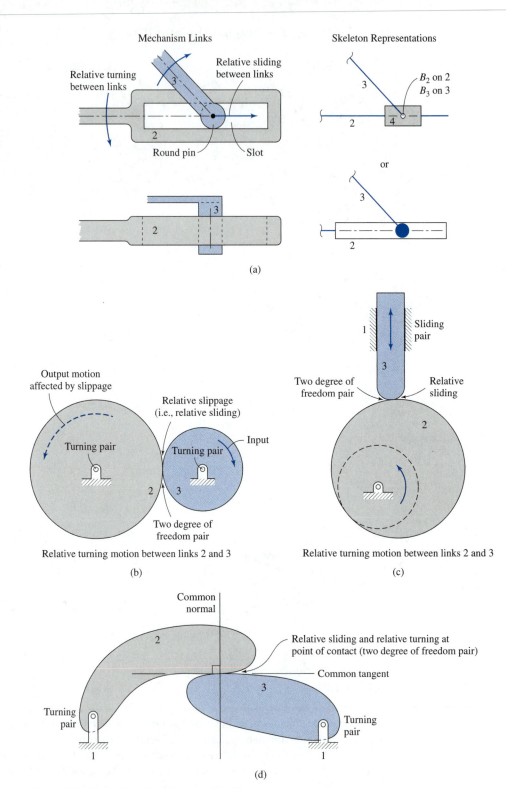

**Figure 1.34** Examples of two degree of freedom pairs.

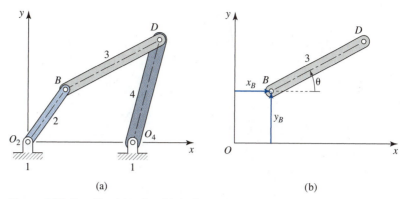

**Figure 1.35** Specification of position of a mechanism link in a plane.

Implementing the above examination, the equation for the mobility, $m$, of a planar mechanism is given as

$$m = 3(n-1) - 2j_1 - j_2$$

(1.5-1)

where

$n$ = number of links in the mechanism
$j_1$ = number of one degree of freedom pairs
$j_2$ = number of two degree of freedom pairs

The coefficient 2, which multiplies $j_1$, corresponds to the two constraints associated with each one degree of freedom pair. Minus signs in front of terms involving the numbers of kinematic pairs relate to a reduction of the mobility brought about by the constraints that are introduced.

## 1.5.1 Examples of Mechanism Mobility

As an example of calculating mobility of a mechanism using Equation (1.5-1), refer to the four-bar mechanism shown in Figure 1.36(a). The mechanism has four links, consisting of the base link and three moving links, and four kinematic pairs, all of which are turning pairs. There are no two degree of freedom pairs for this mechanism. In summary

$$n = 4; \qquad j_1 = 4; \qquad j_2 = 0$$

Substituting the above values in Equation (1.5-1) yields

$$m = 3(n-1) - 2j_1 - j_2 = 3(4-1) - 2 \times 4 - 0 = 1$$

That is, one input value needs to be specified to allow the position of all links to be determined. If $\theta_2$ is specified, then the other variable quantities (i.e., $\theta_3$ and $\theta_4$) of the mechanism may be calculated.

As a second example, consider the five-bar mechanism shown in Figure 1.36(b). In this instance, there are five links ($n = 5$) and a total of five one degree of freedom pairs, all of which are turning pairs ($j_1 = 5$). There are no two degree of freedom pairs ($j_2 = 0$). Using Equation (1.5-1), we find $m = 2$. Thus, in order to determine the geometry of the mechanism, it is necessary to specify two independent parameters. As shown in Figure 1.36(b), this may be accomplished by specifying the values of both $\theta_2$ and $\theta_5$.

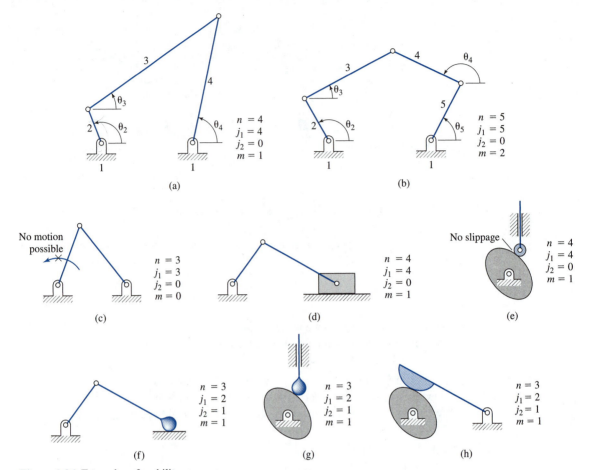

**Figure 1.36** Examples of mobility.

Six more examples are provided in Figure 1.36. The number of links, the numbers of one and two degree of freedom pairs, and the value of the mobility calculated using Equation (1.5-1) are listed for each example.

In calculating mechanism mobility using Equation (1.5-1), it is necessary to modify the calculated result whenever more than one kinematic pair provides the same constraint. For instance, consider the pair of friction gears illustrated in Figure 1.37(a). Links 2 and 3 contact at $A$. If there is no slippage at $A$, then links 2 and 3 are connected by a rolling pair. Links 2 and 3 are each connected to the base link through a turning pair. For this mechanism

$$n = 3; \qquad j_1 = 3; \qquad j_2 = 0$$

Substituting in Equation (1.5-1) yields

$$m = 3(n - 1) - 2j_1 - j_2 = 3(3 - 1) - 2 \times 3 - 0 = 0$$

indicating that the mechanism cannot move. However, by inspection, for a given rotation of link 2, link 3 undergoes a definite rotation, and therefore we should have $m = 1$. In this case, the result obtained using Equation (1.5-1) is incorrect. This is because the constraint

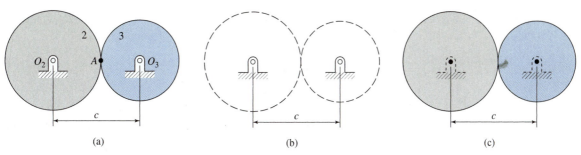

**Figure 1.37** Friction gears.

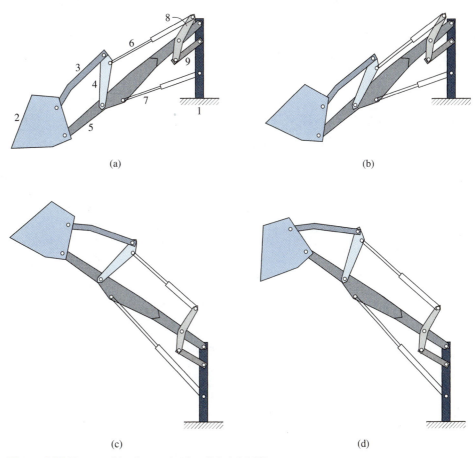

**Figure 1.38** Front-end loader mechanism [Model 1.38].

Model 1.38
Front-End
Loader
Mechanism

of the fixed distance, $c$, between points $O_2$ and $O_3$ is accounted for *twice* by the kinematic pairs. This constraint is satisfied by the two turning pairs (Figure 1.37(b)); however, the *same* constraint is accounted for by considering the rolling pair (Figure 1.37(c)).

Consider the example of a front-end loader, given in Figure 1.38. This mechanism is driven by means of two actuators, components 6 and 7. If the lengths of these two actuators are kept constant, they can be considered as links, and in this case $n = 9$. The total number of kinematic pairs (all turning pairs) is 12. Therefore

$$m = 3(n - 1) - 2j_1 - j_2$$
$$= 3(9 - 1) - 2 \times 12 - 0 = 0$$

This result indicates that if the lengths of two actuators are kept constant, then the front-end loader is not permitted to move.

   The mechanism in fact has a mobility equal to two by allowing the operator to adjust the lengths of the two actuators. Typical positions are shown in Figure 1.38. The starting position is shown in Figure 1.38(a). Then the bucket is rolled back (Figure 1.38(b)) by reducing the length of component 6, also known as the *dump actuator*. This is followed by extending component 7, also known as the *lift actuator*, to raise the load (Figure 1.38(c)). Finally, the dump actuator is extended to empty the bucket (Figure 1.38(d)).

## 1.6 MECHANISM INVERSION

Every mechanism has one stationary base link. All other links may move relative to the fixed base link.

   An *inverted mechanism* is obtained by making the originally fixed link into a moving link and selecting an originally moving link to be the fixed link. As an illustration of the inversion of a mechanism, Figure 1.39 shows four mechanisms that were created from the same kinematic chain. In each case, a different link of the kinematic chain is held fixed. The slider crank mechanism is illustrated in Figure 1.39(a), where link 1 is the

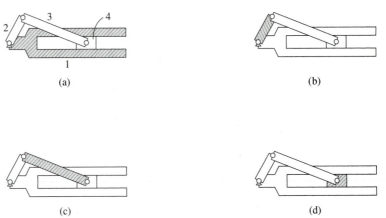

**Figure 1.39** Slider crank mechanism and its three inversions [Model 1.39]: (a) slider crank mechanism (link 1 fixed), (b) inversion #1 (link 2 fixed), (c) inversion #2 (link 3 fixed), (d) inversion #3 (link 4 fixed).

base link. Figures 1.39(b), 1.39(c), and 1.39(d) respectively correspond to the cases where links 2, 3, and 4 are held fixed, and link 1 can move. These three mechanisms are inversions of that given in Figure 1.39(a). All four mechanisms shown in Figure 1.39 have a driving motor between links 1 and 2. Sets of accumulated images throughout a cycle of motion are given in Figure 1.40. In each mechanism, *absolute* motions of the links with respect to the base link are distinct. However, since all mechanisms have the same link dimensions, and all have a driving motor located at the same position in the kinematic chain, the *relative* motions between links are identical for all of these mechanisms.

Another example of the inversion of a mechanism is shown in Figure 1.41. We start with the four-bar mechanism shown in Figure 1.41(a), in which link 1 is held fixed. The three inversions of this mechanism are shown in Figures 1.41(b), 1.41(c), and 1.41(d), in which links 2, 3, and 4, respectively, are held fixed. Sets of accumulated images throughout a cycle of motion are given in Figure 1.42.

In general, a mechanism having $n$ links can have $n - 1$ inversions.

**Model 1.41**
Four-Bar Mechanism and Its Three Inversions

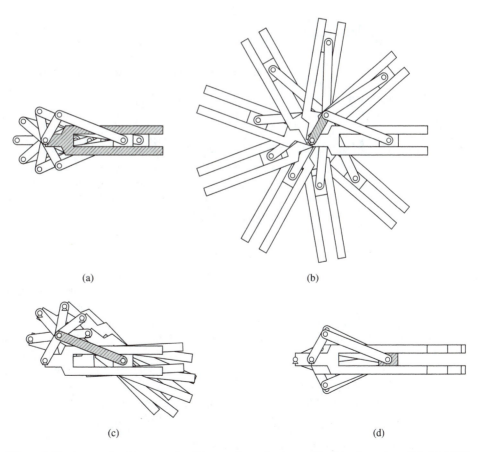

(a)

(b)

(c)

(d)

**Figure 1.40** Accumulated images of a slider crank mechanism and its three inversions [Model 1.39].

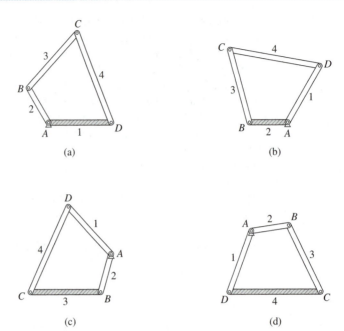

**Figure 1.41** Four-bar mechanism and its three inversions [Model 1.41]: (a) link 1 fixed, (b) link 2 fixed, (c) link 3 fixed, (d) link 4 fixed.

## 1.7 TYPES OF FOUR-BAR AND SLIDER CRANK MECHANISMS

Four-bar and slider crank mechanisms may be classified according to *type*. A type is characterized by the number of links that are able to undergo full rotation, and which are connected to the base link at base pivots. This section provides the method of determining the type of mechanism based on the lengths of the links.

### 1.7.1 Four-Bar Mechanism—Grashof's Criterion

Figures 1.41 and 1.42 illustrate three different types of four-bar mechanisms. For Figures 1.41(a) and 1.41(c), link 2, the *crank*, is able to make a full rotation, whereas link 4, the *rocker*, oscillates between two limit positions. Such four-bar mechanisms are called *crank rockers*. In Figure 1.41(b), both links 1 and 3 are able to make a full rotation, and it is called a *drag link*. For Figure 1.41(d), neither link 1 nor link 3 is able to make a full rotation, and it is referred to as a *rocker-rocker*. Another illustration of these three types of four-bar mechanism is given in Figure 1.43. For the crank rocker mechanism shown in Figure 1.43(a), if link 2 makes complete rotations, then link 4 rocks between two limit positions. The amplitude of the rocking motion is $\Delta\theta_4$.

Four-bar mechanisms may be studied by distinguishing the link lengths as follows:

- $s$: the length of the shortest link
- $l$: the length of the longest link
- $p, q$: the lengths of the other two links

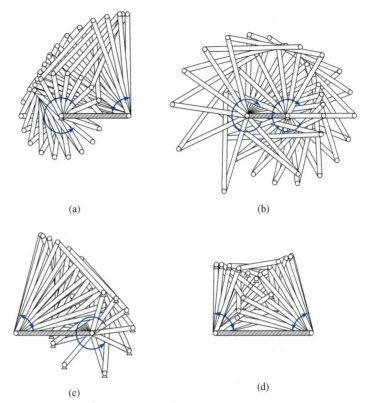

(a)                    (b)

(c)                    (d)

**Figure 1.42** Accumulated images of a four-bar mechanism and its three inversions [Model 1.41]: (a) crank rocker, (b) drag link, (c) crank rocker, (d) rocker-rocker.

To assemble the kinematic chain it is necessary that

$$s + p + q \geq l.$$

When the two sides of the above expression are equal, all links are constrained to remain collinear, and no motion is permitted. Only when the sides of the expression are not equal can there be rotations of the links.

The type of a four-bar mechanism may be determined using *Grashof's Criterion* [2]. Using the above designations of link lengths, Grashof's Criterion is given in Table 1.1.

In Table 1.1, two classes of kinematic chains are identified. Only for a Class I kinematic chain is it possible to obtain all three types of a four-bar mechanism.

As an illustration, consider the four-bar mechanisms illustrated in Figure 1.41. In this instance

$$s = 1.9 \text{ cm}; \qquad p = 3.2 \text{ cm}; \qquad q = 3.8 \text{ cm}; \qquad l = 4.8 \text{ cm}$$

Since

$$s + l = 6.7 \text{ cm} < p + q = 7.0 \text{ cm}$$

the mechanisms are made from a Class I kinematic chain, and thus through inversion it is possible to obtain the different types of four-bar mechanisms, as illustrated in Figure 1.42.

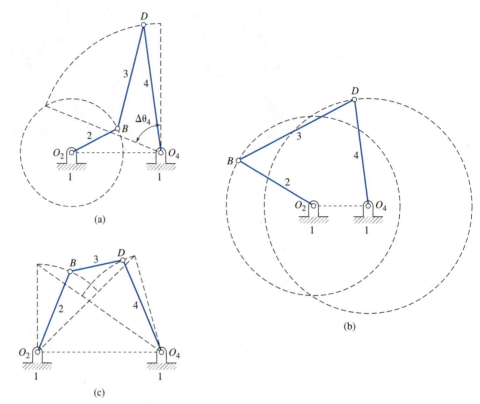

**Figure 1.43** Types of four-bar mechanisms: (a) crank rocker, (b) drag link, (c) rocker-rocker.

**TABLE 1.1** Classification of Four-Bar Mechanisms

| Class I Kinematic Chain $s + l < p + q$ | Class II Kinematic Chain $s + l > p + q$ |
|---|---|
| If $s$ is the input link, then the mechanism is a crank rocker. | The mechanism is a rocker-rocker. |
| If $s$ is the base link, then the mechanism is a drag link. | |
| If otherwise, then the mechanism is a rocker-rocker. | |

The case where

$$s + l = p + q$$

is not covered in Table 1.1. Such a mechanism, referred to as a *change point mechanism*, may be brought into a geometry for which all links are collinear. Figure 1.44 shows an example of a change point mechanism.

A special case of a change point mechanism occurs when

$$r_1 = r_3 \quad \text{and} \quad r_2 = r_4$$

This is referred to as a *parallelogram four-bar mechanism*. Figure 1.45 shows a skeleton representation of such a mechanism. For this mechanism, link 1 always remains parallel to

**Model 1.45**
Parallelogram
Four-Bar
Mechanism

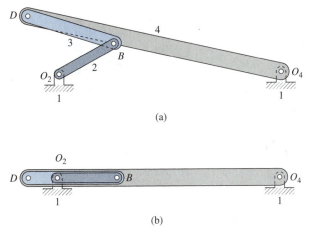

(a)

(b)

**Figure 1.44** Change point mechanism.

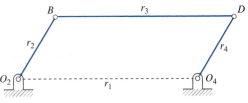

**Figure 1.45** Parallelogram four-bar mechanism [Model 1.45].

![Figure 1.46 automobile jack photo]

**Figure 1.46** Example application of a parallelogram four-bar mechanism.

link 3, and link 2 always remains parallel to link 4. Figure 1.46 shows an automobile jack that incorporates a parallelogram four-bar mechanism. The automobile will always remain parallel to the ground (base link) during a lift.

**Model 1.47**
Variable Base
Link Four-Bar
Mechanism

With **[Model 1.47]** it is possible to adjust the length of the base link and view the resulting motion. The lengths of the three moving links are

$$r_2 = 2.0 \text{ cm}; \qquad r_3 = 4.0 \text{ cm}; \qquad r_4 = 5.0 \text{ cm}$$

Figure 1.47 shows two cases where the length of the base link takes on values

$$r_1 = 6.0 \text{ cm} \qquad \text{and} \qquad r_1 = 1.5 \text{ cm}$$

which correspond to crank rocker and rocker-rocker mechanisms, respectively.

Corresponding to the mechanism shown in Figure 1.47, Table 1.2 lists the ranges of values of $r_1$ and the corresponding types of four-bar mechanisms. The range of link lengths represents the theoretical range of permissible values. However, the practical limits in a design are somewhat restrictive, and it may sometimes be difficult to transmit motions between links. This is described further through examples presented in Chapters 2 and 11.

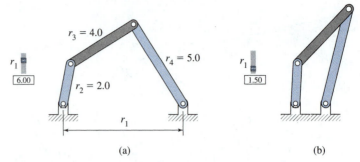

**Figure 1.47** Four-bar mechanisms [Model 1.47]: (a) crank rocker, (b) rocker-rocker.

**TABLE 1.2** Examples of Types of Four-Bar Mechanisms

| $r_1$ (cm) | Type of Four-Bar Mechanism |
|---|---|
| $0.0 < r_1 < 1.0$ | Drag link |
| $r_1 = 1.0$ | Change point |
| $1.0 < r_1 < 3.0$ | Rocker-rocker |
| $r_1 = 3.0$ | Change point |
| $3.0 < r_1 < 7.0$ | Crank rocker |
| $r_1 = 7.0$ | Change point |
| $7.0 < r_1 < 11.0$ | Rocker-rocker |
| $r_1 = 11.0$ | All links collinear, no motion possible |
| $r_1 > 11.0 = r_2 + r_3 + r_4$ | Impossible to assemble mechanism |

Note: $r_2 = 2.0$ cm; $r_3 = 4.0$ cm; $r_4 = 5.0$ cm

### 1.7.2 Slider Crank Mechanism

Consider the slider crank mechanism shown in Figure 1.7. The offset is designated by $r_1$, and the lengths of the crank and coupler are $r_2$ and $r_3$, respectively. In order for the crank to have full rotation, it is necessary that

$$r_2 < r_3 \quad \text{and} \quad r_1 \leq r_3 - r_2$$

When

$$r_1 > r_2 + r_3$$

it is impossible to assemble the mechanism.

As an illustration of the above, Figure 1.48 shows a slider crank mechanism for which

$$r_2 = 1.5 \text{ cm}; \quad r_3 = 2.5 \text{ cm}$$

**Model 1.48**
Variable-Offset
Slider Crank
Mechanism

With **[Model 1.48]** it is possible to adjust the offset and view the resulting motion. Figures 1.48(a), 1.48(b), and 1.48(c) show offsets of 0.0, 0.9, and 1.1 cm, respectively. For the mechanisms shown in Figures 1.48(a) and 1.48(b), the crank is able to make full rotation. However, for the mechanism shown in Figure 1.48(c), since

$$r_1 > r_3 - r_2$$

the crank has a very limited range of permissible motion.

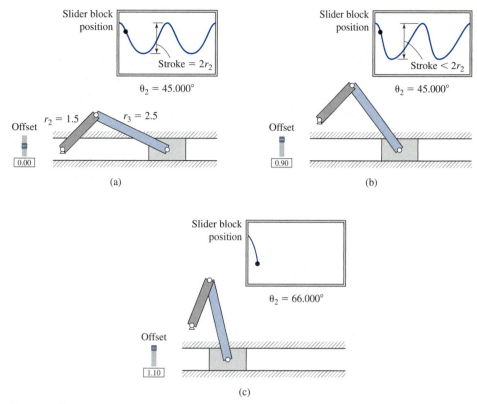

**Figure 1.48** Slider crank mechanisms [Model 1.48].

For the mechanisms shown in Figures 1.48(a) and 1.48(b), the slider moves between two limit positions. The distance between the two limit positions of the slider is referred to as the *stroke*. The offset influences the value of the stroke. When the offset is zero (Figure 1.48(a)),

$$\text{stroke} = 2r_2 = 3.0 \text{ cm}$$

and for a nonzero offset, such as shown in Figure 1.48(b)

$$\text{stroke} < 2r_2$$

## 1.8 COGNATES OF A MECHANISM

Associated with every four-bar mechanism and slider crank mechanism incorporating a coupler point is at least one other mechanism that will generate the identical path of the coupler point. These associated mechanisms are referred to as *cognates*. In a case where a mechanism has been designed to generate a suitable path of the coupler point, but the links of the mechanism have inappropriate locations, it may be possible to implement instead a cognate mechanism. This section presents the method of constructing cognate mechanisms of four-bar and slider crank mechanisms.

### 1.8.1 Four-Bar Mechanism

A four-bar mechanism with a coupler point has two associated cognates. The four-bar mechanism shown in Figure 1.49 has two base pivots, $O_2$ and $O_4$. The moving pivots are designated as $B$ and $D$, and the coupler point is labelled as $C$.

Figure 1.50(a) shows two additional four-bar mechanisms added to the original mechanism. Their four-bar kinematic chains include points $O_{10}HGO_8$ and $O_5EFO_7$. As indicated, the following points coincide: $O_2$ and $O_{10}$; $O_4$ and $O_5$; $O_7$ and $O_8$. All three mechanisms share the same coupler point $C$.

For the three four-bar mechanisms illustrated in Figure 1.50(a) it is necessary that

- the triangles formed by the following points are similar: $BDC$, $HCG$, and $CEF$
- the following points form parallelograms: $O_2BCH$, $DO_4EC$, and $CFO_7G$
- the base pivots, $O_2$, $O_4$, and $O_7$, form a triangle that is similar to that formed by points $BDC$

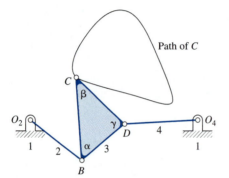

**Figure 1.49** Four-bar mechanism with a coupler point.

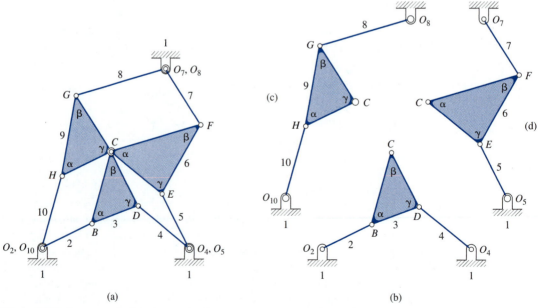

(a)                                        (b)

**Figure 1.50** Cognates of a four-bar mechanism.

**Model 1.51**
Cognates of a
Four-Bar
Mechanism

Figures 1.50(b), 1.50(c), and 1.50(d) show the same three four-bar mechanisms separated from one another. Figure 1.51 shows three four-bar mechanisms that can all move together while attached, along with three separated mechanisms. All mechanisms can move in unison to trace out the same coupler point path.

## 1.8.2 Slider Crank Mechanism

A slider crank mechanism has one associated cognate. Consider the slider crank mechanism shown in Figure 1.52. As the crank rotates, coupler point $C$ traces out the path

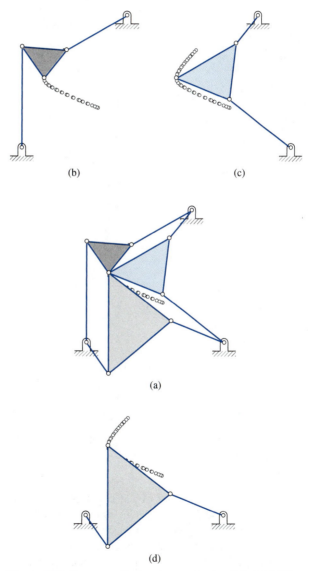

(b)                                   (c)

(a)

(d)

**Figure 1.51** Cognates of a four-bar mechanism [Model 1.51].

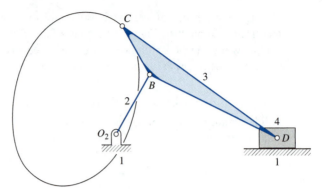

**Figure 1.52** Slider crank mechanism with a coupler point.

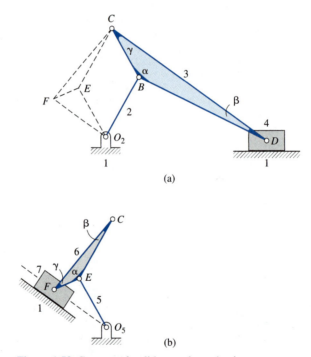

(a)

(b)

**Figure 1.53** Cognate of a slider crank mechanism.

**Model 1.54**

Cognate of a
Slider Crank
Mechanism

indicated. Construction of the cognate mechanism is shown in Figure 1.53(a). For the dashed line indicated, $O_2BCE$ forms a parallelogram. Also, $BDC$ and $ECF$ form similar triangles. The cognate mechanism is illustrated in Figure 1.53(b). Figure 1.54 shows two slider crank mechanisms that can move together while attached, along with two separated mechanisms. All mechanisms can trace out the same coupler point path.

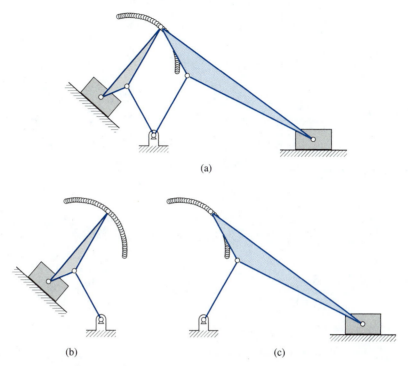

(a)

(b)                                              (c)

**Figure 1.54** Cognate of a slider crank mechanism [Model 1.54].

# PROBLEMS

**P1.1** For each of the mechanisms shown in Figure P1.1, specify the number of links. List the types of kinematic pairs present in each of the mechanisms, and the number of each type. Calculate the mobility.

**P1.2** For each of the mechanisms shown in Figure P1.2, determine the mobility.

**P1.3** For each of the mechanisms shown in Figure P1.3, list the types of kinematic pairs present and the number of each type. Calculate the mobility.

**P1.4** For each of the mechanisms shown in Figure P1.4, list the types of kinematic pairs present and the number of each type. Calculate the mobility.

**P1.5** For each of the mechanisms shown in Figure P1.5, list the types of kinematic pairs present and the number of each type. Calculate the mobility.

**P1.6** For each set of link lengths given in Table P1.6 (see Figure 1.9).

(i) determine whether or not the links can actually form a mechanism

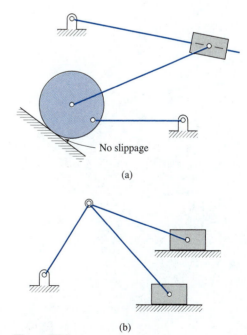

No slippage

(a)

(b)

**Figure P1.1**

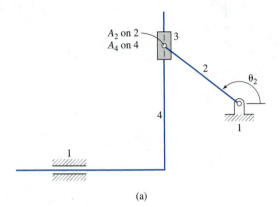

$A_2$ on 2
$A_4$ on 4

(a)

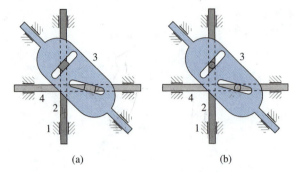

(a)

(b)

**Figure P1.4**

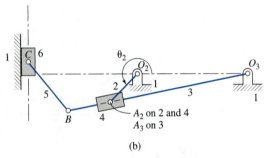

$\theta_2$

$A_2$ on 2 and 4
$A_3$ on 3

(b)

**Figure P1.2**

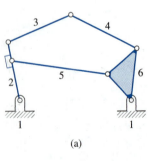

(a)

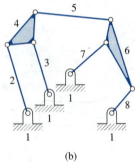

(b)

**Figure P1.3**

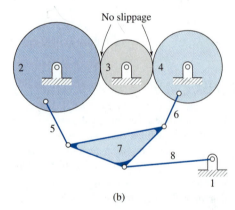

No slippage

(a)

No slippage

(b)

**Figure P1.5**

**TABLE P1.6**

|     | $r_1$ (cm) | $r_2$ (cm) | $r_3$ (cm) | $r_4$ (cm) |
|-----|-----------|-----------|-----------|-----------|
| (a) | 2.0       | 6.5       | 3.0       | 7.0       |
| (b) | 2.0       | 8.0       | 3.0       | 9.0       |
| (c) | 2.5       | 1.0       | 2.5       | 2.0       |
| (d) | 2.5       | 3.0       | 1.0       | 2.0       |
| (e) | 2.0       | 4.5       | 1.5       | 9.0       |
| (f) | 1.5       | 3.0       | 2.5       | 6.0       |

(ii) if a mechanism exists, determine the type of four-bar mechanism

**P1.7** Given the prescribed lengths of three links of a four-bar mechanism (see Figure 1.9), determine the range of values of the length of link 2 so that the mechanism will become a

(a) crank rocker mechanism
(b) drag link mechanism
(c) change point mechanism
(d) rocker-rocker mechanism

$$r_1 = 1.0 \text{ cm}; \quad r_3 = 2.5 \text{ cm}; \quad r_4 = 2.0 \text{ cm}$$

**P1.8** Given the prescribed lengths of three links of a four-bar mechanism (see Figure 1.9), determine the range of values of the length of link 4 so that the mechanism will become a

(a) crank rocker mechanism
(b) drag link mechanism
(c) change point mechanism
(d) rocker-rocker mechanism

$$r_1 = 1.0 \text{ cm}; \quad r_2 = 3.0 \text{ cm}; \quad r_3 = 2.5 \text{ cm}$$

**P1.9** Given the dimensions of a slider crank mechanism (see Figure 1.7), determine the range of values of the length of link 2 so that

(a) link 2 can make a full rotation
(b) the mechanism can be assembled

$$r_1 = 1.0 \text{ cm}; \quad r_3 = 2.5 \text{ cm}$$

**P1.10** Given the prescribed link dimensions of a slider crank mechanism (see Figure 1.7), determine the range of values of the length of link 3 so that

(a) link 2 can make a full rotation
(b) the mechanism can be assembled

$$r_1 = 1.0 \text{ cm}; \quad r_2 = 2.5 \text{ cm}$$

**P1.11** The nomenclature for this group of problems is given in Figure P1.11, and the dimensions and data are given in Table P1.11. For each, draw the two cognates of the mechanism.

**P1.12** The nomenclature for this group of problems is given in Figure P1.12, and the dimensions and data are given in Table P1.12. For each, draw the cognate of the mechanism.

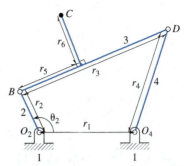

**Figure P1.11**

**TABLE P1.11**

| | $r_1$ (cm) | $r_2$ (cm) | $r_3$ (cm) | $r_4$ (cm) | $r_5$ (cm) | $r_6$ (cm) | $\theta_2$ (degrees) |
|---|---|---|---|---|---|---|---|
| (a) | 1.0 | 0.2 | 0.4 | 1.0 | 0.2 | 0.5 | 30 |
| (b) | 5.0 | 1.0 | 4.0 | 3.0 | −2.0 | 0 | 60 |
| (c) | 8.0 | 2.0 | 10.0 | 6.0 | 6.0 | −3.0 | 120 |
| (d) | 0.2 | 1.0 | 0.4 | 0.9 | 0.2 | 0.4 | 210 |

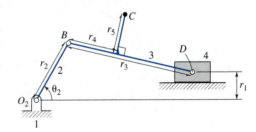

**Figure P1.12**

**TABLE P1.12**

| | $r_1$ (cm) | $r_2$ (cm) | $r_3$ (cm) | $r_4$ (cm) | $r_5$ (cm) | $\theta_2$ (degrees) |
|---|---|---|---|---|---|---|
| (a) | 0 | 5.0 | 10.0 | −4.0 | 0 | 30 |
| (b) | 0 | 2.0 | 6.0 | 2.0 | 4.0 | 60 |
| (c) | 2.0 | 4.0 | 8.0 | 12.0 | 0 | 120 |
| (d) | −2.0 | 4.0 | 7.0 | 5.0 | −3.0 | 315 |

# 2 Mechanics of Rigid Bodies

## 2.1 INTRODUCTION

In this chapter, expressions are presented for the kinematics and kinetics of rigid bodies. These expressions are applicable to the analysis of links of planar mechanisms.

Analyses will be based on the radial-transverse coordinate system. In Appendix D, expressions are presented for determining the position, velocity, and acceleration of a point with respect to a fixed point using this system. A more general case considers one moving point with respect to another moving point. Figure 2.1 shows two moving points, $A$ and $B$, and their paths of motion. The position of point $B$ relative to point $A$ is described by the *relative position vector*, $\bar{r}_{BA}$, which may be expressed as

$$\bar{r}_{BA} = r_{BA}\bar{i}_{r_{BA}}$$

(2.1-1)

where $\bar{i}_{r_{BA}}$ is the *unit relative radial vector*. Also shown in Figure 2.1 is $\bar{i}_{\theta_{BA}}$, the *unit relative transverse vector*, which points 90° counterclockwise from the direction of $\bar{i}_{r_{BA}}$.

## 2.2 RELATIVE VELOCITY BETWEEN TWO POINTS UNDERGOING PLANAR MOTION

The relative velocity between two moving points, $A$ and $B$, is found by differentiating Equation (2.1-1) with respect to time, yielding

$$\bar{v}_{BA} = \dot{r}_{BA}\bar{i}_{r_{BA}} + r_{BA}\frac{d}{dt}(\bar{i}_{r_{BA}})$$

(2.2-1)

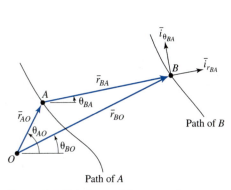

Figure 2.1 Relative position vector between two moving points.

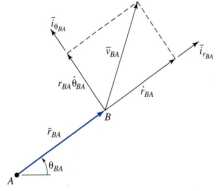

Figure 2.2 Relative velocity components between two points.

An expression for the time derivative of the unit relative radial vector may be found using a procedure similar to that presented in Appendix D. The result is

$$\frac{d}{dt}(\bar{i}_{r_{BA}}) = \dot{\theta}_{BA}\bar{i}_{\theta_{BA}} \tag{2.2-2}$$

where $\dot{\theta}_{BA}$ is the *angular velocity* of the line segment joining $A$ and $B$.

Substituting Equation (2.2-2) in Equation (2.2-1) yields

$$\boxed{\bar{v}_{BA} = \dot{r}_{BA}\bar{i}_{r_{BA}} + r_{BA}\dot{\theta}_{BA}\bar{i}_{\theta_{BA}}} \tag{2.2-3}$$

Components of Equation (2.2-3) are illustrated in Figure 2.2. Special cases of Equation (2.2-3) are presented in Section 2.3.

## 2.3 SPECIAL CASES OF RELATIVE VELOCITY EXPRESSION

### 2.3.1 Two Points Separated by a Fixed Distance

If points $A$ and $B$ are separated by a fixed distance, then

$$\dot{r}_{BA} = 0 \tag{2.3-1}$$

and the relative velocity expressed by Equation (2.2-3) reduces to

$$\boxed{\bar{v}_{BA} = r_{BA}\dot{\theta}_{BA}\bar{i}_{\theta_{BA}}} \tag{2.3-2}$$

For Equation (2.3-2), the relative velocity is in the transverse direction as defined by the line segment joining $A$ and $B$. It is therefore perpendicular to the relative position vector.

In summary, for two points separated by a fixed distance, the relative velocity must always be in a direction perpendicular to the line segment joining the points.

As an illustration, consider the four-bar mechanism shown in Figure 2.3(a). A separate drawing of link 2 is shown in Figure 2.3(b). Points $B$ and $O_2$ are on the same link and thus

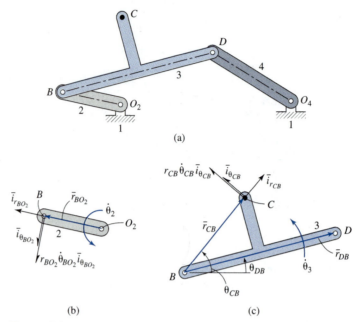

(a)

(b)                                    (c)

**Figure 2.3** Relative velocities between points on mechanism links:
(a) mechanism, (b) link 2, (c) link 3.

are separated by a fixed distance. Therefore

$$\bar{v}_{BO_2} = r_{BO_2}\dot{\theta}_{BO_2}\bar{i}_{\theta BO_2} \tag{2.3-3}$$

Since $O_2$ is fixed, the relative velocity given in Equation (2.3-3) also represents the absolute velocity, and

$$\bar{v}_{BO_2} = \bar{v}_B = r_{BO_2}\dot{\theta}_{BO_2}\bar{i}_{\theta BO_2} \tag{2.3-4}$$

Figure 2.3(c) shows a drawing of link 3. Points $B$ and $C$ are on the same link and separated by a fixed distance. The relative velocity between the two points is

$$\bar{v}_{CB} = r_{CB}\dot{\theta}_{CB}\bar{i}_{\theta CB} \tag{2.3-5}$$

The absolute velocity of point $C$ may be expressed as

$$\bar{v}_C = \bar{v}_B + \bar{v}_{CB} \tag{2.3-6}$$

Substituting Equations (2.3-4) and (2.3-5) in Equation (2.3-6) yields

$$\bar{v}_C = r_{BO_2}\dot{\theta}_{BO_2}\bar{i}_{\theta BO_2} + r_{CB}\dot{\theta}_{CB}\bar{i}_{\theta CB} \tag{2.3-7}$$

A moving link has a single value of angular velocity. Considering link 3 shown in Figure 2.3(c),

$$\dot{\theta}_{CB} = \dot{\theta}_{DB} = \dot{\theta}_3 \tag{2.3-8}$$

### 2.3.2  Slider on a Straight Slide

As a second special case of Equation (2.2-3), consider the movement of a slider on a straight slide, as shown in Figure 2.4. Point $A$ is fixed. Point $B_2$ is on the slide, link 2, and

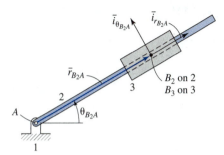

**Figure 2.4** Slider on a straight slide.

point $B_3$ is on the slider, link 3. For convenience in this analysis, points $B_2$ and $B_3$ coincide for the instant at which the relative motion is being considered.

Since point $B_2$ remains a fixed distance from fixed point $A$, its velocity may be found by considering the special case presented in Section 2.3.1. The result is

$$\bar{v}_{B_2A} = \bar{v}_{B_2} = r_{B_2A}\dot{\theta}_{B_2A}\bar{i}_{\theta_{B_2A}} \tag{2.3-9}$$

For point $B_3$, the associated position vector equation is

$$\bar{r}_{B_3A} = \bar{r}_{B_2A} + \bar{r}_{B_3B_2} \tag{2.3-10}$$

and the relative velocity equation is

$$\bar{v}_{B_3A} = \bar{v}_{B_2A} + \bar{v}_{B_3B_2} \tag{2.3-11}$$

or

$$\bar{v}_{B_3} = \bar{v}_{B_2} + \bar{v}_{B_3B_2} \tag{2.3-12}$$

where, using Equation (2.2-3),

$$\bar{v}_{B_3B_2} = \dot{r}_{B_3B_2}\bar{i}_{r_{B_3B_2}} + r_{B_3B_2}\dot{\theta}_{B_3B_2}\bar{i}_{\theta_{B_3B_2}} \tag{2.3-13}$$

However

$$r_{B_3B_2} = 0 \tag{2.3-14}$$

Also, even though $B_2$ and $B_3$ coincide at the instant under consideration, if the slider were to move with respect to the slide, it would be along a line in the same direction as line segment $B_2A$, and therefore we are permitted to write

$$\bar{i}_{r_{B_3B_2}} = \bar{i}_{r_{B_2A}} \tag{2.3-15}$$

Combining Equations (2.3-9) and (2.3-12)–(2.3-15) yields

$$\bar{v}_{B_3} = \bar{v}_{B_2} + \bar{v}_{B_3B_2} = r_{B_2A}\dot{\theta}_{B_2A}\bar{i}_{\theta_{B_2A}} + \dot{r}_{B_3B_2}\bar{i}_{r_{B_2A}} \tag{2.3-16}$$

Comparing expressions for velocities of $B_2$ and $B_3$, given in Equations (2.3-9) and (2.3-16), we find

- $B_2$ and $B_3$ share the *same* transverse velocity component
- $B_2$ and $B_3$ can have *different* radial components

## 2.4 RELATIVE ACCELERATION BETWEEN TWO POINTS UNDERGOING PLANAR MOTION

An expression for the acceleration of a point, using the radial-transverse coordinate system, is presented in Appendix D. Equation (D.3-12) is an expression of the acceleration of a moving point with respect to a fixed point. Similar to the approach taken for the analysis of velocity presented in Section 2.2, we may consider the acceleration of point $B$ with respect to another moving point, point $A$, as illustrated in Figure 2.1. The result is

$$\bar{a}_{BA} = \frac{d}{dt}(\bar{v}_{BA}) = (\ddot{r}_{BA} - r_{BA}\dot{\theta}_{BA}^2)\bar{i}_{r_{BA}} + (r_{BA}\ddot{\theta}_{BA} + 2\dot{r}_{BA}\dot{\theta}_{BA})\bar{i}_{\theta_{BA}} \qquad (2.4\text{-}1)$$

The above expression for acceleration is made up of two components in the radial direction, and two in the transverse direction. This is illustrated in Figure 2.5. Each of these four components is discussed separately below.

### 2.4.1 Sliding Acceleration

Relative *sliding acceleration*, illustrated in Figure 2.6(a), is designated as

$$a_{BA}^S = \ddot{r}_{BA} \qquad (2.4\text{-}2)$$

where point $A$ is on the slide, and point $B$ is on the slider.

In this case, the relative acceleration is in the radial direction defined by the locations of points $A$ and $B$. The direction of sliding acceleration must be in a positive or negative radial direction.

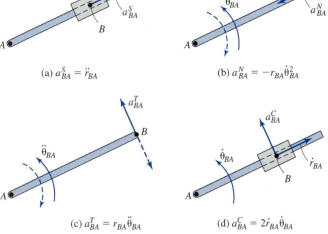

(a) $a_{BA}^S = \ddot{r}_{BA}$

(b) $a_{BA}^N = -r_{BA}\dot{\theta}_{BA}^2$

(c) $a_{BA}^T = r_{BA}\ddot{\theta}_{BA}$

(d) $a_{BA}^C = 2\dot{r}_{BA}\dot{\theta}_{BA}$

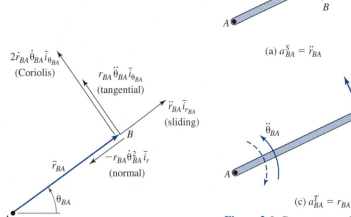

**Figure 2.5** Relative acceleration components between two points.

**Figure 2.6** Components of relative acceleration expression: (a) sliding component, (b) normal component, (c) tangential component, (d) Coriolis component.

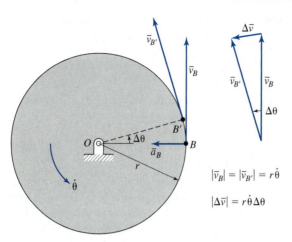

**Figure 2.7** Circular motion.

$$|\bar{v}_B| = |\bar{v}_{B'}| = r\dot{\theta}$$

$$|\Delta\bar{v}| = r\dot{\theta}\Delta\theta$$

### 2.4.2 Normal Acceleration

Relative *normal acceleration*, illustrated in Figure 2.6(b), is designated as

$$a_{BA}^N = -r_{BA}\dot{\theta}_{BA}^2 \qquad (2.4\text{-}3)$$

where the negative sign indicates that normal acceleration is always directed in the negative radial direction. The direction of normal acceleration is independent of the direction of angular velocity.

Normal acceleration occurs in circular motion. Consider the disc with radius $r$ shown in Figure 2.7. It is rotating about base pivot $O$ at a constant rate of $\dot{\theta}$. During a short time interval, point $B$ on the periphery of the disc moves to point $B'$. The velocity of a point may be expressed as a vector in the complex plane (refer to Appendix C, Section C.4). The velocities at points $B$ and $B'$ are

$$\bar{v}_B = r\dot{\theta}e^{i\pi/2}; \qquad \bar{v}_{B'} = r\dot{\theta}e^{i(\pi/2+\Delta\theta)} \qquad (2.4\text{-}4)$$

Point $B$ is moving at constant speed. However, the direction of its velocity is changing. Hence, it undergoes acceleration expressed as

$$\bar{a}_B = \lim_{\Delta t \to 0} \frac{\bar{v}_{B'} - \bar{v}_B}{\Delta t} \qquad (2.4\text{-}5)$$

Substituting Equation (2.4-4) in Equation (2.4-5) and simplifying yields

$$\bar{a}_B = \lim_{\Delta t \to 0} \frac{\bar{v}_{B'} - \bar{v}_B}{\Delta t} = \lim_{\Delta t \to 0} \frac{\Delta\bar{v}_B}{\Delta t} = \lim_{\Delta t \to 0} -\frac{r\dot{\theta}\Delta\theta}{\Delta t} = -r\dot{\theta}^2 \qquad (2.4\text{-}6)$$

The negative sign in Equation (2.4-6) indicates that the acceleration is in the negative radial direction.

### 2.4.3 Tangential Acceleration

Relative *tangential acceleration*, illustrated in Figure 2.6(c), is designated as

$$a_{BA}^T = r_{BA}\ddot{\theta}_{BA} \qquad (2.4\text{-}7)$$

This relative acceleration is perpendicular to the line segment joining $A$ and $B$. It can be in either the positive or negative transverse direction, depending on the positive or negative sign depicting the *angular acceleration*, $\ddot{\theta}_{BA}$.

### 2.4.4 Coriolis Acceleration

Relative *Coriolis acceleration*, illustrated in Figure 2.6(d), is designated as

$$a_{BA}^{C} = 2\dot{r}_{BA}\dot{\theta}_{BA} \tag{2.4-8}$$

where point $A$ is on the slide and point $B$ is on the slider.

In order for Coriolis acceleration to be present, relative sliding (i.e., $\dot{r}_{BA}$) must take place in a frame of reference that has angular velocity (i.e., $\dot{\theta}_{BA}$). This would occur when a slider block moves along a slide, while at the same time the slide has rotational motion. Figure 1.39(b) presents a typical example. The direction of the Coriolis acceleration is dependent upon both the sign of the sliding velocity and the sign of the angular velocity. In this textbook (following the usual convention for analyses of planar systems) positive angular displacements, angular velocities, and angular accelerations are typically taken to be in the counterclockwise direction. The positive value of the sliding velocity corresponds to motion in the positive radial direction defined by the two points under consideration. Figure 2.6(d) shows the combination of positive sliding velocity and positive angular velocity. The resultant Coriolis acceleration is then in the positive transverse direction.

Figure 2.8 shows the four possible combinations of positive and negative angular velocity and sliding velocity. The relative Coriolis acceleration component has been drawn for each combination. All Coriolis acceleration components have a magnitude of

$$\left|a_{BA}^{C}\right| = \left|2\dot{r}_{BA}\dot{\theta}_{BA}\right| \tag{2.4-9}$$

Another illustration of Coriolis acceleration is shown in Figure 2.9. A rotating disc is turning about base pivot $O$. A radial line is painted on the disc between $O$ and point $P$ on the periphery. Let us say there is an ant on the surface of the disc at point $B$. It moves radially along the line from point $B$ toward point $P$ at speed $\dot{r}$. During a short time interval, the ant moves from point $B$ to point $B'$, and the disc rotates $\Delta\theta$. At points $B$ and $B'$, the velocities are

$$\bar{v}_{B} = \dot{r} + r\dot{\theta}e^{i\pi/2}; \qquad \bar{v}_{B'} = \dot{r}e^{i\Delta\theta} + (r + \Delta r)\dot{\theta}e^{i(\pi/2+\Delta\theta)} \tag{2.4-10}$$

Although there is a constant rate of radial movement of the ant with respect to the center of the disc, it experiences an increase in the component of velocity perpendicular to

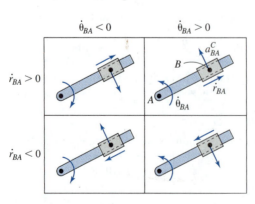

**Figure 2.8** Coriolis acceleration of a slider ($B$) with respect to a slide ($A$).

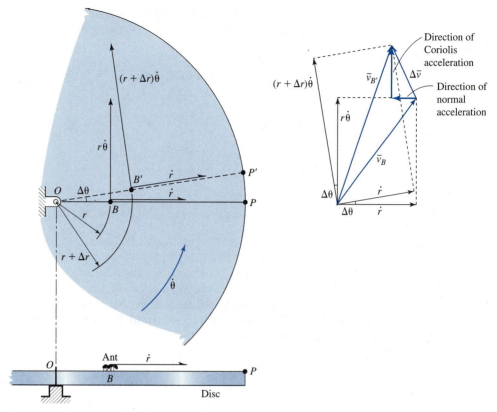

**Figure 2.9** Example of Coriolis acceleration.

the direction of the relative motion. Therefore, the ant undergoes acceleration in that direction. The acceleration of the ant at point $B$ is

$$\overline{a}_B = \lim_{\Delta t \to 0} \frac{\overline{v}_{B'} - \overline{v}_B}{\Delta t} \qquad (2.4\text{-}11)$$

Substituting Equation (2.4-10) in Equation (2.4-11) and simplifying yields

$$\overline{a}_B = -r\dot{\theta}^2 + i2\dot{r}\dot{\theta} \qquad (2.4\text{-}12)$$

Equation (2.4-12) indicates there is a normal component of acceleration directed back toward the base pivot, and a Coriolis component perpendicular to the direction of relative motion.

## 2.5  SPECIAL CASES OF THE RELATIVE ACCELERATION EQUATION

Combining Equations (2.4-1)–(2.4-5), we may express the relative acceleration between points $A$ and $B$ as

$$\overline{a}_{BA} = \left(a_{BA}^S + a_{BA}^N\right)\overline{i}_{r_{BA}} + \left(a_{BA}^T + a_{BA}^C\right)\overline{i}_{\theta_{BA}} \qquad (2.5\text{-}1)$$

The components are illustrated in Figure 2.5. In Subsections 2.5.1–2.5.3, we examine a few special cases of this expression.

## 2.5.1 Two Points Separated by a Fixed Distance

When two points are separated by a fixed distance,

$$\dot{r}_{BA} = 0; \qquad \ddot{r}_{BA} = 0 \tag{2.5-2}$$

Then from Section 2.4,

$$a^C_{BA} = 0; \qquad a^S_{BA} = 0 \tag{2.5-3}$$

and Equation (2.5-1) reduces to

$$\bar{a}_{BA} = \left(-r_{BA}\dot{\theta}^2_{BA}\right)\bar{i}_{r_{BA}} + \left(r_{BA}\ddot{\theta}_{BA}\right)\bar{i}_{\theta_{BA}}$$
$$= \left(a^N_{BA}\right)\bar{i}_{r_{BA}} + \left(a^T_{BA}\right)\bar{i}_{\theta_{BA}} \tag{2.5-4}$$

which indicates that between two points separated by a fixed distance there can only be normal and tangential relative accelerations.

As an illustration, consider the four-bar mechanism shown in Figure 2.10(a). A separate drawing of link 2 is shown in Figure 2.10(b). Points $B$ and $O_2$ are on the same link and separated by a fixed distance. Therefore

$$\bar{a}_{BO_2} = \left(-r_{BO_2}\dot{\theta}^2_{BO_2}\right)\bar{i}_{r_{BO_2}} + \left(r_{BO_2}\ddot{\theta}_{BO_2}\right)\bar{i}_{\theta_{BO_2}} \tag{2.5-5}$$

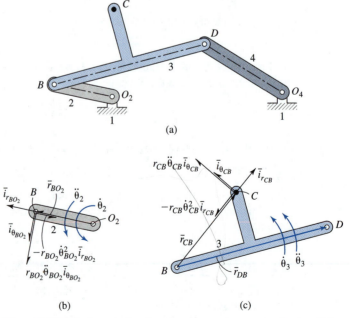

Figure 2.10 Relative accelerations between points on mechanism links: (a) mechanism, (b) link 2, (c) link 3.

Since $O_2$ is fixed, the relative acceleration given in Equation (2.5-5) also represents the absolute acceleration, and

$$\bar{a}_{BO_2} = \bar{a}_B = \left(-r_{BO_2}\dot{\theta}^2_{BO_2}\right)\bar{i}_{r_{BO_2}} + \left(r_{BO_2}\ddot{\theta}_{BO_2}\right)\bar{i}_{\theta_{BO_2}} \qquad (2.5\text{-}6)$$

Furthermore, if link 2 of the mechanism is driven at a constant rate, then

$$\dot{\theta}_{BO_2} = \text{constant}; \qquad \ddot{\theta}_{BO_2} = 0$$

and Equation (2.5-6) reduces to

$$\bar{a}_B = \left(-r_{BO_2}\dot{\theta}^2_{BO_2}\right)\bar{i}_{r_{BO_2}} \qquad (2.5\text{-}7)$$

Figure 2.10(c) illustrates link 3. Since points $B$ and $C$ are on the same link, they are separated by a fixed distance. Therefore

$$\bar{a}_{CB} = \left(-r_{CB}\dot{\theta}^2_{CB}\right)\bar{i}_{r_{CB}} + \left(r_{CB}\ddot{\theta}_{CB}\right)\bar{i}_{\theta_{CB}} \qquad (2.5\text{-}8)$$

The absolute acceleration of point $C$ may be expressed as

$$\bar{a}_C = \bar{a}_B + \bar{a}_{CB} \qquad (2.5\text{-}9)$$

Substituting Equations (2.5-7) and (2.5-8) in Equation (2.5-9) yields

$$\bar{a}_C = \left(-r_{BO_2}\dot{\theta}^2_{BO_2}\right)\bar{i}_{r_{BO_2}} + \left(-r_{CB}\dot{\theta}^2_{CB}\right)\bar{i}_{r_{CB}} + \left(r_{CB}\ddot{\theta}_{CB}\right)\bar{i}_{\theta_{CB}} \qquad (2.5\text{-}10)$$

A moving link can have only one value of angular acceleration. Thus, for link 3 shown in Figure 2.10(c)

$$\ddot{\theta}_{CB} = \ddot{\theta}_{DB} = \ddot{\theta}_3 \qquad (2.5\text{-}11)$$

### 2.5.2 Slider on a Straight Slide
We now consider the special case of a slider moving along a straight slide as illustrated in Figure 2.4. Considering points $B_2$ and $B_3$ for which

$$r_{B_2 B_3} = 0 \qquad (2.5\text{-}12)$$

then from Section 2.4,

$$a^N_{B_2 B_3} = 0; \qquad a^T_{B_2 B_3} = 0 \qquad (2.5\text{-}13)$$

and Equation (2.5-1) reduces to

$$\boxed{\begin{aligned}\bar{a}_{B_3 B_2} &= \left(\ddot{r}_{B_3 B_2}\right)\bar{i}_{r_{B_2 A}} + \left(2\dot{r}_{B_3 B_2}\dot{\theta}_{B_2 A}\right)\bar{i}_{\theta_{B_2 A}} \\ &= \left(a^S_{B_3 B_2}\right)\bar{i}_{r_{B_2 A}} + \left(a^C_{B_3 B_2}\right)\bar{i}_{\theta_{B_2 A}}\end{aligned}} \qquad (2.5\text{-}14)$$

That is, for two points that momentarily coincide, it is possible to have relative sliding acceleration and relative Coriolis acceleration occurring between the two points.

### 2.5.3 Slider on a Curved Slide
Figure 2.11(a) illustrates a slider on a curved slide. The slide has a radius of curvature $\rho$, angular velocity $\dot{\theta}_2$, and angular acceleration $\ddot{\theta}_2$. The velocity of the slider with respect to the slide is given as $\bar{v}_{B_3 B_2}$. Given these motions, Figure 2.11(b) shows two components of acceleration of point $B_2$, relative to point $A$ on the slide. These consist of normal and

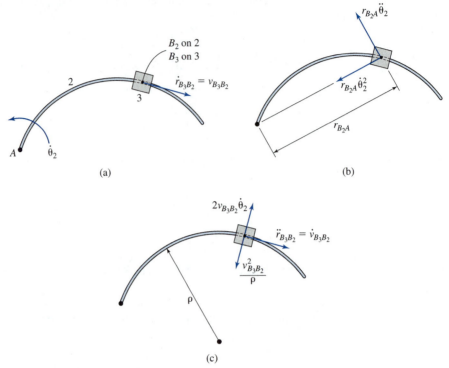

**Figure 2.11**  Slider on a curved slide.

tangential components. Components of acceleration of the slider, represented by point $B_3$, relative to point $B_2$ on the slide, are shown in Figure 2.11(c). Similar to the case of a slider on a straight slide, there are relative sliding and Coriolis components that are tangential and perpendicular to the sliding motion. In addition, there is a component of acceleration directed toward the center of curvature of the slide. The magnitude of this component is

$$\frac{v_{B_2 B_3}^2}{\rho} \tag{2.5-15}$$

Figure 2.11 illustrates a slide in the shape of a circular arc, and hence the radius of its curvature is constant. Equation (2.5-15) is applicable in analyses involving noncircular arcs by applying the value of radius of curvature that corresponds to the position under consideration.

## 2.6  LIMIT POSITIONS AND TIME RATIO OF MECHANISMS

Consider the slider crank mechanism shown in Figure 2.12. The crank is able to make full rotations, and the slider moves between two *limit positions*. Limit positions define the *stroke* of the slider.

In this section, the limit positions of some common mechanisms are presented. We will restrict ourselves to conditions where the input link of the mechanism is able to make full rotation, and the output link moves between two limit positions.

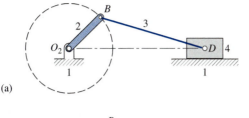

**Figure 2.12** Limit positions of a slider crank mechanism: (a) mechanism, (b) geometry of first limit position, (c) geometry of second limit position.

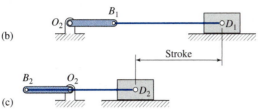

### 2.6.1 Slider Crank Mechanism

In instances when the crank is able to make a complete rotation and is turning at a constant speed, we define the *time ratio* as the time for the slider to move in one direction between the limit positions, divided by the time it takes to move in the opposite direction between the same limit positions. The time ratio is dimensionless.

Figure 2.13(a) shows a slider crank mechanism. The crank is rotating in the counter-clockwise direction. Also shown are the two limit positions. For each limit position, the crank and coupler are collinear. The rotation of the crank required to move the slider from the right limit position, $D_1$, to that on the left, $D_2$, is designated as $\Delta\theta_2$.

Simplified outlines of the limit positions are shown in Figures 2.13(b) and 2.13(c). Comparing Figures 2.13(a), 2.13(b), and 2.13(c),

$$\Delta\theta_2 = 180° + \alpha_2 - \alpha_1 \qquad (2.6\text{-}1)$$

where

$$\alpha_1 = \sin^{-1}\left(\frac{r_1}{r_2 + r_3}\right); \qquad \alpha_2 = \sin^{-1}\left(\frac{r_1}{r_3 - r_2}\right) \qquad (2.6\text{-}2)$$

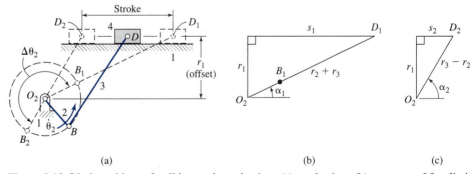

**Figure 2.13** Limit positions of a slider crank mechanism: (a) mechanism, (b) geometry of first limit position, (c) geometry of second limit position.

Also

$$s_1 = \left[(r_2+r_3)^2 - r_1^2\right]^{1/2}; \qquad s_2 = \left[(r_3-r_2)^2 - r_1^2\right]^{1/2} \tag{2.6-3}$$

and

$$\boxed{\text{stroke} = s_1 - s_2 = \left[(r_2+r_3)^2 - r_1^2\right]^{1/2} - \left[(r_3-r_2)^2 - r_1^2\right]^{1/2}} \tag{2.6-4}$$

If the crank turns at a constant rate in the counterclockwise direction, then the time taken for the slider to move to the left, from one limit position to the other is

$$\Delta t_1 = \frac{\Delta \theta_2}{\dot\theta_2} \tag{2.6-5}$$

and the time taken for the slider to move to the right between the limit positions is

$$\Delta t_2 = \frac{2\pi - \Delta\theta_2}{\dot\theta_2} \tag{2.6-6}$$

The time ratio of the mechanism is

$$\boxed{T_R = \frac{\Delta t_1}{\Delta t_2} = \frac{\Delta\theta_2}{2\pi - \Delta\theta_2}} \tag{2.6-7}$$

Having established the stroke and the elapsed times required to move the slider between the limit positions, it is possible to determine the *average* velocity of the slider moving in each direction. The average velocity of the slider to the left is

$$(v_{4,\text{avg}})_{\text{left}} = \frac{\text{stroke}}{\Delta t_1} = \frac{\text{stroke}}{\left(\dfrac{\Delta\theta_2}{\dot\theta_2}\right)} \tag{2.6-8}$$

and the average velocity of the slider to the right is

$$(v_{4,\text{avg}})_{\text{right}} = \frac{\text{stroke}}{\Delta t_2} = \frac{\text{stroke}}{\left(\dfrac{2\pi - \Delta\theta_2}{\dot\theta_2}\right)} \tag{2.6-9}$$

For the case where the offset is zero, the time ratio is unity, and the average velocities of the slider to the left and to the right are equal.

In Chapters 3 and 4 methods are presented whereby *instantaneous* values of the speed of the slider may be determined.

## 2.6.2  Crank Rocker Four-Bar Mechanism

Figure 2.14(a) shows a crank rocker four-bar mechanism. The crank is rotating in the counterclockwise direction. Figures 2.14(b) and 2.14(c) show simplified outlines of the mechanism in its limit geometries. Similar to the analysis of a slider crank mechanism, the counterclockwise angular swing of the crank required to move the rocker from the first to the second limit position indicated is

$$\Delta\theta_2 = 180° + \alpha_2 - \alpha_1 \tag{2.6-10}$$

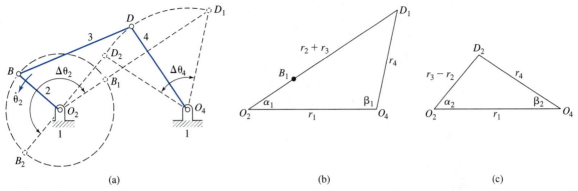

**Figure 2.14** Limit positions of a four-bar mechanism: (a) mechanism, (b) geometry of first limit position, (c) geometry of second limit position.

where expressions for $\alpha_1$ and $\alpha_2$ are determined using the cosine law (refer to Trigonometric Identities, page 494). From Figure 2.14(b)

$$r_4^2 = r_1^2 + (r_2 + r_3)^2 - 2(r_2 + r_3)r_1 \cos \alpha_1 \qquad (2.6\text{-}11)$$

or

$$\alpha_1 = \cos^{-1}\left[\frac{r_1^2 + (r_2 + r_3)^2 - r_4^2}{2(r_2 + r_3)r_1}\right] \qquad (2.6\text{-}12)$$

Similarly, the other angles shown in Figures 2.14(b) and 2.14(c) may be expressed in terms of the link lengths as

$$\alpha_2 = \cos^{-1}\left[\frac{r_1^2 + (r_3 - r_2)^2 - r_4^2}{2(r_3 - r_2)r_1}\right] \qquad (2.6\text{-}13)$$

$$\beta_1 = \cos^{-1}\left[\frac{r_1^2 - (r_2 + r_3)^2 + r_4^2}{2r_1 r_4}\right] \qquad (2.6\text{-}14)$$

$$\beta_2 = \cos^{-1}\left[\frac{r_1^2 - (r_3 - r_2)^2 + r_4^2}{2r_1 r_4}\right] \qquad (2.6\text{-}15)$$

The amplitude of motion of the rocker may be expressed as

$$\Delta\theta_4 = \beta_1 - \beta_2 \qquad (2.6\text{-}16)$$

Equations (2.6-5)–(2.6-7), developed for a slider crank mechanism, may also be applied to the present case. Similar to the analysis of a slider crank mechanism, the average rotational speeds of the rocker in the clockwise and counterclockwise directions are

$$(\dot{\theta}_{4,\text{avg}})_{\text{CW}} = \frac{\Delta\theta_4}{\left(\dfrac{2\pi - \Delta\theta_2}{\dot{\theta}_2}\right)}; \qquad (\dot{\theta}_{4,\text{avg}})_{\text{CCW}} = \frac{\Delta\theta_4}{\left(\dfrac{\Delta\theta_2}{\dot{\theta}_2}\right)} \qquad (2.6\text{-}17)$$

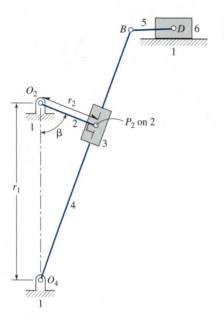

**Figure 2.15** Limit position of a quick-return mechanism.

### 2.6.3  Quick-Return Mechanism

Figure 2.15 shows a quick-return mechanism (see Appendix B, Section B.7) in one of its two limit positions. We will assume that link 2 is driven at a constant rate in the counterclockwise direction. From Figure 2.15, points $O_2$, $P_2$, and $O_4$ form a right triangle. Therefore

$$\beta = \cos^{-1}\left(\frac{r_2}{r_1}\right) \tag{2.6-18}$$

where

$$r_1 = r_{O_2 O_4}; \qquad r_2 = r_{O_2 P_2} \tag{2.6-19}$$

Due to symmetry about the vertical axis (see Appendix B, Figure B.14), the rotation of link 2 required to move link 6 to the right, from the second limit position to the first, is $2\beta$. The time required to execute this motion is

$$\Delta t_1 = \frac{2\beta}{\dot{\theta}_2} = \frac{2\cos^{-1}\left(\frac{r_2}{r_1}\right)}{\dot{\theta}_2} \tag{2.6-20}$$

and the time required to move link 6 to the left, from the first limit position to the second, is

$$\Delta t_2 = \frac{2\pi - 2\cos^{-1}\left(\frac{r_2}{r_1}\right)}{\dot{\theta}_2} \tag{2.6-21}$$

The time ratio of the motions of link 6 to the right and to the left is

$$T_R = \frac{\Delta t_1}{\Delta t_2} \tag{2.6-22}$$

# EXAMPLE 2.1  Average Velocities of a Slider Crank Mechanism

For the slider crank mechanism shown in Figure 2.13(a) with dimensions and crank angular speed

$$r_1 = 2.0\,\text{cm}; \qquad r_2 = 3.5\,\text{cm}; \qquad r_3 = 10.0\,\text{cm}$$

$$\dot\theta_2 = 20.0\,\text{rad/sec CCW}$$

determine the average speed of the slider to the right, and to the left.

### SOLUTION

From Equation (2.6-2)

$$\alpha_1 = \sin^{-1}\left(\frac{r_1}{r_2 + r_3}\right) = \sin^{-1}\left(\frac{2.0}{3.5 + 10.0}\right) = 8.52°$$

$$\alpha_2 = 17.92°$$

From Equation (2.6-1)

$$\Delta\theta_2 = 180° + \alpha_2 - \alpha_1 = 189.40° = 3.31\,\text{rad}$$

Employing Equation (2.6-4), the stroke of the slider is

$$\text{stroke} = \left[(r_2 + r_3)^2 - r_1^2\right]^{1/2} - \left[(r_3 - r_2)^2 - r_1^2\right]^{1/2} = 7.17\,\text{cm}$$

Employing Equations (2.6-8) and (2.6-9) yields

$$(v_{4,\text{avg}})_{\text{left}} = \frac{\text{stroke}}{\left(\dfrac{\Delta\theta_2}{\dot\theta_2}\right)} = \frac{7.17}{\left(\dfrac{3.31}{20.0}\right)} = 43.35\,\text{cm/sec}$$

$$(v_{4,\text{avg}})_{\text{right}} = 48.14\,\text{cm/sec}$$

## 2.7  TRANSMISSION ANGLE

In Section 1.7, expressions were provided for determining the lengths of links required in order to generate a particular type of mechanism. The limits of link lengths are theoretical. However, additional considerations must be given regarding the quality of the transmission of motion between the links for any practical design. One common consideration is the *transmission angle* of the mechanism.

The transmission angle, φ, of a four-bar mechanism and a slider crank mechanism is illustrated in Figures 2.16 and 2.17, respectively. The ideal value of transmission angle is 90°. In this instance, the line of action of the interactive force between links 3 and 4 matches the line of action of motion of the kinematic pair between the two links. This ideal value cannot be maintained in a moving mechanism. However, it is generally acceptable if transmission angles fall within the range

$$45° < \phi < 135° \tag{2.7-1}$$

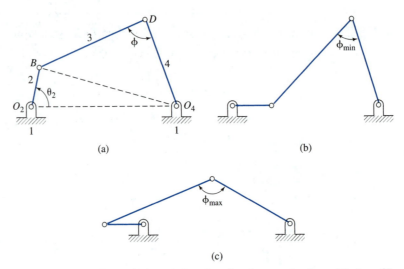

**Figure 2.16** (a) Transmission angle of a four-bar mechanism. (b) $\theta_2 = 0°$. (c) $\theta_2 = 180°$.

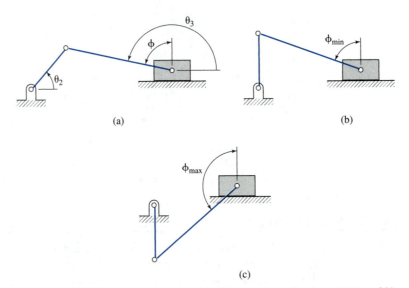

**Figure 2.17** (a) Transmission angle of a slider crank mechanism. (b) $\theta_2 = 90°$. (c) $\theta_2 = 270°$.

Values outside this range result in inefficient transmission of motion. Figure 2.18(a) shows the configuration of a four-bar mechanism having a small transmission angle, outside the acceptable range. Driving torque $M_{12}$ is applied to link 2. As shown in Figure 2.18(b), the direction of force transmitted from link 3 on link 4 results in a small torque $M_{14}$ about point $O_4$, but a high bearing force at the same point. For transmission angles close to zero or 180°, there is a tendency for a mechanism to bind.

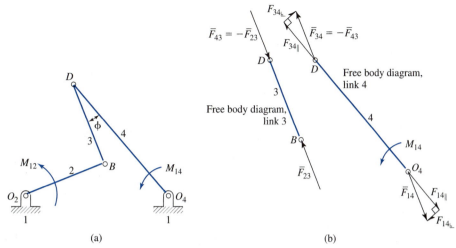

**Figure 2.18** Configuration of a four-bar mechanism with a small transmission angle.

Using Figure 2.16(a), we may determine an expression for the transmission angle of a four-bar mechanism. Applying the cosine law for the triangle formed by points $O_2 O_4 B$,

$$r_{BO_4}^2 = r_1^2 + r_2^2 - 2r_1 r_2 \cos \theta_2 \tag{2.7-2}$$

Using the cosine law again for the triangle formed by points $DO_4 B$,

$$r_{BO_4}^2 = r_3^2 + r_4^2 - 2r_3 r_4 \cos \phi \tag{2.7-3}$$

or

$$\phi = \cos^{-1} \left( \frac{r_3^2 + r_4^2 - r_{BO_4}^2}{2r_3 r_4} \right) \tag{2.7-4}$$

Thus, by combining Equations (2.7-2) and (2.7-4), we may obtain the transmission angle for a particular angular displacement of link 2.

For a four-bar mechanism, extremum values of the transmission angle occur when

$$\theta_2 = 0 \quad \text{and} \quad \theta_2 = 180° \tag{2.7-5}$$

Figures 2.16(b) and 2.16(c) show configurations of a four-bar mechanism where its transmission angle is minimum and maximum.

## EXAMPLE 2.2 Transmission Angles of a Four-Bar Mechanism

For a four-bar mechanism with link lengths

$$r_1 = 8.0 \, \text{cm}; \quad r_2 = 3.5 \, \text{cm}; \quad r_3 = 10.0 \, \text{cm}; \quad r_4 = 9.0 \, \text{cm}$$

determine the transmission angle, $\phi$, as a function of the angular displacement of link 2.

## SOLUTION

Using a typical angle of

$$\theta_2 = 30°$$

then from Equation (2.7-2),

$$r_{BO_4}^2 = r_1^2 + r_2^2 - 2r_1 r_2 \cos \theta_2$$

$$= 8.0^2 + 3.5^2 - 2 \times 8.0 \times 3.5 \times \cos 30° = 27.75 \, \text{cm}^2$$

Using Equation (2.7-4), the transmission angle is

$$\phi = \cos^{-1} \left( \frac{r_3^2 + r_4^2 - r_{BO_4}^2}{2r_3 r_4} \right)$$

$$= \cos^{-1} \left( \frac{10.0^2 + 9.0^2 - 27.75}{2 \times 10.0 \times 9.0} \right) = 31.64°$$

which is outside of the generally acceptable range of values (Equation (2.7-1)).

Based on calculations for other values of $\theta_2$, Figure 2.19 shows a plot of transmission angle as a function of the input angular displacement.

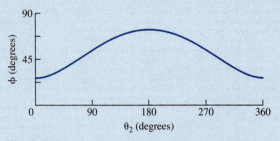

**Figure 2.19** Transmission angle of a four-bar mechanism.

Referring to Figure 2.17(a), an expression for the transmission angle for a slider crank mechanism is

$$\phi = \theta_3 - 90° \tag{2.7-6}$$

An analytical expression of $\theta_3$ will be developed in Chapter 4 in terms of the angular displacement of link 2. When this is combined with Equation (2.7-6) it is possible to determine the transmission angle.

Extremum values of the transmission angle for a slider crank mechanism occur when

$$\theta_2 = 90° \qquad \text{and} \qquad \theta_2 = 270° \tag{2.7-7}$$

Figures 2.17(b) and 2.17(c) show the configurations of a slider crank mechanism where the transmission angle is minimum and maximum.

## 2.8 INSTANTANEOUS CENTER OF VELOCITY

Consider two links, designated as $i$ and $j$, and at least one is undergoing planar motion. The *instantaneous center of velocity*, $P_{i,j}$, associated with these two links has the following properties:

- Two points, one on each of the two links, are coincident for at least an instant in time.
- Linear absolute velocities of the two points are identical. That is, about an instantaneous center of velocity, one link will at most have pure rotational motion relative to the other.

As an illustration, consider two links shown in Figure 2.20. Points $A_2$ and $A_3$ are on links 2 and 3, respectively. Both are coincident with $P_{2,3}$. At the instant shown

$$\bar{v}_{A_2} = \bar{v}_{A_3} = \bar{v}_{P_{2,3}} \tag{2.8-1}$$

Instantaneous centers of velocity will hereafter be referred to simply as *instantaneous centers*. They may be used to complete velocity analyses of planar mechanisms presented in Chapter 3.

Given the velocities of two points on a moving rigid body, it is possible to find the rigid body's instantaneous center with respect to the base link. Figure 2.21 is an illustration of link 2 undergoing planar motion. At the instant illustrated, the *directions* of velocities $\bar{v}_A$ and $\bar{v}_B$ on link 2 are defined. We now apply the condition of relative velocity between two points separated by a fixed distance presented in Section 2.3.1. From such an analysis, we can conclude that point $A$ moves with respect to a stationary point located somewhere on the line that extends through point $A$ and is perpendicular to $\bar{v}_A$. Likewise, point $B$ moves about a stationary point on a line that extends through $B$ and is perpendicular to $\bar{v}_B$. Therefore, the point of zero velocity on link 2, $P_{1,2}$, must be at the intersection of these two lines, where subscripts 1 and 2 indicate the base link and moving link, respectively.

Having located instantaneous center $P_{1,2}$, we can consider at that instant all points on link 2 are rotating about $P_{1,2}$. Using the analysis presented in Section 2.3.1, we have

$$|\bar{v}_A| = r_{P_{1,2}A}|\dot{\theta}_2|; \qquad |\bar{v}_B| = r_{P_{1,2}B}|\dot{\theta}_2| \tag{2.8-2}$$

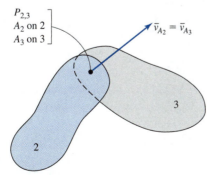

**Figure 2.20** Example of instantaneous center.

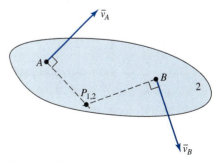

**Figure 2.21** Locating an instantaneous center.

or

$$|\dot{\theta}_2| = \frac{|\bar{v}_A|}{r_{P_{1,2}A}} = \frac{|\bar{v}_B|}{r_{P_{1,2}B}} \tag{2.8-3}$$

Thus, if only $\bar{v}_A$ and the line of action of $\bar{v}_B$ are known, then a scale drawing may be used to provide $r_{P_{1,2}A}$ and $r_{P_{1,2}B}$, which in turn can be used to determine the magnitude of $\bar{v}_B$ and $\dot{\theta}_2$.

For the above case, the instantaneous center lies within the physical boundaries of link 2. This is not always the case. As illustrated in Figure 2.22(a), it is possible to find an instantaneous center that is outside the physical boundaries of a link. Here, we can consider an imaginary extension of the physical boundaries of the link as shown in Figure 2.22(b).

When velocities of two points on a link are parallel, as shown in Figure 2.23(a), the instantaneous center is found by applying Equation (2.8-3). However, in this case,

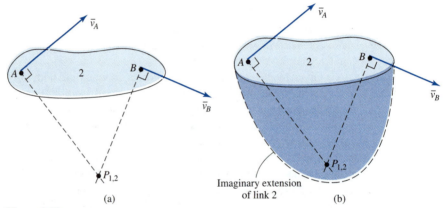

(a)

(b)

**Figure 2.22**

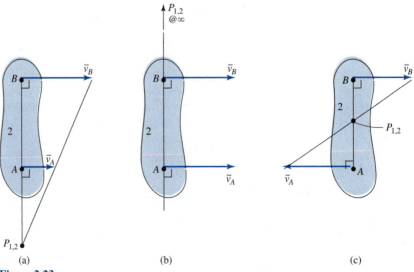

(a)                    (b)                    (c)

**Figure 2.23**

both velocities must be known. If $\bar{v}_A$ and $\bar{v}_B$ are equal in magnitude and direction (Figure 2.23(b)), then the link is undergoing pure translation, and the instantaneous center is at an infinite distance in a direction perpendicular to the motion. If $\bar{v}_A$ and $\bar{v}_B$ are parallel and opposite in direction (Figure 2.23(c)), then the instantaneous center lies between $A$ and $B$.

For two links connected through a kinematic pair, the location of the instantaneous center is summarized in Table 2.1.

**TABLE 2.1** Locations of Instantaneous Centers for Various Types of Kinematic Pairs

| Type of Kinematic Pair | Location of Instantaneous Center |
|---|---|
| Turning pair | The instantaneous center coincides with the turning pair. |
| Sliding pair (curved slide) | The slider is in curvilinear motion with respect to the slide. The instantaneous center is at the center of curvature of the slide. |
| Sliding pair (straight slide) | The slider is in rectilinear motion with respect to the slide. A straight slide is equivalent to a curved slide of infinite radius. Thus, the instantaneous center is at an infinite distance in a direction perpendicular to the sliding motion. |
| Rolling pair | The instantaneous center is at the point of contact, this being the only point where the two links have the same velocity. |
| Two degree of freedom pair <br> ($P_{i,j}$ on common normal)    ($P_{i,j}$ on common normal) | Relative motion takes place along the *common tangent* between the two links at the point of contact. Consequently, the instantaneous center lies on the *common normal*. Its position on the common normal depends on the ratio of the sliding and angular velocities. |

It is possible that each link of a mechanism can have relative motion with respect to every other link. Consequently, the total number of instantaneous centers, $p$, in a mechanism having $n$ links is

$$p = \frac{n(n-1)}{2} \tag{2.8-4}$$

The factor $\frac{1}{2}$ in Equation (2.8-4) takes into account the fact that instantaneous centers $P_{i,j}$ and $P_{j,i}$ are the same. In the four-bar mechanism illustrated in Figure 2.24(a), where $n = 4$, the number of instantaneous centers is

$$p = \frac{n(n-1)}{2} = \frac{4(4-1)}{2} = 6 \tag{2.8-5}$$

The complete set of instantaneous centers is: $P_{1,2}$, $P_{1,3}$, $P_{1,4}$, $P_{2,3}$, $P_{2,4}$, and $P_{3,4}$. Some of the instantaneous centers are found by examining the kinematic pairs and using Table 2.1. These are $P_{1,2}$, $P_{2,3}$, $P_{3,4}$, and $P_{1,4}$, which are shown in Figure 2.24(a).

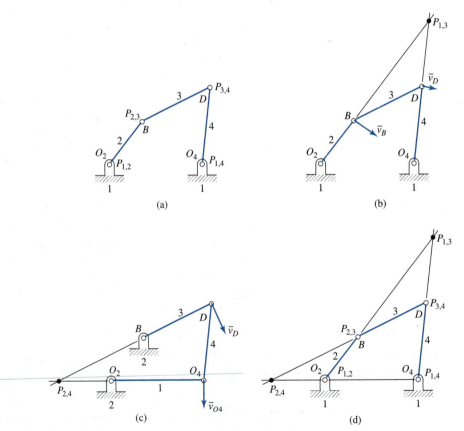

**Figure 2.24** Instantaneous centers of a four-bar mechanism.

The two other instantaneous centers are $P_{1,3}$ and $P_{2,4}$. Instantaneous center $P_{1,3}$ is found by an analysis similar to that illustrated in Figure 2.21. Lines are drawn perpendicular to the velocities at points $B$ and $D$. The intersection of these two lines is the instantaneous center $P_{1,3}$ (Figure 2.24(b)).

The remaining instantaneous center is $P_{2,4}$. Its location may be found by inverting the mechanism and making link 2 the fixed link, as shown in Figure 2.24(c). Using the same reasoning as for the mechanism shown in Figure 2.24(b), the instantaneous center is at the intersection of the two lines that are perpendicular to the velocities at $D$ and $O_4$. Since relative motions for the mechanisms shown in Figures 2.24(b) and 2.24(c) are the same (see Section 1.6), then all six instantaneous centers that have been located apply to both the original and the inverted mechanisms. The complete set of instantaneous centers is shown in Figure 2.24(d).

## 2.9 KENNEDY'S THEOREM

From the analysis of the four-bar mechanism shown in Figure 2.24, the three instantaneous centers associated with any three of the links are collinear. That is, each of the following sets of instantaneous centers lie on the same straight line:

- $P_{1,2}$, $P_{2,3}$, $P_{1,3}$
- $P_{1,4}$, $P_{3,4}$, $P_{1,3}$
- $P_{2,4}$, $P_{1,2}$, $P_{1,4}$
- $P_{2,4}$, $P_{2,3}$, $P_{3,4}$

This result conforms to *Kennedy's Theorem*, which may be applied to any planar mechanism. The theorem states that when considering three links, designated as $i$, $j$, and $k$, undergoing motion relative to one another, the three associated instantaneous centers, $P_{i,j}$, $P_{i,k}$, and $P_{j,k}$, lie on the same straight line [3].

In determining the locations of instantaneous centers of a mechanism, some can be readily located using the guidelines provided in Table 2.1. The remainder may then be found by successive applications of Kennedy's Theorem.

One may identify all instantaneous centers by drawing an *auxiliary polygon* that relates to the mechanism under study. Corresponding to the mechanism and instantaneous centers shown in Figure 2.25(a), an auxiliary polygon is illustrated in Figure 2.25(b). The number of sides and vertices in the polygon must equal the number of links in the mechanism. Each vertex of the polygon represents a link and is accordingly assigned a number.

In an auxiliary polygon, every side and diagonal is associated with two links. Each side and diagonal corresponds to an instantaneous center. Also, an auxiliary polygon illustrates the requirements of Kennedy's Theorem. Each triangle formed by sides and diagonals is associated with three links. The three corresponding instantaneous centers must lie on the same straight line. Figure 2.25(c) shows four auxiliary polygons, each with a shaded triangle. Beside each is listed the corresponding three instantaneous centers lying on a straight line.

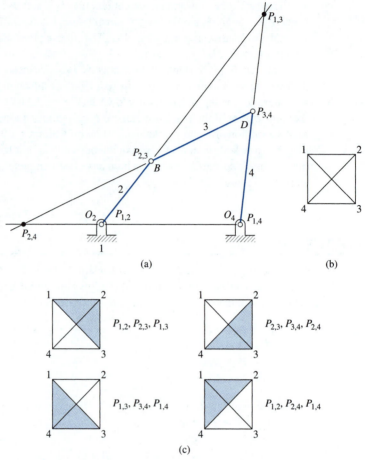

Figure 2.25 Instantaneous centers of a four-bar mechanism.

## EXAMPLE 2.3   Instantaneous Centers of a Slider Crank Mechanism

For the slider crank mechanism shown in Figure 2.26(a), determine all of the instantaneous centers for the position shown.

### SOLUTION

The mechanism has four links ($n = 4$) and, using Equation (2.8-4), six instantaneous centers ($p = 6$). They are: $P_{1,2}$, $P_{1,3}$, $P_{1,4}$, $P_{2,3}$, $P_{2,4}$, and $P_{3,4}$.

The following instantaneous centers are located by examining the kinematic pairs (Table 2.1): $P_{1,2}$, $P_{2,3}$, $P_{3,4}$, and $P_{1,4}$.

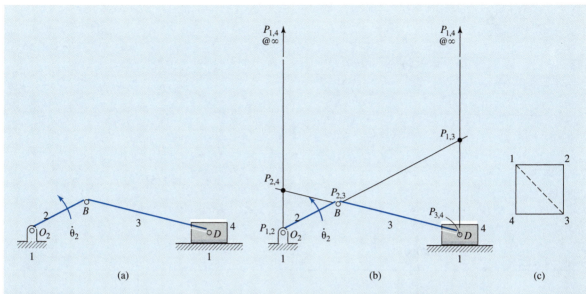

**Figure 2.26**

To locate $P_{1,3}$, we employ the auxiliary polygon. As shown in Figure 2.26(c), the diagonal between vertex 1 and vertex 3 is part of triangles 1-2-3 and 1-3-4. This indicates that $P_{1,3}$ lies on the same straight lines as $P_{1,2}P_{2,3}$ and $P_{1,4}P_{3,4}$. It must therefore lie at the intersection of these two lines. Instantaneous center $P_{2,4}$ is found in a similar manner. All six instantaneous centers are shown in Figure 2.26(b).

## EXAMPLE 2.4 Instantaneous Centers of an Inverted Slider Crank Mechanism

For the inverted slider crank mechanism shown in Figure 2.27(a), determine all of the instantaneous centers for the position shown.

### SOLUTION

This mechanism has four links and six instantaneous centers. The following instantaneous centers are determined by examining the kinematic pairs: $P_{1,2}, P_{1,3}, P_{2,4}, P_{3,4}$. Then, by using an auxiliary polygon (i.e., Kennedy's Theorem), we locate

- $P_{1,4}$ at the intersection of lines $P_{1,2}P_{2,4}$ and $P_{1,3}P_{3,4}$
- $P_{2,3}$ at the intersection of lines $P_{1,2}P_{1,3}$ and $P_{2,4}P_{3,4}$

All of the instantaneous centers are illustrated in Figure 2.27(b).

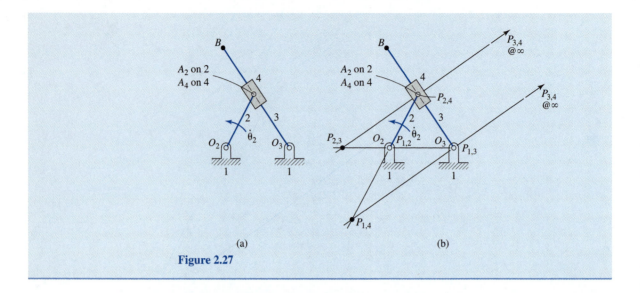

**Figure 2.27**

## EXAMPLE 2.5 Instantaneous Centers of a Mechanism Incorporating a Two Degree of Freedom Pair

For the mechanism shown in Figure 2.28(a), determine all of the instantaneous centers.

### SOLUTION

This mechanism has three links. Using Equation (2.8-4), there are only three instantaneous centers. Instantaneous centers $P_{1,2}$ and $P_{1,3}$ coincide with the turning pairs. Based on Kennedy's Theorem, the third instantaneous center, $P_{2,3}$, lies on the same straight line as $P_{1,2}P_{1,3}$. In addition, from Table 2.1, this instantaneous center lies on the common normal at the point of contact between links 2 and 3. Thus, instantaneous center $P_{2,3}$ is located as shown in Figure 2.28(b).

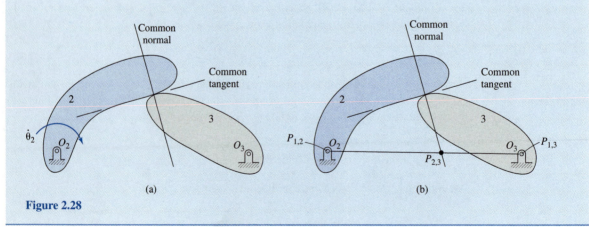

**Figure 2.28**

## EXAMPLE 2.6  Instantaneous Centers of a Six-Link Mechanism

For the mechanism shown in Figure 2.29(a), determine all of the instantaneous centers.

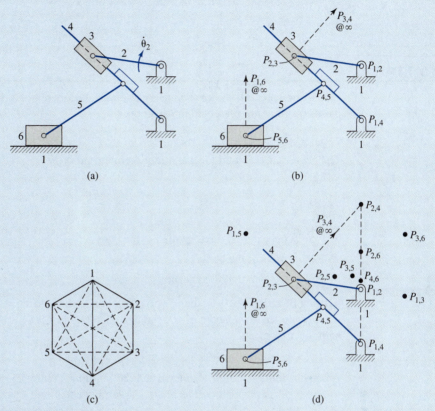

**Figure 2.29**

### SOLUTION

This mechanism has six links and, by employing Equation (2.8-4), 15 instantaneous centers. The following instantaneous centers are determined by examining the kinematic pairs: $P_{1,2}$, $P_{1,4}$, $P_{1,6}$, $P_{2,3}$, $P_{3,4}$, $P_{4,5}$, $P_{5,6}$. These are indicated in Figure 2.29(b). The corresponding auxiliary polygon is shown in Figure 2.29(c). Instantaneous centers that have been determined by examining the kinematic pairs are depicted in the auxiliary polygon as solid lines, while the remaining instantaneous centers are shown as dashed lines.

The location of all of the remaining instantaneous centers are determined by recognizing

- $P_{2,4}$ at the intersection of lines $P_{1,2}$, $P_{1,4}$ and $P_{2,3}P_{3,4}$ (see Figure 2.29(d))
- $P_{4,6}$ at the intersection of lines $P_{1,4}P_{1,6}$ and $P_{4,5}P_{5,6}$
- $P_{1,3}$ at the intersection of lines $P_{1,2}P_{2,3}$ and $P_{1,4}P_{3,4}$
- $P_{1,5}$ at the intersection of lines $P_{1,4}P_{4,5}$ and $P_{1,6}P_{5,6}$
- $P_{3,5}$ at the intersection of lines $P_{1,3}P_{1,5}$ and $P_{3,4}P_{4,5}$

- $P_{2,5}$ at the intersection of lines $P_{1,2}P_{1,5}$ and $P_{2,4}P_{4,5}$
- $P_{2,6}$ at the intersection of lines $P_{1,2}P_{1,6}$ and $P_{2,5}P_{5,6}$
- $P_{3,6}$ at the intersection of lines $P_{2,3}P_{2,6}$ and $P_{3,5}P_{5,6}$

Figure 2.29(d) shows all of the instantaneous centers.

## 2.10 KINETICS

This section presents the governing equations of the kinetics of a rigid body. Derivations of these equations are presented in Appendix D.

### 2.10.1 Linear Motion

The governing equation for linear motion of a rigid body is

$$\boxed{\overline{F} = m\overline{a}_G}$$ (2.10-1)

where

    $\overline{F}$ = net external force applied to the body

    $m$ = mass of the body

    $\overline{a}_G$ = acceleration of the center of mass

In using Equation (2.10-1), there is no restriction as to the location of the point at which the force is applied to the body. Also, as indicated in Figure 2.30, the body may be subjected to additional moments.

### 2.10.2 Angular Motion

A body of finite size may also have angular motions. Figure 2.31 shows a rigid body with its center of mass, point $G$. The related governing equation for angular motion is

$$\boxed{M_O = I_G\ddot{\theta} + m(x_G\ddot{y}_G - \ddot{x}_Gy_G)}$$ (2.10-2)

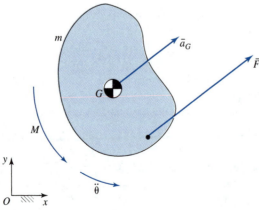

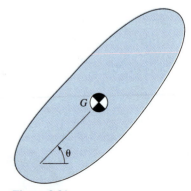

**Figure 2.30** Body subjected to externally applied load and moment.

**Figure 2.31**

where

$M_O$ = net external moment applied to the body measured about point $O$

$\ddot{\theta}$ = angular acceleration of the body

$x_G$, $y_G$ = coordinates of the center of mass with respect to a reference coordinate system

$\ddot{x}_G$, $\ddot{y}_G$ = components of acceleration of the center of mass

$I_G$ = *polar mass moment of inertia* of the body with respect to its center of mass (see Appendix D, Section D.9)

In this book, both the net moment applied to the body and the angular acceleration of the body are considered positive in the counterclockwise direction.

If the origin of the coordinate system is placed at the center of mass, then

$$x_G = 0 = y_G \tag{2.10-3}$$

and Equation (2.10-2) becomes

$$M_G = M_O = I_G\ddot{\theta} \tag{2.10-4}$$

Commonly employed sets of units associated with Equations (2.10-1) and (2.10-2) are presented in Section 2.11.

### 2.10.3 Fixed-Axis Rotation of a Rigid Body

The rigid body shown in Figure 2.32 is allowed to move about a fixed axis at point $O$. For this type of motion, it is convenient to place the origin of the coordinate system at point $O$. In this case

$$x_G = r \cos\theta; \qquad y_G = r \sin\theta \tag{2.10-5}$$

where $r$ is the radial distance from point $O$ to the center of gravity $G$.

Substituting Equations (2.10-5) in Equation (2.10-2) yields

$$
\begin{aligned}
M_O &= I_G\ddot{\theta} + m(x_G\ddot{y}_G - \ddot{x}_G y_G) \\
&= I_G\ddot{\theta} + mr\cos\theta(-r\dot{\theta}^2\sin\theta + r\ddot{\theta}\cos\theta) - mr\sin\theta(-r\dot{\theta}^2\cos\theta - r\ddot{\theta}\sin\theta) \\
&= I_O\ddot{\theta}
\end{aligned}
$$

$$\tag{2.10-6}$$

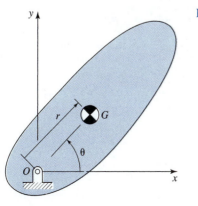

**Figure 2.32**

where after simplification,

$$I_O = I_G + mr^2$$

(2.10-7)

is the polar mass moment of inertia about point $O$. Equation (2.10-7) is referred to as the *parallel-axis theorem*.

## 2.11  SYSTEMS OF UNITS

There are multiple systems of units employed in engineering. They are generally classified into two groups: *International System* (*SI* or *metric*) and *U.S. Customary*. Consistent sets of units for the governing equations of motion of a rigid body are presented in this section.

To accommodate these different sets, we make the following substitutions for the mass and polar mass moment of inertia:

$$m = \frac{m^*}{g_c}; \qquad I_G = \frac{I_G^*}{g_c}$$

(2.11-1)

where quantities $m^*$ and $I_G^*$, like $m$ and $I_G$, are referred to as the mass and polar mass moment of inertia, respectively, and $g_c$ is a *gravitational constant*. The value and units of $g_c$ depend on the units of the other components in the governing equations. Substituting Equations (2.11-1) into Equations (2.10-1) and (2.10-2) yields

$$\overline{F} = \frac{m^*}{g_c}\overline{a}_G$$

(2.11-2)

and

$$M_O = \frac{I_G^*}{g_c}\ddot{\theta} + \frac{m^*}{g_c}(x_G\ddot{y}_G - \ddot{x}_G y_G)$$

(2.11-3)

Tables 2.2 and 2.3 provide units for each of the components of Equations (2.11-2) and (2.11-3). For each table, a consistent set of units is obtained by employing one column in a table.

For the cases in Tables 2.2 and 2.3 for which $g_c = 1$, then from Equations (2.11-1),

$$m = m^*; \qquad I_G = I_G^*$$

(2.11-4)

Employing Equation (2.11-2) and Table 2.2, forces for the SI system may be defined as

$$1\,\text{Newton} = 1\,\text{N} = 1\,\text{kg} \times 1\,\text{m/sec}^2$$

and

$$1\,\text{dyne} = 1\,\text{gr} \times 1\,\text{cm/sec}^2$$

**TABLE 2.2** Consistent Sets of Units Employed from the SI System for Equations (2.11-2) and (2.11-3)

|                                   | 1          | 2          |
|-----------------------------------|------------|------------|
| $\bar{F}$                         | dyne       | N          |
| $M_O$                             | dyne-cm    | N-m        |
| $m^*$                             | gr         | kg         |
| $I_G^*$                           | gr-cm$^2$  | kg-cm$^2$  |
| $x, y$                            | cm         | m          |
| $\bar{a}_G, \ddot{x}, \ddot{y}$   | cm/sec$^2$ | m/sec$^2$  |
| $\ddot{\theta}$                   | rad/sec$^2$| rad/sec$^2$|
| $g_c$                             | 1          | 1          |

**TABLE 2.3** Consistent Sets of Units Employed from the U.S. Customary System for Equations (2.11-2) and (2.11-3)

|                                  | 1                        | 2                       | 3                         | 4                       |
|----------------------------------|--------------------------|-------------------------|---------------------------|-------------------------|
| $\bar{F}$                        | lb                       | lb                      | lb                        | lb                      |
| $M_O$                            | ft-lb                    | ft-lb                   | in-lb                     | in-lb                   |
| $m^*$                            | lbm                      | lb-sec$^2$/ft (slug)    | lbm                       | lb-sec$^2$/in           |
| $I_G^*$                          | lbm-ft$^2$               | lb-ft-sec$^2$           | lbm-in$^2$                | lb-in-sec$^2$           |
| $x, y$                           | ft                       | ft                      | in                        | in                      |
| $\bar{a}_G, \ddot{x}, \ddot{y}$  | ft/sec$^2$               | ft/sec$^2$              | in/sec$^2$                | in/sec$^2$              |
| $\ddot{\theta}$                  | rad/sec$^2$              | rad/sec$^2$             | rad/sec$^2$               | rad/sec$^2$             |
| $g_c$                            | 32.2 lbm-ft/lb-sec$^2$   | 1                       | 386 lbm-in/lb-sec$^2$     | 1                       |

A simple illustration on the use of Tables 2.2 and 2.3 is provided below:

## EXAMPLE 2.7 Body Subjected to an External Force

For the body shown in Figure 2.33, determine the magnitude of acceleration of the center of mass if

$$|\bar{F}| = 12\,\text{lb}; \qquad m^* = 2\,\text{lbm}$$

### SOLUTION

Referring to the units of the given quantities, we employ a set of units listed in Table 2.3. We may use either the column numbered 1 or 3. If we use column 3, then

$$g_c = 386\,\text{lbm-in/lb-sec}^2$$

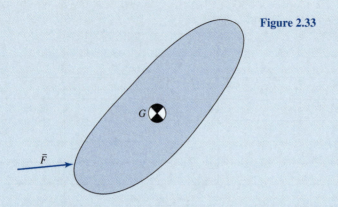

**Figure 2.33**

and from Equation (2.11-2)

$$|\bar{a}_G| = \frac{|\bar{F}|}{\left(\dfrac{m^*}{g_c}\right)} = \frac{12\,\text{lb}}{\left[\dfrac{2\,\text{lbm}}{386\,\text{lbm-in/lb-sec}^2}\right]} = 2{,}320\,\text{in/sec}^2$$

We generally consider the *weight* of a body as the gravitational force exerted on it by the earth at its surface. Weight, being a force, is a vector quantity. The direction of this vector is toward the center of the earth.

When a body is allowed to fall freely under the influence of gravity, the magnitude of acceleration is

$$g = |\bar{a}_G| = 9.81\,\text{m/sec}^2 = 32.2\,\text{ft/sec}^2 = 386\,\text{in/sec}^2$$

The magnitude of this acceleration is independent of the mass of the body. Using Equations (2.11-1) and (2.11-2), the magnitude of the weight, $W$, of a body of mass $m$ is

$$W = mg \tag{2.11-5}$$

A mass of $m^* = 1$ lbm weighs

$$W = mg = \frac{m^*}{g_c}g = \frac{1\,\text{lbm}}{(32.2\,\text{lbm-ft/lb-sec}^2)}32.2\,\text{ft/sec}^2 = 1\,\text{lb}$$

and a mass of $m^* = 1$ kg weighs

$$W = \frac{m^*}{g_c}g = \frac{1\,\text{kg}}{1} \times 9.81\,\text{m/sec}^2 = 9.81\,\text{kg-m/sec}^2 = 9.81\,\text{N}$$

To keep an object from free-falling, an upward force must be applied of the same magnitude as the weight so that there is zero net force on the object.

In some instances, gravity may be neglected in an analysis, such as when the prescribed accelerated motions of a body require forces that are far greater than those caused by gravity. Alternatively, gravity may be neglected if it acts out of the plane of motion of the body.

# EXAMPLE 2.8 Physical Pendulum

The rigid body illustrated in Figure 2.34 is pivoted about point $O$ at a distance $l$ from the center of mass $G$. Gravity acts downward. The equilibrium position corresponds to points $O$ and $G$ in a vertical alignment. If the body is pivoted a small angle $\theta$ from the equilibrium position and released, show that the motion is harmonic and determine the period of one oscillation.

### SOLUTION

Employing Equation (2.10-6),

$$M_O = -mgl\sin\theta = I_O\ddot{\theta} \qquad (2.11\text{-}6)$$

Since for small angles,

$$\sin\theta \approx \theta \qquad (2.11\text{-}7)$$

then Equation (2.11-6) becomes

$$\ddot{\theta} + \omega_n^2\theta = 0 \qquad (2.11\text{-}8)$$

where the *natural frequency* of the system is defined as

$$\omega_n = \left(\frac{mgl}{I_O}\right)^{1/2} \qquad (2.11\text{-}9)$$

The solution of Equation (2.11-8) is [4]

$$\theta = \Theta\cos(\omega_n t + \phi) \qquad (2.11\text{-}10)$$

where $\Theta$ and $\phi$ are arbitrary constants. Angular motion is thus harmonic in time.

**Figure 2.34** Physical pendulum.

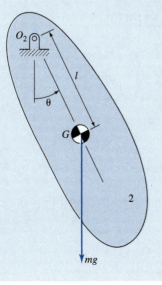

Using Equations (2.11-9) and (2.11-10), the period of time, $\tau$, for one oscillation is found by recognizing that

$$\omega_n \tau = \left(\frac{mgl}{I_O}\right)^{1/2} \quad \tau = 2\pi \tag{2.11-11}$$

from which

$$\tau = 2\pi\left(\frac{I_O}{mgl}\right)^{1/2} \tag{2.11-12}$$

## 2.12 EQUATIONS OF EQUILIBRIUM

### 2.12.1 Static Conditions

For static conditions, where there is no motion,

$$\bar{a}_G = \bar{0}; \qquad \ddot{\theta} = 0 \tag{2.12-1}$$

Then Equations (2.10-1) and (2.10-4) reduce to

$$\boxed{\bar{F} = \bar{0}} \tag{2.12-2}$$

$$\boxed{M_O = 0 = M_G} \tag{2.12-3}$$

Equations (2.12-2) and (2.12-3) are the governing equations of static equilibrium of a body.

## EXAMPLE 2.9  Static Equilibrium of a Pivoting Link

The brake lever shown in Figure 2.35(a) consists of the link pivoted to the base link at point $O_2$. A force of 100 N is applied by the driver's foot, causing tension in the cable. Assuming conditions of static equilibrium, determine

(a) the magnitude of the tension in the cable
(b) the reaction at $O_2$

### SOLUTION

(a) A free body diagram of the link is shown in Figure 2.35(b). Summing moments about $O_2$ yields

$$\sum M_{O_2} = -100 \times 0.20 + |\bar{T}| \times 0.15 = 0$$

from which

$$|\bar{T}| = 133\,\text{N}$$

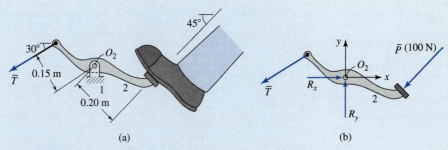

Figure 2.35

(b) Referring to the free body diagram shown in Figure 2.35(b), we require that

$$\sum F_x = 0 = P_x + T_x + R_x = -100 \cos 45° - 133 \cos 30° + R_x$$

$$R_x = 186 \, \text{N}$$

Also

$$\sum F_y = 0 = P_y + T_y + R_y = -100 \sin 45° - 133 \sin 30° + R_y$$

$$R_y = 137 \, \text{N}$$

and therefore

$$\overline{R} = R_x \overline{i} + R_y \overline{j} = 186\overline{i} + 137\overline{j} \, \text{N}$$

## EXAMPLE 2.10

Replace the horizontal 300 N force acting on the lever shown in Figure 2.36(a) by an equivalent system consisting of a force at $O$ and a couple.

### SOLUTION

We apply two equal and opposite 300 N forces at $O$ and identify the counterclockwise couple.

$$M = Fd = 300 \, \text{N} \times 0.15 \, \text{m} \times \sin 60° = 39.0 \, \text{N-m}$$

Thus, the original force is equivalent to the 300 N force at $O$ and the 39.0 N-m couple as shown in the third of the three equivalent figures shown in Figure 2.36(b).

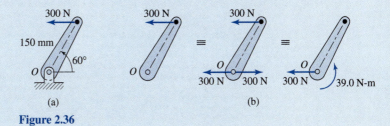

Figure 2.36

## 2.12.2  Dynamic Conditions

Equations (2.10-1) and (2.10-2) can alternatively be expressed as

$$\boxed{\overline{F} + \overline{F}_{IN} = \overline{0}}$$

(2.12-4)

and

$$\boxed{M_O + M_{IN} = 0}$$

(2.12-5)

where using Equations (2.11-1), the *inertia force* is defined as

$$\overline{F}_{IN} = -m\overline{a}_G = -\frac{m^*}{g_c}\overline{a}_G$$

(2.12-6)

and the *inertia moment* or *torque* is

$$M_{IN} = -I_G\ddot{\theta} - m(x_G\ddot{y}_G - \ddot{x}_G y_G)$$
$$= -\frac{I_G^*}{g_c}\ddot{\theta} - \frac{m^*}{g_c}(x_G\ddot{y}_G - \ddot{x}_G y_G)$$

(2.12-7)

Equation (2.12-7) may be simplified if the origin of the coordinate system is placed at the center of mass, and therefore

$$x_G = 0 = y_G$$

(2.12-8)

Then

$$M_{IN} = -I_G\ddot{\theta} = -\frac{I_G^*}{g_c}\ddot{\theta}$$

(2.12-9)

Equations (2.12-4) and (2.12-5) are referred to as *D'Alembert's Principle.* They resemble those employed for static problems.

## EXAMPLE 2.11

The solid disc shown in Figure 2.37(a) is released from rest on an inclined surface. Assuming that the disc rolls without slipping, and that gravity acts downward, use D'Alembert's Principle to determine its angular acceleration.

$$m^* = 2.0\,\text{kg}; \qquad R = 0.3\,\text{m}; \qquad \beta = 20°$$

### SOLUTION

Employing Table 2.2, it is recognized from the units of quantities provided for this problem that

$$g_c = 1; \qquad m = m^*; \qquad I_G = I_G^*$$

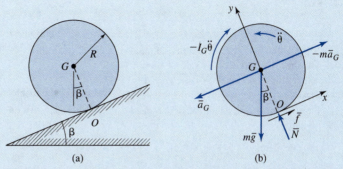

**Figure 2.37** (a) Solid disc on an inclined surface. (b) Free body diagram.

The polar mass moment of inertia of the disc (Appendix D, Section D.9) about its center of mass is

$$I_G^* = \tfrac{1}{2}m^*R^2 = \tfrac{1}{2} \times 2.0\,\text{kg} \times 0.3^2\,\text{m}^2 = 0.090\,\text{kg-m}^2$$

A free body diagram of the cylinder shown in Figure 2.37(b) includes a frictional force, a normal force between the cylinder and incline, an inertia force, and an inertia moment. A Cartesian coordinate system is placed with its origin at the point of contact between the cylinder and the inclined surface. This placement eliminates the need to account for normal and frictional forces when summing moments. Also

$$x_G = 0; \qquad y_G = R$$

Also, by inspection for this problem

$$|\bar{a}_G| = |\ddot{x}_G| = R\ddot{\theta}$$

and

$$M_O = mgR\sin\beta \tag{2.12-10}$$

From Equation (2.12-7) we have

$$\begin{aligned} M_{IN} &= -I_G\ddot{\theta} - m(x_G\ddot{y}_G - \ddot{x}_G y_G) \\ &= -I_G\ddot{\theta} - m(0 - R\ddot{\theta}R) = -(I_G + mR^2)\ddot{\theta} \end{aligned} \tag{2.12-11}$$

Combining Equations (2.12-5), (2.12-10), and (2.12-11) yields

$$mgR\sin\beta - (I_G + mR^2)\ddot{\theta} = 0$$

from which

$$\ddot{\theta} = \frac{mgR\sin\beta}{I_G + mR^2} = \frac{2.0\,\text{kg} \times 9.81\,\text{m/sec}^2 \times 0.3\,\text{m} \times \sin 20°}{0.090\,\text{kg-m}^2 + 2.0\,\text{kg} \times 0.3^2\,\text{m}^2} = 7.46\,\text{rad/sec}^2\,\text{CCW}$$

# PROBLEMS

**P2.1**  For the mechanism shown in Figure P2.1

(a) specify the magnitude of the maximum rotational velocity of link 3 throughout a cycle of motion and the corresponding value(s) of $\theta_2$ when this takes place

(b) specify the value(s) of $\theta_2$ when the rotational speed of link 3 is zero  *Vₐ and V_b have same dir'n.*

$$r_{O_2A} = 3.0 \text{ cm}; \quad r_{AB} = 6.0 \text{ cm}$$

$$\dot{\theta}_2 = 100 \text{ rpm CW}$$

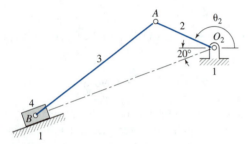

**Figure P2.1**

**P2.2**  For the mechanism shown in Figure P2.2

(a) specify the magnitude of the maximum linear velocity of link 4 throughout a cycle of motion and the corresponding angular position(s) of link 2 where this takes place

(b) specify the magnitude of the maximum sliding velocity of link 3 with respect to link 4 throughout a cycle of motion and the corresponding angular position(s) of link 2 where this takes place

(c) specify the magnitude of the maximum linear acceleration of link 4 throughout a cycle of

motion and the corresponding angular position(s) of link 2 where this takes place

$$r_{O_2A_2} = 5.0 \text{ cm}; \quad \dot{\theta}_2 = 100 \text{ rpm CW}$$

**P2.3**  For the mechanism shown in Figure P2.3

(a) specify the value(s) of $\theta_2$ when the linear speed of link 6 is zero

(b) specify the magnitude of the maximum upward linear speed of link 6 throughout a cycle of motion and the corresponding value(s) of $\theta_2$ when this takes place  *V_b and Vc have same dir'n*

$$r_{O_2O_3} = 5.0 \text{ cm}; \quad r_{O_2A_2} = 3.0 \text{ cm}$$

$$r_{O_3B} = 10.0 \text{ cm}; \quad r_{BC} = 4.0 \text{ cm}$$

$$\dot{\theta}_2 = 10 \text{ rad/sec CCW}$$

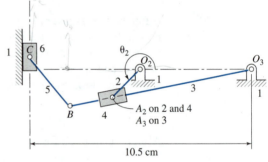

**Figure P2.3**

**P2.4**  Figure P2.4 shows a crank rocker mechanism. Determine

(a) the average angular speed of the rocker in the clockwise direction

(b) the average angular speed of the rocker in the counterclockwise direction

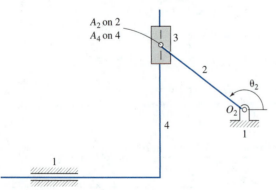

**Figure P2.2**

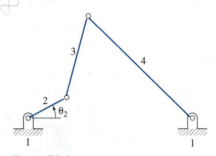

**Figure P2.4**

(c) the time ratio $\dfrac{\Delta\theta_2}{2\pi-\Delta\theta_2}$

$r_1 = 8.0\,\text{cm};\qquad r_2 = 2.0\,\text{cm}$

$r_3 = 4.0\,\text{cm};\qquad r_4 = 7.0\,\text{cm}$

$\dot{\theta}_2 = 40.0\,\text{rad/sec CCW}$

**P2.5** Figure P2.5 shows a slider crank mechanism. Determine

(a) the average speed of the slider to the right
(b) the average speed of the slider to the left
(c) the time ratio

$r_1 = 2.0\,\text{cm};\qquad r_2 = 5.0\,\text{cm};\qquad r_3 = 8.0\,\text{cm}$

$\dot{\theta}_2 = 30.0\,\text{rad/sec CCW}$

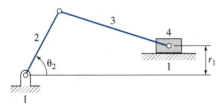

**Figure P2.5**

**P2.6** For the mechanism shown in Figure P2.6

(a) specify the value(s) of $\theta_2$ when the rotational speed of link 3 is zero
(b) specify the value(s) of $\theta_2$ when the linear speed of point $D$ is zero
(c) specify the value(s) of $\theta_2$ when the Coriolis acceleration of $B_1$ with respect to $B_3$ is zero

$r_{O_2A} = 3.0\,\text{cm};\qquad r_{O_2B_1} = 5.0\,\text{cm}$

$r_{AC} = 8.0\,\text{cm};\qquad r_{CD} = 4.0\,\text{cm}$

$\dot{\theta}_2 = 10\,\text{rad/sec CCW (constant)}$

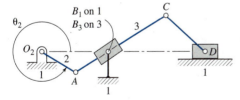

**Figure P2.6**

**P2.7** Figure P2.7 shows an intermittent-motion mechanism known as the Geneva mechanism. The circular peg $P$ is attached to link 2 and slides in the slots of link 3. Specify the magnitude and direction of the maximum rotational speed of link 3 throughout a cycle and the corresponding value of $\theta_2$ when this takes place.

$r_{O_2P} = 4.0\sqrt{2}\,\text{cm};\qquad r_{O_2O_3} = 8.0\,\text{cm}$

$\dot{\theta}_2 = 30\,\text{rad/sec CCW (constant)}$

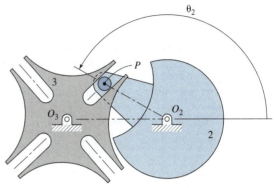

**Figure P2.7**

**P2.8** Prove that the extremum values of the transmission angle of a four-bar mechanism occur when link 2 is parallel to the base link.

**P2.9** Prove that the extremum values of the transmission angle of a slider mechanism occur when the crank is perpendicular to the line of action of the slider. Implement Equation (4.3-77).

**P2.10** Determine

(a) the range of values of the length of link 2 so that the transmission angle remains in the range $45° \le \phi \le 135°$
(b) the type of four-bar mechanism obtained in part (a)

$r_1 = 1.0\,\text{cm};\qquad r_3 = 2.5\,\text{cm};\qquad r_4 = 2.0\,\text{cm}$

**P2.11** Determine (see Figure 1.9)

(a) the range of values of the length of link 2 so that the transmission angle remains in the range $45° \le \phi \le 135°$
(b) the type of four-bar mechanism obtained in part (a)

$r_1 = 1.0\,\text{cm};\qquad r_2 = 3.0\,\text{cm};\qquad r_3 = 2.5\,\text{cm}$

**P2.12** Using the prescribed link dimensions of a slider crank mechanism, determine the range of values of the length of link 2 so that link 2 can make a full rotation, and the transmission angle remains in the range $45° \le \phi \le 135°$. Implement Equation (4.3-77).

$r_1 = 1.0\,\text{cm};\qquad r_3 = 2.5\,\text{cm}$

**P2.13** Using the prescribed link dimensions of a slider crank mechanism, determine the range of values of the length of link 3 so that link 2 can make a full rotation, and the transmission angle remains in the range $45° \le \phi \le 135°$. Implement Equation (4.3-77).

$$r_1 = 1.0\,\text{cm}; \qquad r_2 = 2.5\,\text{cm}$$

**P2.14** For the mechanism in the position shown in Figure P2.14

(a) specify the number of links, the types of kinematic pairs, and the number of each type, and calculate the mobility

(b) determine the positions of all instantaneous centers

$$r_{O_2 B_2} = 4.0\,\text{cm}$$

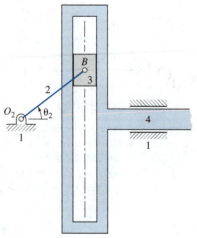

**Figure P2.14**

**P2.15** Locate all instantaneous centers of the mechanism shown in Figure P2.15.

$$\theta_2 = 45°; \qquad r_{O_2 B} = 2.5\,\text{cm}$$

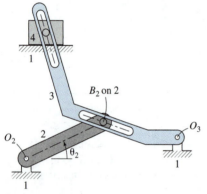

**Figure P2.15**

**P2.16** Locate all instantaneous centers of the mechanism shown in Figure P2.16.

$$r_{O_2 B_2} = 2.0\,\text{in}; \qquad r_{O_2 O_4} = 1.5\,\text{in}$$
$$r_{O_4 C} = 1.0\,\text{in}; \qquad r_{CD} = 3.0\,\text{in}$$
$$\theta_2 = 30°$$

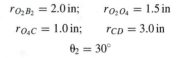
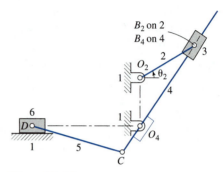

**Figure P2.16**

**P2.17** Locate all instantaneous centers, except $P_{4,5}$, of the mechanism shown in Figure P2.17.

$$r_{O_2 O_3} = 6.0\,\text{cm}; \qquad r_{O_2 B_2} = 4.0\,\text{cm}$$
$$r_{O_3 C} = 12.0\,\text{cm}; \qquad r_{CD} = 7.0\,\text{cm}$$
$$\theta_2 = 30°$$

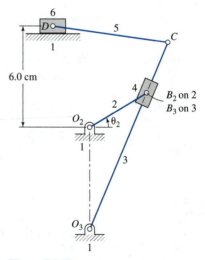

**Figure P2.17**

**P2.18** For the crane shown in Figure P2.18, calculate the moment about point $A$ exerted by the 120 kN force supported by the hoisting cable of the crane for the position shown.

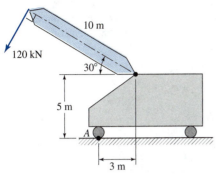

**Figure P2.18**

**P2.19** In raising the massless pole from the position shown in Figure P2.19, the tension $T$ in the cable must supply a moment about $O$ of 72 kN-m. Determine $T$.

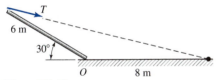

**Figure P2.19**

**P2.20** For the link shown in Figure P2.20, replace the couple and force shown by a single force $\overline{F}$ applied at point $D$. Locate $D$ by determining distance $b$.

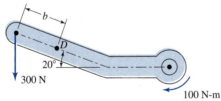

**Figure P2.20**

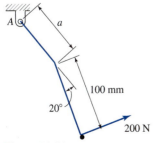

**Figure P2.21**

**P2.21** The lever shown in Figure P2.21 is to be designed to operate with a 200 N force. The force is to provide a counterclockwise couple of 80 N-m at $A$. Determine dimension $a$.

**P2.22** A box weighing 10 N rests on the floor of an elevator as shown in Figure P2.22. Determine the force exerted by the box on the floor of the elevator if
(a) the elevator starts up with an acceleration of 3 m/sec$^2$
(b) the elevator starts down with an acceleration of 3 m/sec$^2$

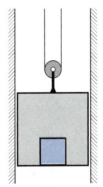

**Figure P2.22**

**P2.23** Determine the constant force required to accelerate an automobile with mass 1,000 kg on a level road from rest to 30 m/sec in 5 seconds. Gravity acts vertically downward. (Figure P2.23)

**Figure P2.23**

**P2.24** Determine the constant force required to accelerate an automobile with mass 1,000 kg on a road with a 10° incline from rest to 30 m/sec in 5 seconds. Gravity acts vertically downward. (Figure P2.24)

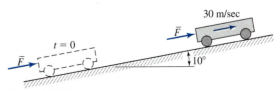

**Figure P2.24**

**P2.25** A freight elevator contains three crates as shown in Figure P2.25. The mass of the cage of the elevator is 1,000 kg, and the masses of crates *A*, *B*, and *C* are 200, 100, and 50 kg, respectively. At one instant, the elevator undergoes an upward acceleration of 5 m/sec². At this instant, determine

(a) the tension in the cable

(b) the force exerted by crate *B* on crate *C*

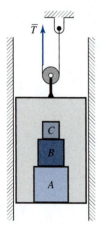

**Figure P2.25**

**P2.26** The solid disc shown in Figure P2.26 has a mass of 20 kg and a radius of 0.3 m. Block *B* has a mass of 5 kg. When a torque of $T = 75$ N-m is applied to the disc, determine the acceleration of block *B* and the tension in the cable. Gravity acts vertically downward.

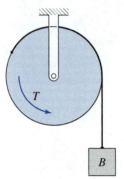

**Figure P2.26**

# 3  Graphical Kinematic Analysis of Planar Mechanisms

## 3.1  INTRODUCTION

In this chapter, graphical velocity and acceleration analyses of planar mechanisms are presented. Each analysis is suitable for one configuration of the mechanism and requires that the mechanism be drawn to scale in the position for which the analysis is to be completed. One method for velocity analysis employs the concept of instantaneous centers (Section 2.8). Another method involves construction of vector polygons of velocity and acceleration. All methods presented are based on material presented in Chapter 2.

## 3.2  VELOCITY ANALYSIS USING INSTANTANEOUS CENTERS

Instantaneous centers of a mechanism may be used to complete a velocity analysis for a single configuration of a planar mechanism. Since the velocity at instantaneous center $P_{i,j}$ is identical for links $i$ and $j$, it can be thought of as a transfer point of motion between links at the instant under consideration.

A velocity analysis of a planar mechanism can be completed by carrying out the following general steps:

1. Draw a diagram of the mechanism to scale, in the configuration under analysis.
2. Obtain the locations of the instantaneous centers, which are required to complete the analysis.
3. Express the velocities at the instantaneous centers based on input motions. For the crank, link 2, of the mechanism shown in Figure 2.26(b), the magnitude of the velocity of instantaneous center $P_{2,3}$ may be expressed in terms of the rotational

speed of link 2 as

$$|\bar{v}_{P_{2,3}}| = r_{P_{1,2}P_{2,3}}|\dot{\theta}_2| \tag{3.2-1a}$$

4. Write the appropriate equations needed in calculating other velocities. When considering the instantaneous center between links $i$ and $j$, we can state

$$|\bar{v}_{P_{i,j}}| = r_{P_{1,i}P_{i,j}}|\dot{\theta}_i| = r_{P_{1,j}P_{i,j}}|\dot{\theta}_j| \tag{3.2-1b}$$

5. Complete the analysis by repeating the process until all required velocities are determined.

For the velocity analysis of a mechanism, it may not be necessary to determine the location of all of the instantaneous centers. This will be demonstrated by examples.

## EXAMPLE 3.1  Velocity Analysis of a Slider Crank Mechanism

For the slider crank mechanism shown in Figure 3.1, determine the velocity of the slider as a function of the rotational velocity of link 2 for the position shown. Use the instantaneous centers.

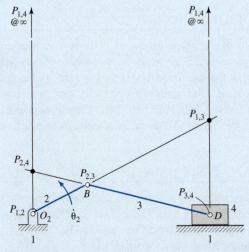

**Figure 3.1**

### SOLUTION

The instantaneous centers for this mechanism were determined in Example 2.3.

The velocity of the slider can be determined by more than one method. For this example, two methods are presented. The first method progresses through the mechanism links and instantaneous centers from the input to the output. The second method determines the final result by considering transfer of motion through one instantaneous center.

**Method #1**

We consider instantaneous center $P_{2,3}$. It is the transfer point of motion between links 2 and 3. The velocity of this point is the same for both links. Considering that link 2 revolves about $P_{1,2}$, and link 3 revolves about $P_{1,3}$,

$$|\bar{v}_{P_{2,3}}| = |\bar{v}_B| = r_{P_{1,2}P_{2,3}}|\dot{\theta}_2| = r_{P_{1,3}P_{2,3}}|\dot{\theta}_3| \tag{3.2-2}$$

which gives

$$|\dot{\theta}_3| = \frac{r_{P_{1,2}P_{2,3}}}{r_{P_{1,3}P_{2,3}}}|\dot{\theta}_2| \tag{3.2-3}$$

Instantaneous center $P_{3,4}$ is the motion transfer point between links 3 and 4. Its magnitude of velocity, considering it as a point on link 3, is

$$|\bar{v}_{P_{3,4}}| = |\bar{v}_D| = r_{P_{1,3}P_{3,4}}|\dot{\theta}_3| \tag{3.2-4}$$

Combining Equations (3.2-3) and (3.2-4) yields

$$|\bar{v}_D| = r_{P_{1,3}P_{3,4}}\frac{r_{P_{1,2}P_{2,3}}}{r_{P_{1,3}P_{2,3}}}|\dot{\theta}_2| \tag{3.2-5}$$

The slider, link 4, is undergoing pure translation. Therefore, velocity $\bar{v}_D$ is the same for any point on this link.

In the above analysis, instantaneous center $P_{2,4}$ was not employed.

**Method #2**

If we employ instantaneous center $P_{2,4}$, then we consider transfer of motion directly between links 2 and 4. The speed of this instantaneous center expressed as a point on link 2 is

$$|\bar{v}_{P_{2,4}}| = r_{P_{1,2}P_{2,4}}|\dot{\theta}_2| \tag{3.2-6}$$

Because the slider, link 4, is in pure translation,

$$|\bar{v}_{P_{2,4}}| = |\bar{v}_D| = r_{P_{1,2}P_{2,4}}|\dot{\theta}_2| \tag{3.2-7}$$

In the above analysis, instantaneous centers $P_{1,3}$ and $P_{1,4}$ were not employed. It was also not required to determine the angular velocity of link 3.

# EXAMPLE 3.2  Velocity Analysis of an Inverted Slider Crank Mechanism

For the inverted slider crank mechanism shown in Figure 3.2, determine the velocity of point $B$ as a function of the rotational speed of link 2 for the position shown. Use the instantaneous centers.

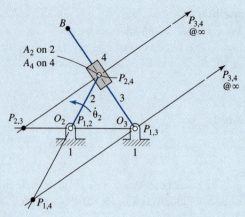

**Figure 3.2**

## SOLUTION

The instantaneous centers for this mechanism were determined in Example 2.4. Two methods for determining the velocity of point $B$ are presented.

### Method #1

Instantaneous center $P_{2,4}$ is the motion transfer point between links 2 and 4. Hence

$$|\bar{v}_{P_{2,4}}| = |\bar{v}_{A_2}| = r_{P_{1,2}P_{2,4}}|\dot{\theta}_2| = r_{P_{1,4}P_{2,4}}|\dot{\theta}_4| \tag{3.2-8}$$

from which

$$|\dot{\theta}_4| = \frac{r_{P_{1,2}P_{2,4}}}{r_{P_{1,4}P_{2,4}}}|\dot{\theta}_2| \tag{3.2-9}$$

By inspection, the rotational speed of link 4 is in the counterclockwise direction.

At the instantaneous center $P_{3,4}$, the velocities of links 3 and 4 are the same, from which we obtain

$$r_{P_{1,3}P_{3,4}}|\dot{\theta}_3| = r_{P_{1,4}P_{3,4}}|\dot{\theta}_4| \tag{3.2-10}$$

However, since $P_{3,4}$ is located at infinity, we conclude that

$$|\dot{\theta}_3| = |\dot{\theta}_4| \tag{3.2-11}$$

The magnitude of the velocity of point $B$ is

$$|\bar{v}_B| = r_{P_{1,3}B}|\dot{\theta}_3| \tag{3.2-12}$$

Combining Equations (3.2-9), (3.2-11), and (3.2-12), the magnitude of the velocity may be expressed as

$$|\bar{v}_B| = r_{P_{1,3}B}|\dot{\theta}_3| = r_{P_{1,3}B}\frac{r_{P_{1,2}P_{2,4}}}{r_{P_{1,4}P_{2,4}}}|\dot{\theta}_2| \tag{3.2-13}$$

By inspection, the velocity is directed downward and to the left.

**Method #2**

Instantaneous center $P_{2,3}$ is the motion transfer point between links 2 and 3. Hence

$$r_{P_{1,2}P_{2,3}}|\dot{\theta}_2| = r_{P_{1,3}P_{2,3}}|\dot{\theta}_3| \tag{3.2-14}$$

Combining Equations (3.2-12) and (3.2-14), the magnitude of the velocity of point $B$ may be expressed as

$$|\bar{v}_B| = r_{P_{1,3}B}|\dot{\theta}_3| = r_{P_{1,3}B}\frac{r_{P_{1,2}P_{2,3}}}{r_{P_{1,3}P_{2,3}}}|\dot{\theta}_2| \tag{3.2-15}$$

## EXAMPLE 3.3  Velocity Analysis of a Mechanism Incorporating a Two Degree of Freedom Pair

For the mechanism shown in Figure 3.3, link 2 has a prescribed rotational speed in the clockwise direction. Determine the rotational speed of link 3. Use the instantaneous centers.

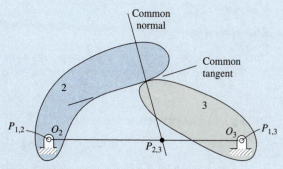

**Figure 3.3**

### SOLUTION

The instantaneous centers for this mechanism were determined in Example 2.5.

The magnitude of the velocity of the instantaneous center $P_{2,3}$ may be expressed as

$$|\bar{v}_{P_{2,3}}| = r_{P_{1,2}P_{2,3}}|\dot{\theta}_2| = r_{P_{1,3}P_{2,3}}|\dot{\theta}_3| \tag{3.2-16}$$

from which

$$|\dot{\theta}_3| = \frac{r_{P_{1,2}P_{2,3}}}{r_{P_{1,3}P_{2,3}}}|\dot{\theta}_2| \tag{3.2-17}$$

By inspection, the direction of the rotational speed of link 3 is counterclockwise, opposite that of link 2.

## EXAMPLE 3.4  Velocity Analysis of a Six-Link Mechanism

For the mechanism shown in Figure 3.4, determine the velocity of link 6 as a function of the rotational speed of link 2 for the position shown. Employ instantaneous centers.

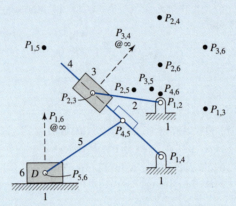

**Figure 3.4** Instantaneous centers of a six-link mechanism.

### SOLUTION

The instantaneous centers for this mechanism were determined in Example 2.6.

The speed of $P_{2,6}$ expressed as a point on input link 2 is

$$|\bar{v}_{P_{2,6}}| = r_{P_{1,2}P_{2,6}}|\dot{\theta}_2| \tag{3.2-18}$$

Because the slider is in pure translation,

$$|\bar{v}_{P_{2,6}}| = |\bar{v}_D| = r_{P_{1,2}P_{2,6}}|\dot{\theta}_2| \tag{3.2-19}$$

In the above analysis, it was not necessary to determine the angular velocities of links 3 through 5, inclusive.

## 3.3  VELOCITY POLYGON ANALYSIS

For a given configuration of a planar mechanism, it is possible to construct an associated velocity polygon. This polygon corresponds to a single configuration of the mechanism and is a graphical representation of all the relative velocities that occur between selected points of interest of the mechanism. Once constructed, it is possible to determine the absolute velocity of any point, the relative velocity between any two points, or the rotational velocity of any link.

To construct a velocity polygon, the following general steps are required:

1. Draw a diagram of the mechanism to scale, in the configuration under analysis.
2. Set up a polar diagram, where every point with zero velocity is at the origin (*pole point*) $O_V$.

3. Starting from the pole point, add velocity vectors that can be readily calculated. For instance, consider those based on prescribed input motions with respect to the base link.
4. Write the appropriate relative velocity equations to aid in calculating and adding other velocity vectors.
5. Proceed through the mechanism until all points of interest are analyzed.

The velocity polygon can then be employed to determine the absolute and relative velocities. This procedure is illustrated in the following examples.

## EXAMPLE 3.5  Velocity Analysis of a Four-Bar Mechanism

For the four-bar mechanism shown in Figure 3.5(a), the link lengths are

$$r_{O_2O_4} = r_1 = 8.0 \, \text{cm}; \qquad r_{O_2B} = r_2 = 3.5 \, \text{cm}$$

$$r_{BD} = r_3 = 10.0 \, \text{cm}; \qquad r_{O_4D} = r_4 = 9.0 \, \text{cm}$$

(Note: Drawings of mechanisms and polygons have been scaled down.)

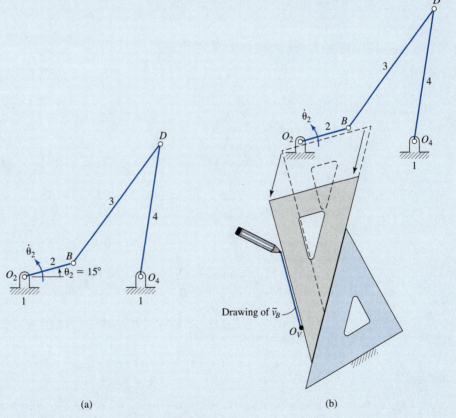

(a)                                         (b)

**Figure 3.5** Mechanism and velocity polygon.

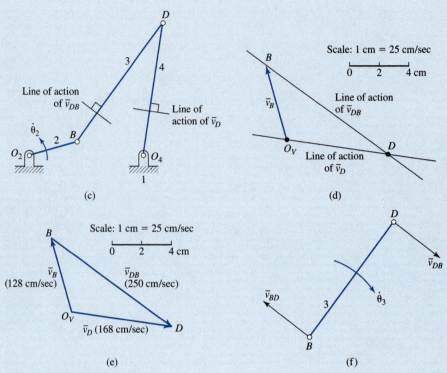

**Figure 3.5** *(Continued)*

The input rotational speed is

$$\dot{\theta}_2 = 350\,\text{rpm CCW} = 350 \times \frac{2\pi}{60}\,\text{rad/sec CCW} = 36.7\,\text{rad/sec CCW}$$

Determine the rotational speeds of links 3 and 4 for the position shown through construction of a velocity polygon.

**SOLUTION**

Base pivots are designated as $O_2$ and $O_4$. The velocities of these points are zero and will therefore be located at the pole point of the velocity polygon. The two moving turning pairs are designated as $B$ and $D$.

Construction of the velocity polygon begins with determining the velocity of point $B$ since it is at the distal end of link 2 and moves relative to point $O_2$. The appropriate relative velocity equation is

$$\bar{v}_B = \bar{v}_{O_2} + \bar{v}_{BO_2} \tag{3.3-1}$$

However, since $O_2$ has zero velocity, the relative velocity between points $B$ and $O_2$ also represents the absolute velocity of point $B$, so the vector is drawn from the pole point. Also, point $B$ is separated from point $O_2$ by a fixed distance, and therefore from Section 2.3.1, the direction of the relative velocity of point $B$ must be perpendicular to line segment $O_2B$ on the mechanism, that is,

$$\bar{v}_B = \bar{v}_{BO_2} = r_2\dot{\theta}_2\bar{i}_{\theta_{BO_2}} \tag{3.3-2}$$

The magnitude of the velocity is

$$|\bar{v}_B| = |r_2\dot{\theta}_2| = 3.5\,\text{cm} \times 36.7\,\text{rad/sec} = 128\,\text{cm/sec}$$

The direction of this velocity is determined by inspecting the direction of the rotational speed of link 2. Using a selected scale of

$$1\,\text{cm} = 25\,\text{cm/sec} \tag{3.3-3}$$

vector $\bar{v}_B$ is added to the polygon, and the head of this vector is labelled as $B$. Figure 3.5(b) illustrates how a pair of drafting triangles is used to draw this velocity vector. One edge of a triangle is placed parallel to line segment $O_2B$ on the mechanism. A second triangle is butted up against another edge of the first triangle as shown. While holding the second triangle stationary, the first triangle is slid into the position whereby an edge touches the pole point. By inspection of Figure 3.5(b), the edge touching the pole point is perpendicular to link 2 and can therefore be used to draw the velocity vector. Triangles can also be easily employed to draw vectors, as required, that are parallel to line segments on the mechanism.

Next, we seek to determine the velocity of point $D$. We recognize that $D$ moves relative to point $O_4$, and

$$\bar{v}_D = \bar{v}_{O_4} + \bar{v}_{DO_4} \tag{3.3-4}$$

Since the distance between $D$ and $O_4$ is constant, we conclude that the relative velocity between these points is perpendicular to the line joining the points on the mechanism (Figure 3.5(c)), that is,

$$\bar{v}_D = \bar{v}_{DO_4} = r_4\dot{\theta}_4\bar{i}_{\theta_{DO_4}} \tag{3.3-5}$$

Since $O_4$ is fixed, we draw in the line of action of $\bar{v}_D$ from the pole point (Figure 3.5(d)). However, we do not yet know the magnitude of this vector. Also, the direction has yet to be determined.

Point $D$ on the mechanism also moves relative to point $B$, and therefore we employ the relation

$$\bar{v}_D = \bar{v}_B + \bar{v}_{DB} \tag{3.3-6}$$

Since $B$ and $D$ are separated by a fixed distance, the corresponding relative velocity is perpendicular to the corresponding direction on the mechanism (Figure 3.5(c)), that is,

$$\bar{v}_{DB} = r_3\dot{\theta}_3\bar{i}_{\theta_{DB}} \tag{3.3-7}$$

Since the velocity of point $D$ in Equation (3.3-6) is expressed with respect to the velocity of point $B$, the line of action of the relative velocity is added to the polygon starting from $B$

(Figure 3.5(d)). Although we know the line of action, we do not know the magnitude and direction of this vector.

The velocity of point $D$ is unique. In order to satisfy the relative velocity expressions, Equations (3.3-5) and (3.3-7), as well as the lines of action of the velocities, the head of the velocity of point $D$ must lie at the intersection of the two lines of action (Figure 3.5(d)). This is designated as point $D$ on the polygon. The relative velocity between points $B$ and $D$ is labelled. The completed velocity polygon is given in Figure 3.5(e). Note also, as illustrated in Figure 3.5(f), that

$$\bar{v}_{DB} = -\bar{v}_{BD} \tag{3.3-8}$$

Now that the velocity polygon has been completed, it is possible to determine the magnitudes and directions of angular speeds of links 3 and 4. To do this, we measure speeds from the polygon employing the scale given by Equation (3.3-3). The results are

$$v_{BD} = |\bar{v}_{BD}| = 250 \,\text{cm/sec}; \qquad v_D = |\bar{v}_D| = 168 \,\text{cm/sec}$$

Considering the magnitudes of Equation (3.3-7) and rearranging,

$$|\dot{\theta}_3| = \frac{v_{DB}}{r_3} = \frac{250}{10.0} = 25.0 \,\text{rad/sec}$$

Similarly, from Equation (3.3-5), the magnitude of the rotational speed of link 4 is

$$|\dot{\theta}_4| = \frac{v_D}{r_4} = \frac{168}{9.0} = 18.7 \,\text{rad/sec}$$

The direction of rotational speed of link 3, with points $B$ and $D$ at its ends, may be found by examining the direction of the relative velocity $\bar{v}_{DB}$. This relative velocity is that of $D$ with respect to $B$. That is, if we were to view point $D$ from $B$ on the mechanism, we would

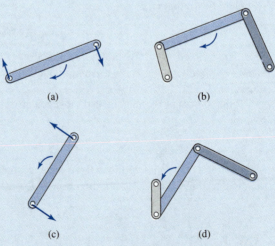

(a)                    (b)

(c)                    (d)

**Figure 3.6** Example of relative velocities [Video 3.6].

see it moving to the right. This corresponds to the link turning in the clockwise direction. This is illustrated in Figure 3.5(f).

We would obtain the same result if we were to use the relative velocity $\bar{v}_{BD}$. In this case, we see point $B$ moving to the right with respect to point $D$, once again leading us to the conclusion that the link is turning in the clockwise direction. This is illustrated in Figure 3.5(f). A similar analysis for link 4 reveals that it is also turning in the clockwise direction. Therefore, the rotational speeds of links 3 and 4 are

$$\dot{\theta}_3 = 25.0 \, \text{rad/sec CW}; \qquad \dot{\theta}_4 = 18.7 \, \text{rad/sec CW}$$

**Video 3.6**
Relative Velocity

As an illustration of relative velocity, consider the four-bar mechanism shown in Figure 3.6. The coupler is isolated in Figure 3.6(a). At the ends of the link, the relative velocities are shown. For the configuration shown in Figure 3.6(b), the rotation of the coupler is in the clockwise direction. The angular arrow indicates the direction of rotational speed. Later in the cycle, as for instance shown in Figures 3.6(c) and 3.6(d), rotation of the coupler is in the counterclockwise direction.

## EXAMPLE 3.6 Velocity Analysis of a Four-Bar Mechanism with a Coupler Point

Consider the same four-bar kinematic chain as in Example 3.5. Coupler point $C$ is added to link 3 as shown in Figure 3.7. Determine the velocity of the point $C$.

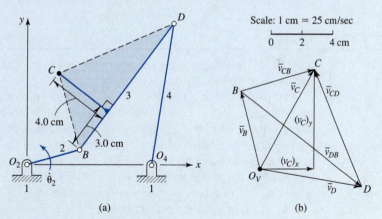

**Figure 3.7** (a) Mechanism. (b) Velocity polygon.

### SOLUTION

We begin by drawing the portion of the velocity polygon that is the same as that for Example 3.5. Then we consider the velocity of the point $C$, first with respect to point $B$. The appropriate relative velocity equation is

$$\bar{v}_C = \bar{v}_B + \bar{v}_{CB} \qquad (3.3\text{-}9)$$

Since points $B$ and $C$ are separated by a fixed distance, the relative velocity between these two points is in a direction perpendicular to the line segment joining these points on the mechanism. Furthermore, since $C$ is moving with respect to $B$, the line of action of its relative velocity vector is drawn starting from point $B$ on the polygon. However, we do not yet know the magnitude and direction of this vector.

In a similar fashion, the velocity of the point $C$ moves relative to the point $D$. The appropriate relative velocity equation is

$$\bar{v}_C = \bar{v}_D + \bar{v}_{CD} \tag{3.3-10}$$

The relative velocity between $C$ and $D$ on the mechanism is perpendicular to the direction of the line segment joining these points. In this instance, the line of action of this relative velocity is drawn from point $D$ of the polygon.

Since the velocity of the point $C$ is unique and must satisfy both Equations (3.3-9) and (3.3-10), the head of the velocity vector of point $C$ must lie at the intersection of the two lines of action of the relative velocities. This is labelled as point $C$ on the polygon. The relative velocities between points $B$, $C$, and $D$ are labelled. The absolute velocity of point $C$ is then found by drawing a vector from the pole point to point $C$. The result is

$$|\bar{v}_C| = 181 \text{ cm/sec}$$

The direction of this velocity is indicated on the polygon in Figure 3.7(b). The velocity of point $C$ may be broken into its Cartesian components, as shown. The results are

$$(v_C)_x = 87 \text{ cm/sec}; \qquad (v_C)_y = 159 \text{ cm/sec}$$

## 3.4  VELOCITY IMAGE

A useful tool for obtaining the velocity of any point on a link, using a velocity polygon, is referred to as the *velocity image*.

All points within a link are separated by fixed distances as the mechanism moves. Therefore, the relative velocities between a pair of points on the same link will be in a direction perpendicular to a line joining those two points on the mechanism. Furthermore, the magnitude of relative velocity equals the distance on the mechanism multiplied by the rotational speed of the link. However, since the rotational speed of a link is unique, the separation of points on the same link in the velocity polygon is proportional to the distance separating the corresponding points on the mechanism.

As an illustration, reconsider Example 3.6. The velocity polygon is redrawn in Figure 3.8(b). Expressions for magnitudes of relative velocities between points $B$, $C$, and $D$ are indicated. Since $\dot{\theta}_3$ is common to all of the expressions, the triangle defined by points $B$, $C$, and $D$ on the mechanism (Figure 3.8(a)) is *similar* to that defined by points $B$, $C$, and $D$ as they appear on the velocity polygon (Figure 3.8(b)).

Once the velocity image of a link in the velocity polygon has been established, it is a simple matter to find the velocity of any point on the link. For instance, point $E$ on the mechanism shown in Figure 3.8(a) is found on its image in the velocity polygon. The absolute velocity of that point then corresponds to a vector drawn from the pole point to point $E$ on the polygon.

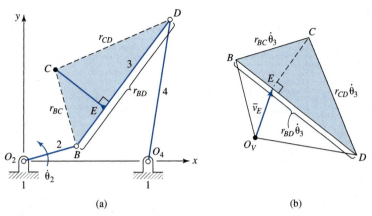

**Figure 3.8** Example of velocity image: (a) mechanism, (b) outline of velocity polygon.

Velocity images of links 2 and 4 also appear in the velocity polygon. The image of link 2 is between the pole point and point $B$ on the polygon, and the image of link 4 is between the pole point and point $D$ on the polygon.

Further examples of velocity polygon analyses that incorporate velocity images are provided below.

## EXAMPLE 3.7  Velocity Analysis of a Slider Crank Mechanism

For the slider crank mechanism shown in Figure 3.9(a), the link lengths are

$$r_{O_2B} = r_2 = 3.5\,\text{in}; \qquad r_{BD} = r_3 = 9.0\,\text{in}$$

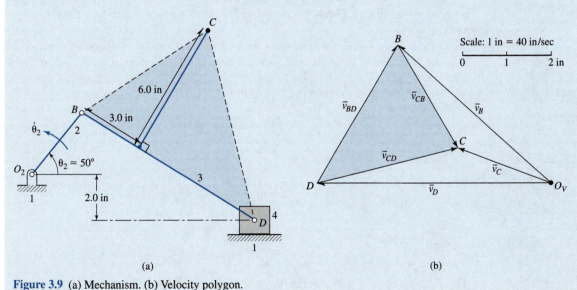

**Figure 3.9** (a) Mechanism. (b) Velocity polygon.

The input rotational speed is

$$\dot{\theta}_2 = 500\,\text{rpm CCW} = 52.4\,\text{rad/sec CCW}$$

Determine the rotational speed of link 3 and the velocity of point $C$, for the position shown.

### SOLUTION

The speed of $B$ is

$$|\bar{v}_B| = |r_2 \dot{\theta}_2| = 3.5\,\text{in} \times 52.4\,\text{rad/sec} = 183\,\text{in/sec}$$

and the vector is added to the polygon starting from the pole point. Point $B$ is labelled on the polygon in Figure 3.9(b).

The slider, designated as point $D$, moves with respect to the base link. Therefore, its line of action of velocity is drawn with respect to the pole point. However, at this instant, we do not know the magnitude and direction of this vector.

Point $D$ on the mechanism also moves relative to point $B$, and therefore we employ the relation

$$\bar{v}_D = \bar{v}_B + \bar{v}_{DB} \qquad (3.4\text{-}1)$$

The line of action of the relative velocity is added to the polygon, starting from $B$.

The intersection of the two lines of action gives point $D$ on the polygon. Vectors $\bar{v}_D$ and $\bar{v}_{DB}$ are labelled.

Using the same procedure as presented in Example 3.6, point $C$ is determined on the polygon. Alternatively, $C$ may be located by means of the velocity image. Then, $\bar{v}_C$, $\bar{v}_{CB}$, and $\bar{v}_{CD}$ are labelled.

Similar to the analysis performed in Example 3.5, by examining the direction of $\bar{v}_{DB}$, the direction of the rotational speed of link 3 is determined to be in the clockwise direction.

The magnitude of the rotational speed of link 3 is found by measuring

$$v_{DB} = 138\,\text{in/sec}$$

and therefore

$$\dot{\theta}_3 = \frac{v_{DB}}{r_3} = \frac{138}{9.0} = 15.3\,\text{rad/sec CW}$$

Also

$$|\bar{v}_C| = 91\,\text{in/sec}$$

The direction of $\bar{v}_C$ is given on the polygon.

## EXAMPLE 3.8  Velocity Analysis of an Inverted Slider Crank Mechanism

For the inverted slider crank mechanism shown in Figure 3.10(a), the link lengths are

$$r_{O_2 O_3} = r_1 = 5.0 \text{ in}; \qquad r_{O_2 B_2} = r_2 = 6.0 \text{ in}$$

The rotational speed of link 2 is

$$\dot\theta_2 = 30.0 \text{ rad/sec CW}$$

Determine the velocity polygon for the mechanism and the rotational speed of link 3 for the given configuration.

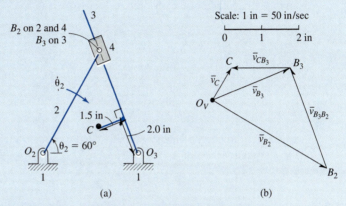

**Figure 3.10**  (a) Mechanism. (b) Velocity polygon.

### SOLUTION

The speed of $B_2$ is

$$|\bar{v}_{B_2}| = |r_2 \dot\theta_2| = 6.0 \text{ in} \times 30 \text{ rad/sec} = 180 \text{ in/sec}$$

The velocity of $B_2$ is drawn on the polygon.

Next, we consider the velocity of point $B_3$ and note that

$$\bar{v}_{B_3} = \bar{v}_{B_3 O_3} = r_{B_3 O_3} \dot\theta_3 \bar{i}_{\theta_{B_3 O_3}} \tag{3.4-2}$$

where

$$r_{B_3 O_3} = 5.57 \text{ in}$$

is measured from the scaled diagram.

The line of action of the velocity of $B_3$ is added to the polygon starting from the pole point. We now examine the motion of $B_3$ with respect to $B_2$, employing

$$\bar{v}_{B_3} = \bar{v}_{B_2} + \bar{v}_{B_3 B_2} \tag{3.4-3}$$

In Equation (3.4-3), $\bar{v}_{B_2}$ is known. In addition, from the special case of a slider moving on a straight slide presented in Section 2.3.2, in this instance we have

$$\bar{v}_{B_3 B_2} = \dot{r}_{B_3 B_2} \bar{i}_{r_{B_3 O_3}} \tag{3.4-4}$$

The line of action of this relative velocity, which is along the direction of the slide, is added to the polygon, starting from $B_2$. The intersection of the two lines of action gives point $B_3$ on the polygon.

Similar analyses as in previous examples leads to the determination of point $C$ on the polygon. All vectors in the polygon can now be labelled, as shown in Figure 3.10(b).

Now that the polygon has been completed, the sliding velocity between points $B_3$ and $B_2$ has a magnitude

$$|\bar{v}_{B_3B_2}| = |\dot{r}_{B_3B_2}| = 140 \text{ in/sec}$$

and the magnitude of the absolute velocity of $B_3$ is

$$|\bar{v}_{B_3}| = v_{B_3O_3} = 113 \text{ in/sec}$$

The rotational speed of link 3 is

$$\dot{\theta}_3 = \frac{v_{B_3O_3}}{r_{B_3O_3}} = \frac{113}{5.57} = 20.3 \text{ rad/sec CW}$$

where the direction of the rotational speed has been found by examining the direction of $\bar{v}_{B_3}$ and the location of $B_3$ with respect to $O_3$.

## EXAMPLE 3.9  Velocity Polygon Analysis

For the mechanism shown in Figure 3.11(a), the link lengths are

$$r_{O_2O_4} = r_1 = 2.0 \text{ in}; \qquad r_{BO_2} = r_2 = 1.0 \text{ in}; \qquad r_{BD} = r_3 = 3.5 \text{ in}$$

$$r_{BC} = 2.0 \text{ in}; \qquad r_{DO_4} = r_4 = 2.0 \text{ in}; \qquad r_{E_5C} = 1.5 \text{ in}$$

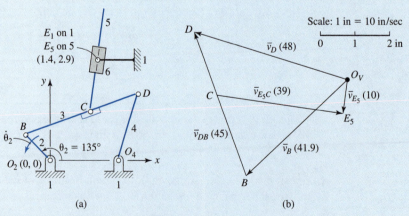

**Figure 3.11** (a) Mechanism. (b) Velocity polygon.

The input rotational speed is

$$\dot\theta_2 = 400 \, \text{rpm CCW} = 41.9 \, \text{rad/sec CCW}$$

Determine the velocity polygon of the mechanism and the rotational speeds of links 3, 4, and 5 for the given configuration.

### SOLUTION

The following procedure is employed:

- Calculate $v_B$:

$$|\bar v_B| = |r_2 \dot\theta_2| = 1.0 \, \text{in} \times 41.9 \, \text{rad/sec} = 41.9 \, \text{in/sec}$$

- Locate $B$ on the polygon.
- Employ

$$\bar v_D = \bar v_B + \bar v_{DB} \qquad (3.4\text{-}5)$$

and draw the lines of action of $\bar v_D$ and $\bar v_{DB}$ to locate $D$ on the polygon.
- Locate $C$ using the velocity image.
- Employ

$$\bar v_{E_5} = \bar v_C + \bar v_{E_5 C} \qquad (3.4\text{-}6)$$

and draw the lines of action of $\bar v_{E_5}$ and $\bar v_{E_5 C}$ to locate $E_5$ on the polygon.
- Label all vectors. The completed velocity polygon is shown in Figure 3.11(b). Magnitudes of the velocities are indicated in parentheses.
- The rotational speeds are

$$\dot\theta_3 = \frac{v_{DB}}{r_3} = \frac{45}{3.5} = 12.9 \, \text{rad/sec CCW}$$

$$\dot\theta_4 = \frac{v_D}{r_4} = \frac{48}{2.0} = 24.0 \, \text{rad/sec CCW}$$

$$\dot\theta_5 = \frac{v_{CE_5}}{r_{CE_5}} = \frac{39}{1.5} = 26.0 \, \text{rad/sec CW}$$

## 3.5 ACCELERATION POLYGON ANALYSIS

The acceleration polygon method has many similarities to the velocity polygon method. However, in order to complete an acceleration polygon, results from the velocity polygon analysis must first be obtained. Values of rotational and sliding velocities are needed to determine the normal and Coriolis accelerations.

To construct an acceleration polygon, the following general steps are required:

1. Draw a diagram of the mechanism to scale, in the configuration under analysis.
2. Set up a polar diagram, where every point with zero acceleration is at the origin (pole point) $O_A$.

3. Starting from the pole point, draw in acceleration vectors that can be readily calculated. For instance, consider those based on prescribed input motions with respect to the base link.
4. Write the appropriate relative acceleration equations to aid in calculating and constructing other acceleration vectors.
5. Proceed through the mechanism until all points of interest are analyzed.

The acceleration polygon can then be employed to determine absolute and relative accelerations. This will be illustrated by the following examples.

## EXAMPLE 3.10 Acceleration Analysis of a Four-Bar Mechanism with Constant Input Speed

Consider the four-bar mechanism presented in Example 3.5 (refer to Figure 3.5). Pertinent information required from the velocity analysis is illustrated in Figure 3.12(a). The rotational speed of link 2 is constant. Determine the angular accelerations of links 3 and 4 for the position shown by constructing an acceleration polygon.

### SOLUTION

For this example, all the kinematic pairs are separated by fixed distances. Therefore, from Section 2.5.1, we need only concern ourselves with normal and tangential components of relative acceleration.

Construction of the acceleration polygon begins by determining the acceleration of point $B$, which is separated from $O_2$ by a fixed distance. Using Equation (2.5-4),

$$\bar{a}_B = \bar{a}_{BO_2} = \left(-r_2\dot{\theta}_2^2\right)\bar{i}_{r_{BO_2}} + \left(r_2\ddot{\theta}_2\right)\bar{i}_{\theta_{BO_2}}$$
$$= \left(a_{BO_2}^N\right)\bar{i}_{r_{BO_2}} + \left(a_{BO_2}^T\right)\bar{i}_{\theta_{BO_2}} \tag{3.5-1}$$

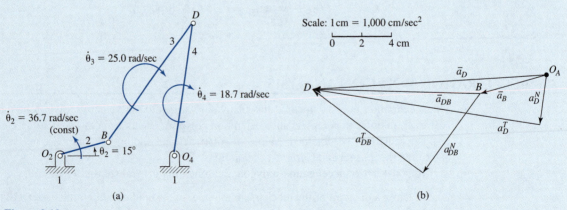

**Figure 3.12** (a) Mechanism. (b) Acceleration polygon.

Furthermore, since

$$\dot{\theta}_2 = \text{constant}; \qquad \ddot{\theta}_2 = 0$$

Equation (3.5-1) reduces to

$$\bar{a}_B = \left(-r_2\dot{\theta}_2^2\right)\bar{i}_{r_{BO_2}} \tag{3.5-2}$$

The direction of the acceleration of $B$ must be in the negative radial direction of the line segment defined by points $O_2$ and $B$. The value is

$$a_B^N = -r_2\dot{\theta}_2^2 = -3.5 \times 36.7^2 = -4{,}710 \text{ cm/sec}^2$$

Using the scale

$$1 \text{ cm} = 1{,}000 \text{ cm/sec}^2 \tag{3.5-3}$$

the acceleration vector of point $B$ is drawn relative to the pole point.

Next, we seek to determine the acceleration of point $D$. We recognize that $D$ moves relative to point $O_4$, so we write the relative acceleration equation

$$\bar{a}_D = \bar{a}_{O_4} + \bar{a}_{DO_4} \tag{3.5-4}$$

Since the distance between $D$ and $O_4$ is constant, the relative acceleration is

$$\begin{aligned}\bar{a}_{DO_4} &= \left(-r_4\dot{\theta}_4^2\right)\bar{i}_{r_{DO_4}} + \left(r_4\ddot{\theta}_4\right)\bar{i}_{\theta_{DO_4}} \\ &= \left(a_{DO_4}^N\right)\bar{i}_{r_{DO_4}} + \left(a_{DO_4}^T\right)\bar{i}_{\theta_{DO_4}}\end{aligned} \tag{3.5-5}$$

Using the value of rotational speed from Example 3.5, the normal acceleration component is

$$a_{DO_4}^N = -r_4\dot{\theta}_4^2 = -9.0 \times 18.7^2 = -3{,}150 \text{ cm/sec}^2$$

From the pole point, we draw the normal component of acceleration, and from the head of the vector for normal acceleration, the line of action of the tangential component. We do not yet know the magnitude of the tangential component. Also, the vector may point in one of two possible directions and still be perpendicular to the line segment defined by points $D$ and $O_4$. This direction has yet to be determined.

Point $D$ on the mechanism also moves relative to point $B$. Also, points $B$ and $D$ are separated by a fixed distance. We employ the following relative acceleration equation:

$$\bar{a}_D = \bar{a}_B + \bar{a}_{DB} \tag{3.5-6}$$

where

$$\begin{aligned}\bar{a}_{DB} &= \left(-r_3\dot{\theta}_3^2\right)\bar{i}_{r_{DB}} + \left(r_3\ddot{\theta}_3\right)\bar{i}_{\theta_{DB}} \\ &= \left(a_{DB}^N\right)\bar{i}_{r_{DB}} + \left(a_{DB}^T\right)\bar{i}_{\theta_{DB}}\end{aligned} \tag{3.5-7}$$

The normal acceleration component is

$$a_{DB}^N = -r_3\dot{\theta}_3^2 = -10.0 \times 25.0^2 = -6{,}250 \text{ cm/sec}^2$$

The relative acceleration $\bar{a}_{DB}$ in Equation (3.5-6) is with respect to point $B$. Therefore, the normal acceleration component and the line of action of the tangential acceleration are added to the polygon starting from the point $B$. For the tangential component, we only know the line of action, not its magnitude and direction.

It must be recognized, however, that the acceleration of point $D$ is unique. In order to satisfy the relative acceleration expressions, Equations (3.5-4) and (3.5-6), as well as the lines of action of the tangential acceleration components, the head of $\bar{a}_D$ must lie at the intersection of the two lines of action. This is designated as point $D$ on the polygon. The completed acceleration polygon is given in Figure 3.12(b).

Now that the acceleration polygon has been completed, it is possible to determine the magnitudes and directions of angular accelerations of links 3 and 4. To do this, we measure the tangential components of acceleration from the polygon employing the scale given by Equation (3.5-3). The result is

$$a_{DB}^T = 9,080 \text{ cm/sec}^2; \qquad a_D^T = 15,300 \text{ cm/sec}^2$$

Considering only the tangential component of Equations (3.5-5) and (3.5-7), expressions for the magnitude of the angular acceleration of links 3 and 4 are

$$|\ddot{\theta}_3| = \frac{a_{DB}^T}{r_3}; \qquad |\ddot{\theta}_4| = \frac{a_D^T}{r_4} \qquad (3.5\text{-}8)$$

Substituting values in Equations (3.5-8) gives

$$|\ddot{\theta}_3| = \frac{a_{DB}^T}{r_3} = \frac{9,080}{10.0} = 908 \text{ rad/sec}^2$$

$$(3.5\text{-}9)$$

$$|\ddot{\theta}_4| = \frac{a_D^T}{r_4} = \frac{15,300}{9.0} = 1,700 \text{ rad/sec}^2$$

The direction of the angular acceleration of link 3, with points $B$ and $D$ at its ends, may be found by examining the direction of $a_{DB}^T$. This relative acceleration is as seen with respect to point $B$. That is, if we were to view $D$ from $B$ on the mechanism, we would see it accelerating upwards and to the left. This corresponds to the link at this instant having an angular acceleration in the counterclockwise direction. This is illustrated in Figure 3.13.

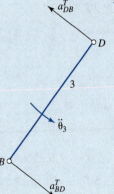

**Figure 3.13** Relative components of tangential acceleration between points $B$ and $D$.

Alternatively, we could obtain the same conclusion regarding the direction of the angular acceleration by employing the relative acceleration $a_{BD}^T$, where

$$a_{BD}^T = -a_{DB}^T \qquad (3.5\text{-}10)$$

as is illustrated in Figure 3.13.

A similar analysis for link 4 reveals that it has an angular acceleration in the counterclockwise direction at the same instant. Therefore, the angular accelerations of links 3 and 4 are

$$\ddot{\theta}_3 = 908 \text{ rad/sec}^2 \text{ CCW}; \qquad \ddot{\theta}_4 = 1{,}700 \text{ rad/sec}^2 \text{ CCW}$$

Since the direction of angular acceleration for both links (i.e., counterclockwise) is opposite to that of the angular velocity (i.e., clockwise), both links in this position are undergoing an angular *deceleration*, or in other words, slowing down. If directions of angular velocity and angular acceleration for a link were the same, the link would be undergoing an angular *acceleration*, or in other words, speeding up.

# EXAMPLE 3.11 Acceleration Analysis of a Four-Bar Mechanism with Variable Input Speed

Consider the four-bar mechanism presented in Example 3.10. The configuration of the mechanism and the instantaneous value of the input rotational speed of link 2 are the same as given in Example 3.10. However, link 2 is decelerating at the rate

$$\ddot{\theta}_2 = 600 \text{ rad/sec}^2 \text{ CW}$$

Determine the angular accelerations of links 3 and 4 for the position shown by constructing an acceleration polygon.

## SOLUTION

Since the configuration of the mechanism and input rotational speed are the same as for Example 3.10, so too are the rotational speeds of links 3 and 4, as well as the normal components of acceleration.

Using Equation (2.5-4), the acceleration of point $B$ is

$$
\begin{aligned}
\overline{a}_B = \overline{a}_{BO_2} &= \left(-r_2\dot{\theta}_2^2\right)\overline{i}_{r_{BO_2}} + (r_2\ddot{\theta}_2)\overline{i}_{\theta_{BO_2}} \\[6pt]
&= (-3.5 \times 36.7^2)\overline{i}_{r_{BO_2}} + [3.5 \times (-600)]\overline{i}_{\theta_{BO_2}} \\[6pt]
&\hspace{8cm} (3.5\text{-}11) \\[-4pt]
&= (-4{,}710)\,\overline{i}_{r_{BO_2}} + (-2{,}100)\,\overline{i}_{\theta_{BO_2}} \\[6pt]
&= \left(a_{BO_2}^N\right)\overline{i}_{r_{BO_2}} + \left(a_{BO_2}^T\right)\overline{i}_{\theta_{BO_2}}
\end{aligned}
$$

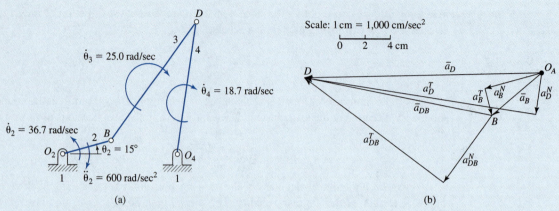

(a)                                                    (b)

**Figure 3.14**  (a) Mechanism. (b) Acceleration polygon.

This acceleration is added to the polygon, starting from the pole point. The remainder of the construction of the acceleration polygon is similar to that presented in Example 3.10. From the polygon shown in Figure 3.14 we measure

$$a_{DB}^T = 13,200 \text{ cm/sec}^2; \qquad a_D^T = 18,000 \text{ cm/sec}^2$$

from which, using Equations (3.5-8), the angular accelerations of links 3 and 4 are

$$\ddot{\theta}_3 = 1,320 \text{ rad/sec}^2 \text{ CCW}; \qquad \ddot{\theta}_4 = 2,000 \text{ rad/sec}^2 \text{ CCW}$$

# EXAMPLE 3.12  Acceleration Analysis of a Four-Bar Mechanism with a Coupler Point

Consider the four-bar mechanism presented in Example 3.6 (Figure 3.7) where coupler point $C$ has been added to link 3. The mechanism is redrawn in Figure 3.15(a). The rotational speed of link 2 is constant. Determine the acceleration of the point $C$.

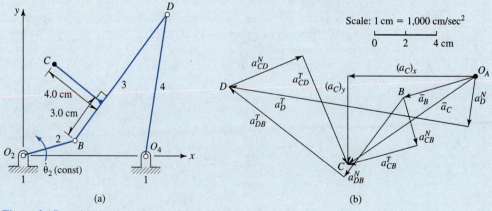

(a)                                                    (b)

**Figure 3.15**  (a) Mechanism. (b) Acceleration polygon.

### SOLUTION

We begin by drawing the portion of the acceleration polygon that is the same as that given in Example 3.10. Then we consider the acceleration of point $C$ first with respect to point $B$. The appropriate relative acceleration equation is

$$\bar{a}_C = \bar{a}_B + \bar{a}_{CB} \tag{3.5-12}$$

Since points $B$ and $C$ are separated by a fixed distance, $\bar{a}_{CB}$ has only normal and tangential components. The normal component is in the negative radial direction of the line segment on the mechanism. The tangential component is in a direction perpendicular to the line segment connecting the same two points on the mechanism. Furthermore, this relative acceleration is with respect to moving point $B$ and will be drawn with respect to point $B$ on the polygon.

We start with the normal component since both its magnitude and direction may be determined. The magnitude of the normal acceleration is

$$a_{CB}^N = -r_{CB}\dot{\theta}_3^2 = -(3.0^2 + 4.0^2)^{1/2} \times 25.0^2$$
$$= -3{,}130 \text{ cm/sec}^2$$

which is drawn on the polygon. Also drawn is the line of action of $a_{CB}^T$. However, at this instant, we do not know its magnitude and direction.

In a similar fashion, the acceleration of point $C$ moves relative to point $D$. The appropriate relative acceleration equation is

$$\bar{a}_C = \bar{a}_D + \bar{a}_{CD} \tag{3.5-13}$$

where the normal component of $\bar{a}_{CD}$ is

$$a_{CD}^N = -r_{CD}\dot{\theta}_3^2 = -8.06 \times 25.0^2$$
$$= -5{,}040 \text{ cm/sec}^2$$

This normal component, along with the line of action of $a_{CD}^T$, are added to the polygon, starting from point $D$.

Since the acceleration of point $C$ is unique and must satisfy both of the requirements as given in Equations (3.5-12) and (3.5-13), point $C$ on the polygon must lie at the intersection of the two lines of action of the relative tangential accelerations. The absolute acceleration of point $C$ is then found by drawing a vector from the pole point to point $C$. The vector drawn gives the direction, and employing the same scale as other vectors in the polygon, the magnitude may be determined. The magnitude is

$$|\bar{a}_C| = 9{,}740 \text{ cm/sec}^2$$

The completed acceleration polygon is shown in Figure 3.15(b). The acceleration of point $C$ may be broken into its Cartesian components, as shown. The results are

$$(a_C)_x = -8{,}050 \text{ cm/sec}^2; \qquad (a_C)_y = -5{,}740 \text{ cm/sec}^2$$

## 3.6 ACCELERATION IMAGE

A useful tool for obtaining the acceleration of any point on a link from an acceleration polygon is referred to as the *acceleration image.*

All points within a link are separated by fixed distances. Therefore, the relative accelerations between a pair of points on the same link can have only normal and tangential components of relative acceleration. That is, there are no sliding or Coriolis components of relative acceleration between points on the same link. The magnitudes of the normal and tangential components of relative acceleration, as given by Equations (2.4-3) and (2.4-7), are proportional to the distances between the points. In addition, the normal acceleration component is dependent on the rotational speed of the link, whereas the tangential component is proportional to the angular acceleration. However, since there are unique values of rotational speed and angular acceleration for a link, the distance separating points on the same link in a mechanism is proportional to the distance of separation in the acceleration polygon.

As an illustration, let us reconsider Example 3.12. The acceleration polygon is re-drawn in Figure 3.16(b). The magnitude of the relative accelerations between points $B$ and $C$ are indicated. The relative acceleration consists of a normal component and a tangential component. The triangle defined by points $B$, $C$, and $D$ in the mechanism (Figure 3.16(a)) is similar to the triangle defined by points $B$, $C$, and $D$ as they appear in the acceleration polygon (Figure 3.16(b)).

Once the acceleration image of a link in the acceleration polygon has been established, it is a simple matter to find the acceleration of any point on the link. For instance, point $E$ on the mechanism shown in Figure 3.16(a) is found on its image in the acceleration polygon. The absolute acceleration of $E$ then corresponds to a vector drawn from the pole point to point $E$ on the polygon.

Acceleration images of links 2 and 4 also appear in the acceleration polygon. The image of link 2 is between the pole point and point $B$ on the polygon, and the image of link 4 is between the pole point and point $D$ on the polygon.

Further examples of acceleration polygon analyses that incorporate acceleration images are provided below.

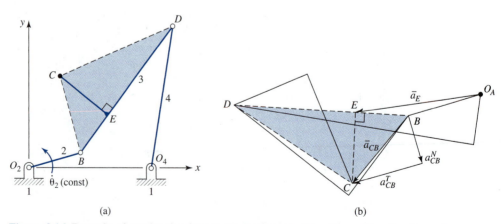

(a)                                          (b)

**Figure 3.16** Example of acceleration image: (a) mechanism, (b) outline of acceleration polygon.

# EXAMPLE 3.13  Acceleration Analysis of a Slider Crank Mechanism

Consider the slider crank mechanism presented in Example 3.7 (refer to Figure 3.9). The mechanism is redrawn in Figure 3.17(a). The rotational speed of link 2 is constant. Determine the acceleration polygon of the mechanism and the angular acceleration of link 3 for the given configuration.

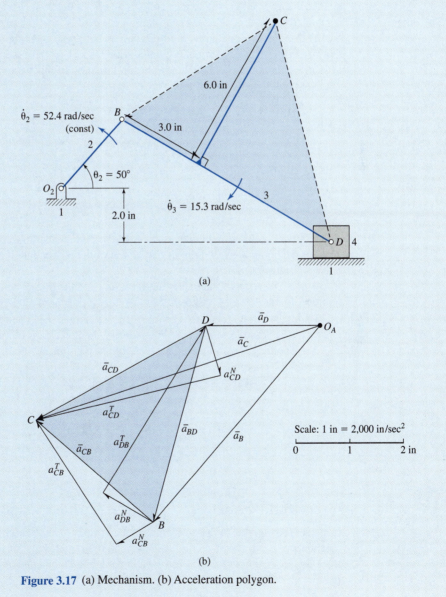

(a)

(b)

**Figure 3.17** (a) Mechanism. (b) Acceleration polygon.

**SOLUTION**

In addition to normal and tangential components of relative acceleration, there can be a sliding component of acceleration of the slider (link 4) with respect to the slide (link 1). However, since the direction of the slide is fixed, no component of Coriolis acceleration exists.

The normal component of the acceleration of $B$ is

$$a_B^N = -r_2\dot{\theta}_2^2 = -3.5 \times 52.4^2 = -9,610 \text{ in/sec}^2$$

and the vector is added to the polygon starting from the pole point. Point $B$ on the polygon is labelled.

The slider, designated as point $D$, moves with respect to the base link. Therefore, its line of action of acceleration is drawn with respect to the pole point. However, at this instant, we do not know the magnitude and direction of this vector.

Next, we employ the relation

$$\bar{a}_D = \bar{a}_B + \bar{a}_{DB} \tag{3.6-1}$$

where

$$\begin{aligned}\bar{a}_{DB} &= \left(-r_3\dot{\theta}_3^2\right)\bar{i}_{r_{DB}} + (r_3\ddot{\theta}_3)\bar{i}_{\theta_{DB}} \\ &= \left(a_{DB}^N\right)\bar{i}_{r_{DB}} + \left(a_{DB}^T\right)\bar{i}_{\theta_{DB}}\end{aligned} \tag{3.6-2}$$

The normal acceleration component is

$$a_{DB}^N = -r_3\dot{\theta}_3^2 = -9.0 \times 15.3^2 = -2,110 \text{ in/sec}^2$$

The relative acceleration between points $B$ and $D$ in Equation (3.6-2) is with respect to $B$. The normal component and the line of action of the tangential component are added to the polygon, starting from $B$.

The intersection of the line of action of $\bar{a}_D$ and the line of action of $a_{DB}^T$ gives point $D$ on the polygon. Vector $\bar{a}_D$ and the tangential component of $\bar{a}_{DB}$ are labelled.

Using the same procedure as presented in Example 3.12, point $C$ is determined. Alternatively, it is possible to locate $C$ by means of the acceleration image. Then, $\bar{a}_C, \bar{a}_{CB}$, and $\bar{a}_{CD}$ are labelled. The completed acceleration polygon is shown in Figure 3.17(b).

From the polygon, we measure

$$a_{DB}^T = 7,330 \text{ in/sec}^2$$

from which

$$\ddot{\theta}_3 = \frac{a_{DB}^T}{r_3} = \frac{7,330}{9.0} = 814 \text{ rad/sec}^2 \text{ CCW}$$

## EXAMPLE 3.14  Acceleration Analysis of an Inverted Slider Crank Mechanism

Consider the inverted slider crank mechanism presented in Example 3.8 (Figure 3.10). The mechanism is redrawn in Figure 3.18(a). The rotational speed of link 2 is constant. Determine the acceleration polygon for the mechanism and the angular acceleration of link 3 in the given configuration.

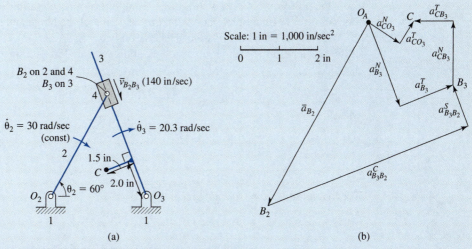

**Figure 3.18** (a) Mechanism. (b) Acceleration polygon.

SOLUTION

For this problem, there can be a sliding component of acceleration of the slider with respect to the slide. However, since the slide rotates, there is also a component of Coriolis acceleration between the slide and slider.

Construction of the acceleration polygon begins with determining the acceleration of point $B_2$. Since $B_2$ and $O_2$ are separated by fixed distance $r_2$,

$$\bar{a}_{B_2} = \bar{a}_{B_2 O_2} = \left(-r_2 \dot{\theta}_2^2\right)\bar{i}_{r_{B_2 O_2}} + \left(r_2 \ddot{\theta}_2\right)\bar{i}_{\theta_{B_2 O_2}} \tag{3.6-3}$$

For constant rotational speed, Equation (3.6-3) reduces to

$$\bar{a}_{B_2} = \left(-r_2 \dot{\theta}_2^2\right)\bar{i}_{r_{B_2 O_2}} = (-6.0 \times 30^2)\bar{i}_{r_{B_2 O_2}} = -5{,}400 \text{ in/sec}^2 \, \bar{i}_{r_{B_2 O_2}}$$

Acceleration vector $\bar{a}_{B_2}$ is drawn relative to the pole point in the negative radial direction defined by points $B_2$ and $O_2$.

Next, we determine the acceleration of point $B_3$. Recognizing that $B_3$ moves relative to point $O_3$, we write the relative acceleration equation

$$\bar{a}_{B_3} = \bar{a}_{O_3} + \bar{a}_{B_3 O_3} \tag{3.6-4}$$

For a fixed distance between $B_3$ and $O_3$

$$\bar{a}_{B_3 O_3} = \left(-r_{B_3 O_3}\dot{\theta}_3^2\right)\bar{i}_{r_{B_3 O_3}} + \left(r_{B_3 O_3}\ddot{\theta}_3\right)\bar{i}_{\theta_{B_3 O_3}}$$

$$= \left(a^N_{B_3 O_3}\right)\bar{i}_{r_{B_3 O_3}} + \left(a^T_{B_3 O_3}\right)\bar{i}_{\theta_{B_3 O_3}} \tag{3.6-5}$$

The normal acceleration component is

$$a^N_{B_3 O_3} = -r_{B_3 O_3}\dot{\theta}_3^2 = -5.57 \times 20.3^2 = -2{,}300 \text{ in/sec}^2$$

From the pole point, we draw the normal component of acceleration, and from the head of the vector for normal acceleration, the line of action of the tangential component.

We now examine the relative acceleration between $B_3$ and $B_2$. Using the special case of the acceleration of a slider on a straight slide presented in Section 2.5.2, we have in this instance

$$\bar{a}_{B_3} = \bar{a}_{B_2} + \bar{a}_{B_3 B_2} \tag{3.6-6}$$

where

$$\begin{aligned}
\bar{a}_{B_3 B_2} &= (\ddot{r}_{B_3 B_2})\bar{i}_{r_{B_3 O_3}} + (2\dot{r}_{B_3 B_2}\dot{\theta}_3)\bar{i}_{\theta_{B_3 O_3}} \\
&= \left(a_{B_3 B_2}^S\right)\bar{i}_{r_{B_2 A}} + \left(a_{B_3 B_2}^C\right)\bar{i}_{\theta_{B_2 A}}
\end{aligned} \tag{3.6-7}$$

The magnitude of the Coriolis acceleration is

$$\left|a_{B_3 B_2}^C\right| = |2\dot{r}_{B_3 B_2}\dot{\theta}_3| = |2 \times 140 \times 20.3| = 5,680 \,\text{in/sec}^2$$

In Equation (3.6-7), the relative Coriolis and sliding acceleration components are of point $B_3$ with respect to point $B_2$. Therefore, these relative acceleration components are added to the polygon starting from point $B_2$. We can begin by adding the Coriolis acceleration, since we now know its magnitude and can determine its direction. At this instant, we only know the line of action of the sliding acceleration.

We currently have the acceleration of the slider (point $B_2$) and will draw a Coriolis acceleration of the slide (point $B_3$) with respect to the slider. Directions of relative Coriolis acceleration are given in Figure 2.8. Using this figure, with the known combination of relative sliding and rotational speed of the slide, we determine that the direction of the relative Coriolis acceleration is downwards and to the left. However, this figure was set up with the acceleration of the slider with respect to the slide. In this instance, we know the acceleration of the slider, and are seeking the acceleration of the slide. Recognizing that

$$a_{B_2 B_3}^C = -a_{B_3 B_2}^C \tag{3.6-8}$$

we reverse the direction of the acceleration as shown in Figure 2.8. This is illustrated in Figure 3.19.

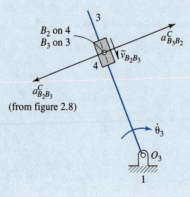

**Figure 3.19**  Directions of Coriolis acceleration.

The Coriolis component is added to the polygon in Figure 3.18(b). At the head of the vector of the Coriolis acceleration, the line of action of the sliding acceleration is added.

The acceleration of the point $B_3$ is unique. Therefore, in order to satisfy the relative acceleration expressions, Equations (3.6-4) and (3.6-6), as well as the lines of action of the tangential and sliding acceleration components, the head of the acceleration of point $B_3$ must lie at the intersection of the two lines of action. This is designated as point $B_3$ on the polygon.

The acceleration of point $C$, located on link 3, is now determined by considering the appropriate normal and tangential components with respect to points $B_3$ and $O_3$, or by using the acceleration image. The completed acceleration polygon is given in Figure 3.18(b).

From this figure

$$a_{B_3 O_3}^T = 1{,}490 \text{ in/sec}^2$$

and therefore the angular acceleration of link 3 is

$$\ddot{\theta}_3 = \frac{a_{B_3 O_3}^T}{r_{B_3 O_3}} = \frac{1{,}490}{5.57} = 268 \text{ rad/sec}^2 \text{ CW}$$

Since this angular acceleration is in the same direction as the rotational speed, link 3 is speeding up.

## EXAMPLE 3.15  Acceleration Polygon Analysis

Consider the mechanism presented in Example 3.9 (Figure 3.11). The mechanism is redrawn in Figure 3.20(a). The rotational speed of link 2 is constant. Determine the acceleration polygon of the mechanism and the angular acceleration of link 5 for the given configuration.

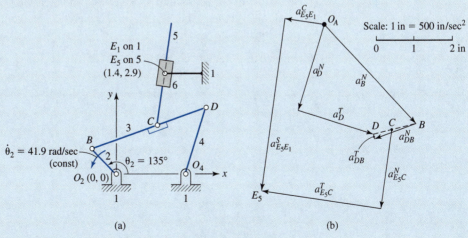

(a)

(b)

**Figure 3.20**  (a) Mechanism. (b) Acceleration polygon.

**SOLUTION**

For this problem, there can be a sliding component of acceleration of the slider (link 6) with respect to the slide (link 5). However, since the slide rotates, a component of Coriolis acceleration is present.

The following procedure is employed:

(a) Calculate the normal acceleration components:

$$a_{BO_2}^N = -r_{BO_2}\dot{\theta}_2^2 = -1.0 \times 41.9^2 = -1{,}760 \text{ in/sec}^2$$

$$a_{DB}^N = -r_{DB}\dot{\theta}_3^2 = -3.5 \times 12.9^2 = -582 \text{ in/sec}^2$$

$$a_{DO_4}^N = -r_{DO_4}\dot{\theta}_4^2 = -2.0 \times 24.0^2 = -1{,}150 \text{ in/sec}^2$$

$$a_{E_5C}^N = -r_{E_5C}\dot{\theta}_5^2 = -1.5 \times 26.0^2 = -1{,}010 \text{ in/sec}^2$$

(b) Calculate the magnitude of the Coriolis acceleration:

$$a_{E_5E_1}^C = 2\dot{r}_{E_5E_1}\dot{\theta}_5 = 2 \times 10 \times 26.0 = 520 \text{ in/sec}^2$$

(c) Construct the acceleration polygon as follows:

- Draw $a_B^N$ and locate $B$.
- Draw $a_D^N$ and $a_{DB}^N$.
- Draw the lines of action of $a_D^T$ and $a_{DB}^T$ and locate $D$.
- Locate $C$ by using the acceleration image.
- Draw $a_{E_5C}^N$ and $a_{E_5E_1}^C$ (refer to Figure 3.21 for the direction of the Coriolis acceleration).
- Draw the lines of action of $a_{E_5C}^T$ and $a_{E_5E_1}^S$ and locate $E_5$.
- Label all vectors. The completed acceleration polygon is shown in Figure 3.20(b).

(d) Determine the angular acceleration of link 5:

$$\ddot{\theta}_5 = \frac{a_{E_5C}^T}{r_{E_5C}} = \frac{1{,}550}{1.5} = 1{,}030 \text{ rad/sec}^2 \text{ CCW}$$

**Figure 3.21** Directions of Coriolis acceleration.

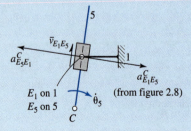

## EXAMPLE 3.16 Velocity and Acceleration Analysis of a Cam Mechanism

Consider the cam mechanism shown in Figure 3.22. The rotational speed of link 2 and link lengths are

$$\dot{\theta}_2 = 3.0 \, \text{rad/sec CCW (constant)}$$

$$r_{O_2 B_2} = 1.3 \, \text{cm}; \qquad r_{O_4 B_4} = 1.8 \, \text{cm}$$

The radii of the circular cam and roller follower are indicated in the figure. Determine the velocity and acceleration polygons for the given configuration.

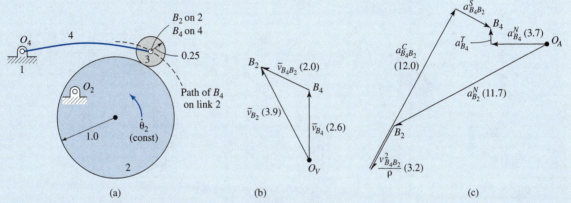

(a)                         (b)                         (c)

**Figure 3.22** (a) Mechanism. (b) Velocity polygon. (c) Acceleration polygon.

### SOLUTION

We consider the center of the roller follower, point $B$, to be composed of two points, $B_2$ and $B_4$. Point $B_2$ is considered as an imaginary physical extension of link 2. The motion between these two points is equivalent to a slider moving on a curved slide. The radius of curvature is equal to the radius of the cam plus the radius of the roller follower. Between points $B_2$ and $B_4$ there can be relative sliding, plus Coriolis and circular components of acceleration.

The calculated values of velocity and acceleration required to complete the polygons are

$$v_{B_2} = r_{B_2 O_2} \dot{\theta}_2 = 1.3 \times 3.0 = 3.9 \, \text{cm/sec}$$

$$\dot{\theta}_4 = \frac{v_{B_4}}{r_{B_4 O_4}} = \frac{2.6}{1.8} = 1.44 \, \text{rad/sec CCW}$$

$$a_{B_2}^N = -r_{B_2 O_2} \dot{\theta}_2^2 = -1.3 \times 3.0^2 = -11.7 \, \text{cm/sec}^2$$

$$a_{B_4}^N = -r_{B_4 O_4} \dot{\theta}_4^2 = -1.8 \times 1.44^2 = -3.7 \, \text{cm/sec}^2$$

$$a_{B_4 B_2}^C = 2 v_{B_4 B_2} \dot{\theta}_2 = 2 \times 2.0 \times 3.0 = 12.0 \, \text{cm/sec}^2$$

$$\frac{v_{B_4 B_2}^2}{\rho} = \frac{v_{B_4 B_2}^2}{r_{C B_2}} = \frac{2.0^2}{(1.0 + 0.25)} = 3.2 \, \text{cm/sec}^2$$

The completed velocity polygon and acceleration polygon are shown in Figures 3.22(b) and 3.22(c), respectively. Magnitudes of the velocities and accelerations are indicated in parentheses.

# PROBLEMS

**P3.1**  For the mechanism shown in Figure P3.1, determine the magnitudes of the linear velocities of points $B$, $C$, and $D$ using instantaneous centers.

$$r_{O_2O_4} = 4.0\,\text{in};    r_{O_2B} = 1.0\,\text{in}$$
$$r_{BD} = 3.0\,\text{in};    r_{O_4D} = 2.5\,\text{in}$$
$$r_{BC} = 2.0\,\text{in};    r_{CD} = 1.5\,\text{in}$$
$$\theta_2 = 45°;    \dot\theta_2 = 60\,\text{rad/sec CCW}$$

$V_c = r_{P_{13C}} |\dot\theta_3|$

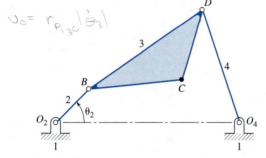

**Figure P3.1**

**P3.2**  For the mechanism shown in Figure P3.2, determine the linear velocity of the follower using instantaneous centers. $\quad \checkmark P_{2,3}$

$$r_{O_2C} = 1.5\,\text{in};    \rho = 0.75\,\text{in}$$
$$\theta_2 = 110°;    \dot\theta_2 = 600\,\text{rpm CW}$$

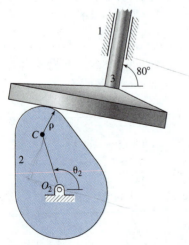

**Figure P3.2**

**P3.3**  For the mechanism shown in Figure P3.3, determine the magnitude of the linear velocity of link 4 using instantaneous centers.

$$r_{O_2B} = 2.5\,\text{cm}$$
$$\theta_2 = 45°;    \dot\theta_2 = 40\,\text{rpm CCW}$$

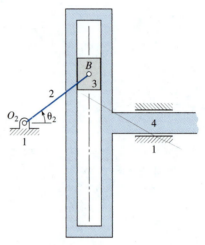

**Figure P3.3**

**P3.4**  For the mechanism shown in Figure P3.4, determine, using instantaneous centers,

(a) the angular velocity of link 4
(b) the linear velocity of point $D$

$$r_{O_2B_2} = 2.0\,\text{in};    r_{O_2O_4} = 1.5\,\text{in}$$
$$r_{O_4C} = 1.0\,\text{in};    r_{CD} = 3.0\,\text{in}$$
$$\theta_2 = 30°;    \dot\theta_2 = 36\,\text{rad/sec CW}$$

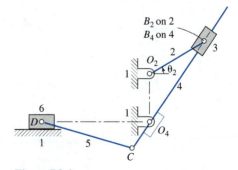

**Figure P3.4**

**P3.5**  For the mechanism shown in Figure P3.5, determine the linear velocity of link 6 using instantaneous centers. $\quad up_{2,6}$

$r_{O_2O_3} = 6.0\,\text{cm};$    $r_{O_2B_2} = 4.0\,\text{cm}$

$r_{O_3C} = 11.0\,\text{cm};$    $r_{CD} = 5.0\,\text{cm}$

$\theta_2 = 30°;$    $\dot{\theta}_2 = 20\,\text{rad/sec CCW}$

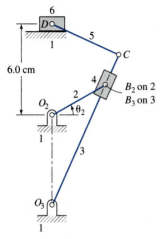

**Figure P3.5**

**P3.6**  For the mechanism shown in Figure P3.6, determine the rotational speed of link 6 using instantaneous centers.

$r_{O_2O_6} = 8.0\,\text{cm};$    $r_{O_2O_3} = 6.0\,\text{cm}$

$r_{O_2B_2} = 3.0\,\text{cm};$    $r_{O_3C} = 12.0\,\text{cm}$

$r_{CD} = 4.0\,\text{cm};$    $r_{O_6D} = 6.0\,\text{cm}$

$\theta_2 = 300°;$    $\dot{\theta}_2 = 5\,\text{rad/sec CCW}$

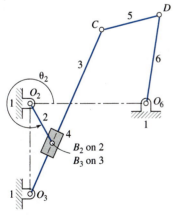

**Figure P3.6**

**P3.7**  For the mechanism shown in Figure P3.7, determine the linear velocity of point $C$ using instantaneous centers.

$r_{O_6A} = r_{O_6C} = 1.0\,\text{in}$

$r_{AD} = 1.5\,\text{in};$    $r_{O_2B_2} = 1.0\,\text{in}$

$\theta_2 = 315°;$    $\dot{\theta}_2 = 3\,\text{rad/sec CW}$

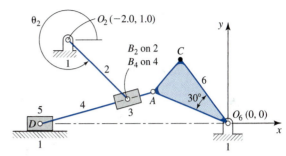

**Figure P3.7**

**P3.8**  For the mechanism shown in Figure P3.8, determine the linear velocity of point $D$ using instantaneous centers.

$r_{O_2O_4} = 1.0\,\text{in};$    $r_{O_2O_6} = 2.5\,\text{in}$

$r_{O_2A} = 1.0\,\text{in};$    $r_{AB} = 2.5\,\text{in}$

$r_{O_4B} = 1.5\,\text{in};$    $r_{O_4C} = 2.0\,\text{in}$

$r_{CD} = 1.0\,\text{in};$    $r_{CE} = 1.5\,\text{in}$

$r_{O_6E} = 2.0\,\text{in}$

$\theta_2 = 45°;$    $\dot{\theta}_2 = 5\,\text{rad/sec CCW}$

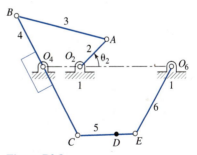

**Figure P3.8**

**P3.9**  For the mechanism in the position shown in Figure P3.9, determine

(a) the rotational speed of link 3
(b) the linear speed of link 4

$r_{O_2B_2} = 4.0\,\text{cm};$    $\dot{\theta}_2 = 75\,\text{rpm CW (constant)}$

**P3.10**  For the mechanism shown in Figure P3.10, the two links are in contact at point $P$. For the position shown, determine

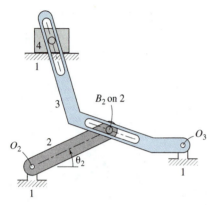

**Figure P3.9**

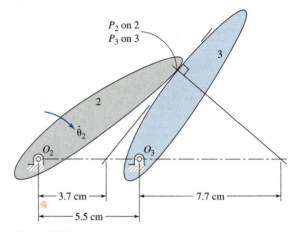

**Figure P3.11**

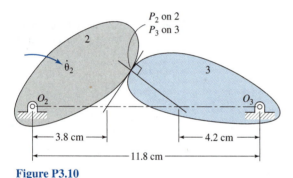

**Figure P3.10**

(a) the absolute velocity of $P$ on 2
(b) the absolute velocity of $P$ on 3
(c) the sliding velocity between links at $P$, if any
(d) the magnitude of the ratio of rotational speeds of links 2 and 3

$$r_{O_2P_2} = 5.3 \text{ cm}; \qquad r_{O_3P_3} = 7.0 \text{ cm}$$
$$\dot{\theta}_2 = 10 \text{ rpm CW}$$

**P3.11** For the mechanism shown in Figure P3.11, the two links are in contact at point $P$. For the position shown, determine

(a) the absolute velocity of $P$ on 2
(b) the absolute velocity of $P$ on 3
(c) the sliding velocity between links at $P$, if any
(d) the magnitude of the ratio of rotational speeds of links 2 and 3

$$r_{O_2P_2} = 8.8 \text{ cm}; \qquad r_{O_3P_3} = 5.1 \text{ cm}$$
$$\dot{\theta}_2 = 20 \text{ rpm CW}$$

**P3.12** For the mechanism shown in Figure P3.12

(a) draw the velocity polygon (employ scale: 1 in = 10 in/sec)
(b) specify
  (i)    the velocity of point $B$
  (ii)   the velocity of point $C$
  (iii)  the velocity of point $D$
  (iv)   the velocity of point $B$ relative to point $D$
  (v)    the angular velocity of link 3

$$r_{O_2B} = 2.0 \text{ in}; \qquad r_{BD} = 3.0 \text{ in}$$
$$r_{BC} = 1.0 \text{ in}$$
$$\theta_2 = 135°; \qquad \dot{\theta}_2 = 120 \text{ rpm CW}$$

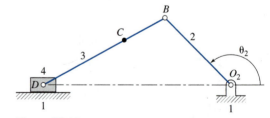

**Figure P3.12**

**P3.13** For the mechanism shown in Figure P3.13

(a) draw the velocity polygon (employ scale: 1 in = 10 in/sec)
(b) specify
  (i)    the velocity of point $B$
  (ii)   the velocity of point $C$
  (iii)  the velocity of point $D$

(iv) the angular velocity of link 3
(v) the angular velocity of link 4

$$r_{O_2B} = 1.5 \text{ in}; \qquad r_{BD} = 3.0 \text{ in}$$
$$r_{O_2O_4} = 4.0 \text{ in}; \qquad r_{O_4D} = 2.5 \text{ in}$$
$$r_{BC} = 3.5 \text{ in}; \qquad r_{CD} = 1.0 \text{ in}$$
$$\theta_2 = 60°; \qquad \dot\theta_2 = 140 \text{ rpm CW}$$

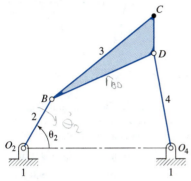

**Figure P3.13**

**P3.14** For the mechanism shown in Figure P3.14

(a) draw the velocity polygon (employ scale: 1 cm = 5 cm/sec)
(b) specify
   (i) the velocity of point $D$
   (ii) the velocity of point $E$
   (iii) the angular velocity of link 3
   (iv) the angular velocity of link 5

$$r_{O_2B} = 4.5 \text{ cm}; \qquad r_{BD} = 6.0 \text{ cm}$$
$$r_{CD} = 1.5 \text{ cm}; \qquad r_{DE} = 3.0 \text{ cm}$$
$$\theta_2 = 135°; \qquad \dot\theta_2 = 6 \text{ rad/sec CW}$$

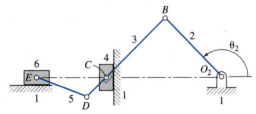

**Figure P3.14**

**P3.15** For the mechanism shown in Figure P3.15

(a) draw the velocity polygon (employ scale: 1 cm = 10 cm/sec)
(b) draw the acceleration polygon (employ scale: 1 cm = 50 cm/sec²)

(c) specify
   (i) the sliding acceleration of point $B_2$ relative to point $B_3$
   (ii) the angular accelerations of links 3 and 5

$$r_{O_2B_2} = 6.0 \text{ cm}; \qquad r_{O_2O_3} = 5.0 \text{ cm}$$
$$r_{O_3C} = 6.0 \text{ cm}; \qquad r_{CD} = 9.0 \text{ cm}$$
$$\theta_2 = 10°; \qquad \dot\theta_2 = 80 \text{ rpm CW (constant)}$$

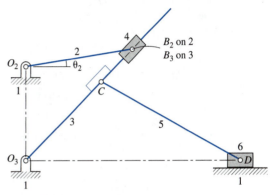

**Figure P3.15**

**P3.16** For the mechanism shown in Figure P3.16

(a) draw the velocity polygon (employ scale: 1 cm = 10 cm/sec)
(b) draw the acceleration polygon (employ scale: 1 cm = 100 cm/sec²)

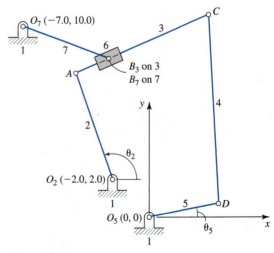

**Figure P3.16**

(c) specify the angular acceleration of link 7

$$r_{O_2A} = 6.0 \, \text{cm}; \qquad r_{AC} = 8.0 \, \text{cm}$$
$$r_{CD} = 10.0 \, \text{cm}; \qquad r_{O_5D} = 4.0 \, \text{cm}$$
$$r_{AB_3} = 2.0 \, \text{cm}; \qquad r_{O_7B_7} = 5.0 \, \text{cm}$$
$$\theta_2 = 110°; \qquad \dot{\theta}_2 = 75 \, \text{rpm CW (constant)}$$
$$\theta_5 = 10°; \qquad \dot{\theta}_5 = 100 \, \text{rpm CW}$$
$$\ddot{\theta}_5 = 75 \, \text{rad/sec}^2 \, \text{CCW}$$

**P3.17** For the mechanism shown in Figure P3.17

(a) draw the velocity polygon (employ scale: 1 cm = 10 cm/sec)

(b) draw the acceleration polygon (employ scale: 1 cm = 100 cm/sec²)

(c) specify the angular acceleration of link 5

$$r_{O_2A} = 3.0 \, \text{cm}; \qquad r_{AD_3} = 10.0 \, \text{cm}$$
$$r_{AB} = 6.0 \, \text{cm}; \qquad r_{BC} = 7.0 \, \text{cm}$$
$$\theta_2 = 100°; \qquad \dot{\theta}_2 = 160 \, \text{rpm CCW (constant)}$$

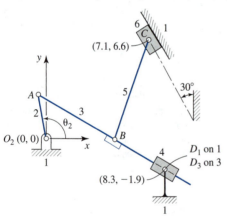

**Figure P3.17**

**P3.18** For the mechanism shown in Figure P3.18

(a) draw the velocity polygon (employ scale: 1 cm = 10 cm/sec)

(b) draw the acceleration polygon (employ scale: 1 cm = 100 cm/sec²)

(c) specify the angular acceleration of link 6

$$r_{O_2A} = 4.0 \, \text{cm}; \qquad r_{AB} = 9.0 \, \text{cm}$$
$$r_{O_4B} = 8.0 \, \text{cm}; \qquad r_{O_4C_4} = 4.0 \, \text{cm}$$
$$r_{O_6C_6} = 6.0 \, \text{cm}$$
$$\theta_2 = 45°; \qquad \dot{\theta}_2 = 140 \, \text{rpm CW (constant)}$$

**P3.19** For the mechanism shown in Figure P3.19

(a) draw the velocity polygon (employ scale: 1 cm = 10 cm/sec)

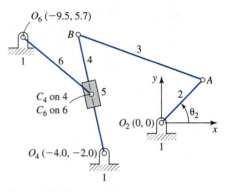

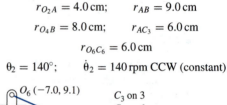

**Figure P3.18**

(b) draw the acceleration polygon (employ scale: 1 cm = 100 cm/sec²)

(c) specify the angular acceleration of link 6

$$r_{O_2A} = 4.0 \, \text{cm}; \qquad r_{AB} = 9.0 \, \text{cm}$$
$$r_{O_4B} = 8.0 \, \text{cm}; \qquad r_{AC_3} = 6.0 \, \text{cm}$$
$$r_{O_6C_6} = 6.0 \, \text{cm}$$
$$\theta_2 = 140°; \qquad \dot{\theta}_2 = 140 \, \text{rpm CCW (constant)}$$

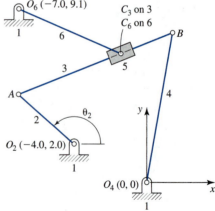

**Figure P3.19**

**P3.20** For the scaled drawing of a mechanism shown in Figure P3.20

(a) draw the velocity polygon (employ scale: 1 cm = 10 cm/sec)

(b) draw the acceleration polygon (employ scale: 1 cm = 100 cm/sec²)

(c) specify

(i) the acceleration of point $B_6$

(ii) the angular accelerations of links 3 and 6

(d) if the angular velocity of link 2 is changed to a constant rate of 80 rpm CCW, specify the new angular acceleration for link 6

$$r_{O_2A} = 5.0\,\text{cm}; \qquad r_{AC} = 13.0\,\text{cm}$$

$$r_{AB_3} = 5.0\,\text{cm}; \qquad r_{O_6B_6} = 8.0\,\text{cm}$$

$$\theta_2 = 135°; \qquad \dot{\theta}_2 = 130\,\text{rpm CCW (constant)}$$

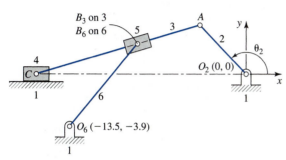

**Figure P3.20**

**P3.21** For the mechanism shown in Figure P3.21

(a) draw the velocity polygon (employ scale: 1 cm = 10 cm/sec)
(b) draw the acceleration polygon (employ scale: 1 cm = 200 cm/sec$^2$)
(c) specify
    (i) the angular acceleration of link 4
    (ii) the acceleration of link 6

$$r_{O_2A_2} = 3.0\,\text{cm}; \qquad r_{O_4B} = 1.0\,\text{cm}$$

$$r_{O_2O_4} = 5.0\,\text{cm}; \qquad r_{BC} = 4.5\,\text{cm}$$

$$\theta_2 = 150°; \qquad \dot{\theta}_2 = 180\,\text{rpm CW (constant)}$$

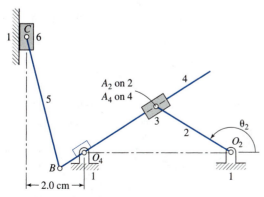

**Figure P3.21**

**P3.22** For the mechanism shown in Figure P3.22

(a) draw the velocity polygon (employ scale: 1 in = 10 in/sec)
(b) draw the acceleration polygon (employ scale: 1 in = 500 in/sec$^2$)
(c) specify
    (i) the sliding acceleration between points $B_2$ and $B_4$
    (ii) the angular acceleration of link 2

$$r_{O_2B_2} = 4.0\,\text{in}; \qquad r_{CB_4} = 0.8\,\text{in}$$

$$r_{O_2D} = 4.25\,\text{in}; \qquad \rho = 1.5\,\text{in}$$

$$\dot{\theta}_4 = 100\,\text{rpm CW (constant)}$$

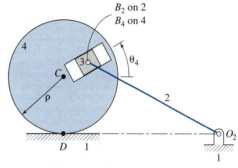

**Figure P3.22**

**P3.23** For the mechanism shown in Figure P3.23

(a) draw the velocity polygon for the limit position shown (employ scale: 1 cm = 10 cm/sec)
(b) repeat part (a) for the other limit position
(c) draw the acceleration polygon for the limit position shown (employ scale: 1 cm = 200 cm/sec$^2$)
(d) for part (c), specify the angular acceleration of links 3 and 4

$$r_{O_2O_4} = 8.0\,\text{cm}; \qquad r_{O_2B} = 2.5\,\text{cm}$$

$$r_{BD} = 9.5\,\text{cm}; \qquad r_{O_4D} = 6.0\,\text{cm}$$

$$\dot{\theta}_2 = 20\,\text{rad/sec CCW (constant)}$$

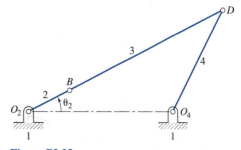

**Figure P3.23**

**P3.24** For the mechanism shown in Figure P3.24

(a) draw the velocity polygon (employ scale: 1 cm = 10 cm/sec)
(b) draw the acceleration polygon (employ scale: 1 cm = 100 cm/sec$^2$)

$$r_{O_2B} = 3.0 \text{ cm}; \qquad r_{BC} = 7.0 \text{ cm}$$

$$r_{O_2O_4} = 8.0 \text{ cm}; \qquad r_{O_4C} = 4.0 \text{ cm}$$

$$r_{O_4D} = 2.0 \text{ cm}; \qquad r_{DE} = 8.0 \text{ cm}$$

$$\theta_2 = 45°; \qquad \dot{\theta}_2 = 20 \text{ rad/sec CCW (constant)}$$

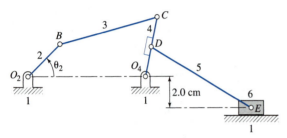

**Figure P3.24**

**P3.25** For the mechanism shown in Figure P3.25

(a) draw the velocity polygon (employ scale: 1 in = 5 in/sec)
(b) draw the acceleration polygon (employ scale: 1 in = 50 in/sec$^2$)

$$r_{O_2C} = 0.75 \text{ in}; \qquad \rho = 1.5 \text{ in}$$

$$\theta_2 = 315°; \qquad \dot{\theta}_2 = 10 \text{ rad/sec CW (constant)}$$

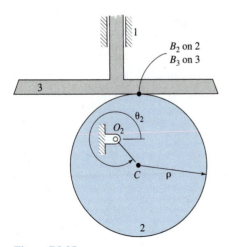

**Figure P3.25**

**P3.26** For the mechanism shown in Figure P3.26

(a) draw the velocity polygon (employ scale: 1 cm = 10 cm/sec)
(b) specify
   (i) the rotational speeds of links 3 and 4
   (ii) the velocities of the midpoints of links 2, 3, and 4 (i.e., points $D$, $E$, and $F$)
(c) draw the acceleration polygon (employ scale: 1 cm = 400 cm/sec$^2$)
(d) specify
   (i) the angular accelerations of links 3 and 4
   (ii) the accelerations of the midpoints of links 2, 3, and 4

$$r_{O_2O_4} = 12.0 \text{ cm}; \qquad r_{O_2B} = 4.0 \text{ cm}$$

$$r_{BC} = 14.0 \text{ cm}; \qquad r_{O_4C} = 8.0 \text{ cm}$$

$$\theta_2 = 45°; \qquad \dot{\theta}_2 = 30 \text{ rad/sec CW (constant)}$$

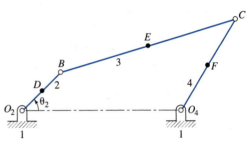

**Figure P3.26**

**P3.27** For the mechanism shown in Figure P3.27

(a) draw the velocity polygon (employ scale: 1 cm = 10 cm/sec)
(b) draw the acceleration polygon (employ scale: 1 cm = 100 cm/sec$^2$)
(c) specify the angular acceleration of link 3

$$r_{O_2O_3} = 6.0 \text{ cm}; \qquad r_{O_2B_2} = 3.0 \text{ cm}$$

$$r_{O_3C} = 12.0 \text{ cm}; \qquad r_{CD} = 4.5 \text{ cm}$$

$$\theta_2 = 30°; \qquad \dot{\theta}_2 = 20 \text{ rad/sec CCW (constant)}$$

**P3.28** For the scaled mechanism shown in Figure P3.28

(a) draw the velocity polygon (employ scale: 1 cm = 10 cm/sec)
(b) draw the acceleration polygon (employ scale: 1 cm = 100 cm/sec$^2$)
(c) specify
   (i) the angular accelerations of links 3 and 5

# Problems

**119**

(ii) the acceleration of link 6

$$r_{O_2A} = 3.0 \text{ cm}; \quad r_{BA} = 4.0 \text{ cm}$$
$$r_{AC_3} = 8.0 \text{ cm}; \quad r_{BD} = 5.0 \text{ cm}$$
$$\theta_2 = 130°; \quad \dot{\theta}_2 = 180 \text{ rpm CCW (constant)}$$

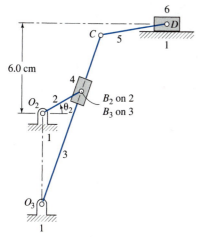

**Figure P3.27**

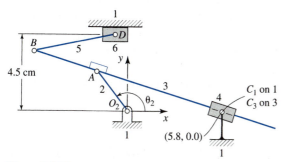

**Figure P3.28**

**P3.29** For the mechanism shown in Figure P3.29

(a) draw the velocity polygon (employ scale: 1 cm = 10 cm/sec)
(b) specify
   (i) the velocity of point $G$
   (ii) the Coriolis acceleration of $E_5$ with respect to $E_1$

$$r_{O_2B} = 4.0 \text{ cm}; \quad r_{BC} = 4.0 \text{ cm}$$
$$r_{BD} = 10.0 \text{ cm}; \quad r_{CE_5} = 6.0 \text{ cm}$$
$$r_{CG} = 9.0 \text{ cm}$$
$$\theta_2 = 30°$$
$$\dot{\theta}_2 = 160 \text{ rpm CCW}$$
$$\ddot{\theta}_2 = 50 \text{ rad/sec}^2 \text{ CW}$$

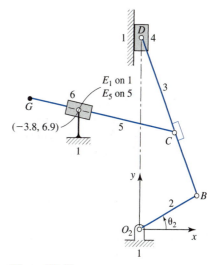

**Figure P3.29**

# 4   Analytical Kinematic Analysis of Planar Mechanisms

## 4.1 INTRODUCTION

This chapter presents a method of generating analytical expressions for kinematic quantities of planar mechanisms. These expressions may be programmed on a computer to determine numerical values for numerous mechanism positions more efficiently than using the graphical techniques presented in Chapter 3.

The method described herein is based on the use of complex numbers. A brief overview of complex numbers is included in Appendix C.

## 4.2 LOOP CLOSURE EQUATION

A *loop* of a mechanism involves traversing along links and through kinematic pairs, to arrive back at the starting point. Mechanisms may be classified according to the number of loops required to cover all links and kinematic pairs.

Figure 4.1 illustrates two mechanisms, each having only one loop. In each case, a single loop covers all links and kinematic pairs of the mechanism. Figure 4.2 illustrates two mechanisms where each requires two loops.

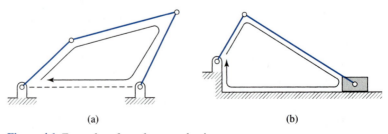

(a)                                    (b)

**Figure 4.1** Examples of one-loop mechanisms.

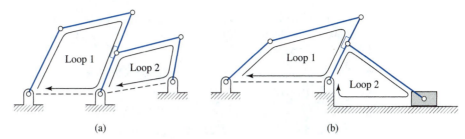

**Figure 4.2** Examples of two-loop mechanisms.

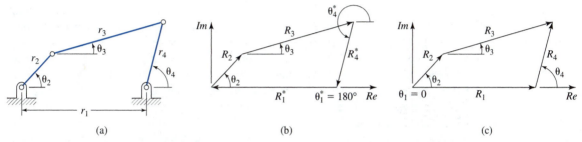

**Figure 4.3** Complex vector representation of a four-bar mechanism: (a) mechanism, (b) vector representation, version 1, (c) vector representation, version 2.

Vectors in a complex plane may be combined to represent a mechanism loop. However, no vector representation is unique. Figures 4.3(b) and 4.3(c) illustrate two different complex vector representations of the mechanism shown in Figure 4.3(a). The representation shown in Figure 4.3(b) has all vectors placed head to tail. The corresponding angular displacements of the vectors are also specified. The representation shown in Figure 4.3(c) has two vectors, $R_1$ and $R_4$, pointing in opposite directions to those shown in Figure 4.3(b). Hence, the corresponding arguments (i.e., angles) are different.

In constructing a vector loop, one proceeds by assigning a direction for each vector in the loop. In comparing the two versions presented, one would probably select the second version, since the arguments of the vectors correspond to those commonly employed in depicting angular displacements of the links of the mechanism.

It is possible to write a *loop closure equation* for each vector loop to specify the condition that the loop must remain closed as the links of the mechanism move. The loop closure equation corresponding to the vectors shown in Figure 4.3(b) is

$$R_1^* + R_2 + R_3 + R_4^* = 0 \tag{4.2-1}$$

and that associated with Figure 4.3(c) is

$$R_1 - R_2 - R_3 + R_4 = 0 \tag{4.2-2}$$

where

$$R_1^* = -R_1; \qquad R_4^* = -R_4 \tag{4.2-3}$$

There are *independent variables* and *dependent variables* that apply to a mechanism. Inputs to a mechanism are considered to be independent variables. Dependent variables are produced as a result of the inputs.

- The number of independent variables is equal to the mobility of the mechanism (see Section 1.5).
- There are two dependent variables for each mechanism loop.

The dependent variables may be either magnitudes or arguments (i.e., angular displacements) of the vectors. If we consider the case of the four-bar mechanism shown in Figure 4.3(c), and the independent variable is the angle of link 2, then the two dependent variables are $\theta_3$ and $\theta_4$.

As a second example, consider the complex vector representation of a slider crank mechanism as shown in Figure 4.4(a). The corresponding complex vector representation is shown in Figure 4.4(b). If the independent variable is selected as the angular displacement of link 2, then the dependent variables are the angular displacement of link 3 and the magnitude of vector $R_4$. Vector $R_4$ is along the line of action of the slide. Vector $R_1$ corresponds to the slider offset, which is fixed, and neither its length nor angle will change as the mechanism operates. Therefore, neither the magnitude nor argument of vector $R_1$ is variable.

Figure 4.5 shows a five-bar mechanism. The mobility of this mechanism is two. If the two independent variables are $\theta_2$ and $\theta_5$, then the two remaining variable quantities are $\theta_3$ and $\theta_4$, which are the dependent variables.

Figure 4.6 shows a six-bar mechanism. The mobility of this mechanism is one. The single independent variable is taken as $\theta_2$. This mechanism has two loops, and thus there are a total of four dependent variables (two for each loop). Complex vectors are superimposed on the mechanism links in Figure 4.6 to identify more easily these variables. The dependent

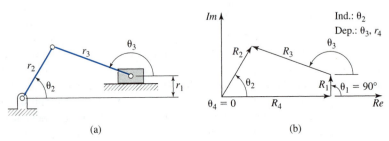

(a)                                          (b)

**Figure 4.4** (a) Slider crank mechanism. (b) Complex vector representation.

**Figure 4.5** Five-bar mechanism.

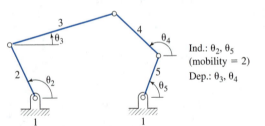

Ind.: $\theta_2$, $\theta_5$
(mobility = 2)
Dep.: $\theta_3$, $\theta_4$

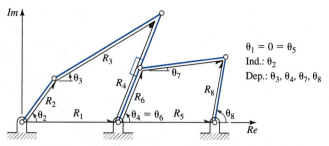

**Figure 4.6** Complex vector representation of a six-bar mechanism.

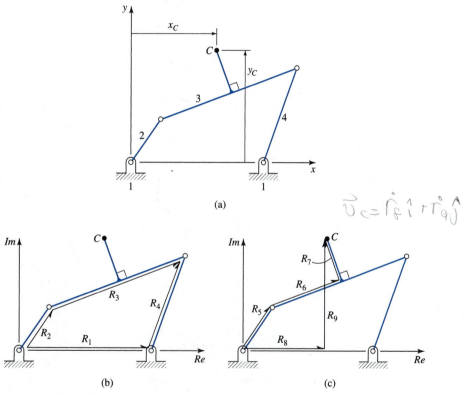

(a)

(b)             (c)

**Figure 4.7** Complex vector representation of a four-bar mechanism with a coupler point:
(a) mechanism, (b) first vector loop, (c) second vector loop.

variables are $\theta_3$, $\theta_4$, $\theta_7$, and $\theta_8$. Also

$$\theta_6 = \theta_4$$

and therefore, once the analysis of the first loop has been completed, $\theta_6$ can be considered as a known value in the analysis of the second loop.

Figure 4.7(a) shows a four-bar mechanism with a coupler point. This mechanism may be represented by two vector loops. One vector loop, shown in Figure 4.7(b), is similar to Figure 4.3(c). The second loop, shown in Figure 4.7(c), involves the coupler point. For the

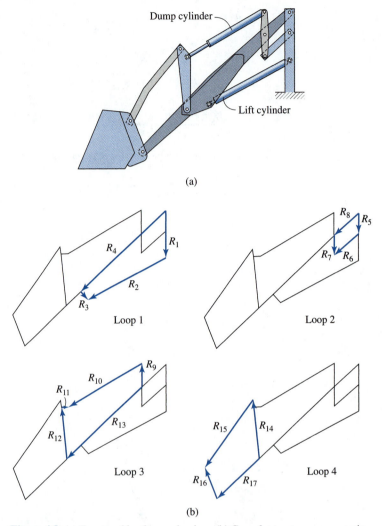

(a)

(b)

**Figure 4.8**  (a) Front-end loader mechanism. (b) Complex vector representation.

second loop, we let

$$\theta_8 = 0; \qquad \theta_9 = 90°$$

The dependent variables associated with the second vector loop are the magnitudes of vectors $R_8$ and $R_9$. Note that angular displacements of vectors for the second vector loop can be easily determined if angles in the first loop are known. The relationships between angles of these two loops are

$$\theta_5 = \theta_2; \qquad \theta_6 = \theta_3; \qquad \theta_7 = \theta_3 + 90°$$

A front-end loader is shown in Figure 4.8(a). The corresponding complex vector representation of this mechanism is given in Figure 4.8(b). The loop closure equations are

- **Loop 1:** $R_1 + R_2 - R_3 - R_4 = 0$
- **Loop 2:** $R_5 + R_6 - R_7 - R_8 = 0$

- **Loop 3:** $R_9 + R_{10} - R_{11} - R_{12} - R_{13} = 0$
- **Loop 4:** $R_{14} + R_{15} - R_{16} - R_{17} = 0$

The mobility of this mechanism is two. The two independent variables are the magnitudes of vectors $R_2$ and $R_{10}$. Since four vector loops are needed to represent this mechanism, eight dependent variables are present.

In the following two sections, analyses are performed for mechanisms that may be represented by one vector loop (Section 4.3), followed by those which require more than one vector loop (Section 4.4).

## 4.3 COMPLEX VECTOR ANALYSIS OF A PLANAR ONE-LOOP MECHANISM

To perform a complex vector analysis, the following general steps are required:

1. Represent the mechanism links, using complex vectors, to form a polygon.
2. Generate a loop closure equation and identify the two dependent variables.
3. Generate two equations, which are real and imaginary parts of the loop closure equation.
4. Solve for the two dependent variables using equations from step 3.
5. Determine the values of the time derivatives of the dependent variables, using one of the following methods:

   (i) *Method #1*: Use the solved expressions for the dependent variables, obtained in step 4, and differentiate them with respect to time.
   (ii) *Method #2*: Differentiate the real and imaginary components of the loop closure equation, obtained in step 3, and solve for the time derivative quantities.

The above procedure requires the use of trigonometric identities. A listing of several such identities is provided on page 494.

### 4.3.1 Scotch Yoke Mechanism

Consider the scotch yoke mechanism shown in Figure 4.9(a), where the input motion is the rotation of link 2.

A complex vector representation of this mechanism is given in Figure 4.9(b), and the corresponding loop closure equation is

$$R_2 - R_3 - R_1 = 0 \qquad (4.3-1)$$

Using

$$R_j = r_j(\cos \theta_j + i \sin \theta_j), \qquad j = 1, 2, \ldots \qquad (4.3-2)$$

we obtain

$$r_2(\cos \theta_2 + i \sin \theta_2) - r_3(\cos \theta_3 + i \sin \theta_3) - r_1(\cos \theta_1 + i \sin \theta_1) = 0 \qquad (4.3-3)$$

Recognizing that $\theta_1 = 0$ and $\theta_3 = 90°$, Equation (4.3-3) may be simplified to

$$r_2 \cos \theta_2 - r_1 + i(r_2 \sin \theta_2 - r_3) = 0 \qquad (4.3-4)$$

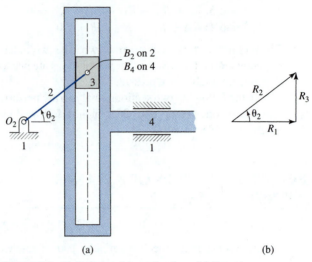

(a)                                         (b)

**Figure 4.9** (a) Scotch yoke mechanism. (b) Complex vector representation.

Considering the real and imaginary components of Equation (4.3-4), we obtain

$$r_1 = r_2 \cos \theta_2 \qquad (4.3\text{-}5)$$

and

$$r_3 = r_2 \sin \theta_2 \qquad (4.3\text{-}6)$$

The independent variable is the angular displacement $\theta_2$, and the dependent variables are $r_1$ and $r_3$. That is, Equations (4.3-5) and (4.3-6) have been arranged to express the dependent variables in terms of the independent variable and a specified link dimension.

Time derivative quantities of the dependent variables may now be determined. Since the two dependent variables are expressed explicitly in terms of known quantities, both methods described above are identical.

Taking time derivatives of Equations (4.3-5) and (4.3-6), where we recognize that $r_2$ is constant, and therefore $\dot{r}_2 = 0$, gives

$$\dot{r}_1 = -r_2 \dot{\theta}_2 \sin \theta_2 \qquad (4.3\text{-}7)$$

and

$$\dot{r}_3 = r_2 \dot{\theta}_2 \cos \theta_2 \qquad (4.3\text{-}8)$$

The second time derivatives are

$$\ddot{r}_1 = -r_2 \dot{\theta}_2^2 \cos \theta_2 - r_2 \ddot{\theta}_2 \sin \theta_2 \qquad (4.3\text{-}9)$$

and

$$\ddot{r}_3 = -r_2\dot{\theta}_2^2 \sin\theta_2 + r_2\ddot{\theta}_2 \cos\theta_2 \qquad (4.3\text{-}10)$$

A special case is obtained for Equations (4.3-9) and (4.3-10) if we consider a constant input rotational speed, by setting

$$\ddot{\theta}_2 = 0$$

It is possible to compare expressions for calculated quantities obtained using the graphical and the analytical techniques. For this example

$$|\dot{r}_1| = |\bar{v}_{B_4}|; \qquad |\dot{r}_3| = |\bar{v}_{B_4B_2}|; \qquad |\ddot{r}_1| = |\bar{a}_{B_4}|; \qquad |\ddot{r}_3| = \left|a_{B_4B_2}^S\right|$$

## 4.3.2 Inverted Slider Crank Mechanism

Consider the inverted slider crank mechanism shown in Figure 4.10(a). The input motion is the rotation of link 2.

A vector representation of this mechanism is given in Figure 4.10(b), and the corresponding loop closure equation is

$$R_2 - R_3 - R_1 = 0 \qquad (4.3\text{-}11)$$

Using Equation (4.3-2), we obtain

$$r_2(\cos\theta_2 + i\sin\theta_2) - r_3(\cos\theta_3 + i\sin\theta_3) - r_1(\cos\theta_1 + i\sin\theta_1) = 0 \qquad (4.3\text{-}12)$$

Recognizing that $\theta_1 = 0$, Equation (4.3-12) becomes

$$r_2\cos\theta_2 - r_3\cos\theta_3 - r_1 + i(r_2\sin\theta_2 - r_3\sin\theta_3) = 0 \qquad (4.3\text{-}13)$$

The real component of Equation (4.3-13) is

$$r_2\cos\theta_2 - r_3\cos\theta_3 - r_1 = 0 \qquad (4.3\text{-}14)$$

and the imaginary component is

$$r_2\sin\theta_2 - r_3\sin\theta_3 = 0 \qquad (4.3\text{-}15)$$

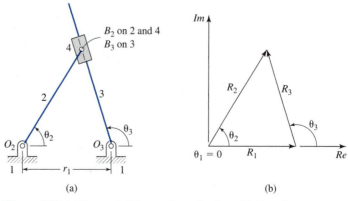

(a)　　　　　　　　　　　　　(b)

**Figure 4.10** (a) Inverted slider crank mechanism. (b) Complex vector representation.

The independent variable is the input angular displacement $\theta_2$, and the dependent variables are $r_3$ and $\theta_3$.

Rearranging Equations (4.3-14) and (4.3-15) to bring these dependent variables onto one side,

$$r_3 \cos \theta_3 = r_2 \cos \theta_2 - r_1 \qquad (4.3\text{-}16)$$

$$r_3 \sin \theta_3 = r_2 \sin \theta_2 \qquad (4.3\text{-}17)$$

Squaring and adding Equations (4.3-16) and (4.3-17), and employing the identity

$$\cos^2 \theta + \sin^2 \theta = 1$$

we obtain

$$r_3^2 = r_1^2 + r_2^2 - 2r_1 r_2 \cos \theta_2 \qquad (4.3\text{-}18)$$

and therefore

$$r_3 = \left( r_1^2 + r_2^2 - 2r_1 r_2 \cos \theta_2 \right)^{1/2} \qquad (4.3\text{-}19)$$

where only the positive value of the square root was retained.

Note that the dependent variable $r_3$ in Equation (4.3-19) is expressed in terms of the independent variable and the given dimensions of links 1 and 2. It is essential to rearrange the equations given in Equations (4.3-16) and (4.3-17) prior to squaring and adding; otherwise, we would end up with additional cross terms, making it impossible to solve for the dependent variable.

Dividing Equation (4.3-17) by Equation (4.3-16) gives

$$\tan \theta_3 = \left( \frac{r_2 \sin \theta_2}{r_2 \cos \theta_2 - r_1} \right) \qquad (4.3\text{-}20)$$

or

$$\theta_3^* = \tan^{-1} \left[ \frac{\sin \theta_2}{\cos \theta_2 - \left( \dfrac{r_1}{r_2} \right)} \right] \qquad (4.3\text{-}21)$$

Equation (4.3-21) is expressed in terms of the independent variable $\theta_2$ and the given link dimensions. An asterisk has been added to the left-hand side of Equation (4.3-21) to signify that the equation includes an inverse trigonometric function, and when calculating this expression using a calculator or computer, only the principal values of angles are supplied. The ranges of principal values of inverse trigonometric functions are Arcsine, $-90°$ to $90°$; Arccosine, 0 to $180°$; Arctangent, $-90°$ to $90°$. As a result, the value obtained using Equation (4.3-21) may not correspond to the actual angular displacement of the link. Therefore, it becomes necessary to check the value obtained using Equation (4.3-21) against a known result. For instance, if we select

$$r_1 = 1.0 \text{ cm}; \qquad r_2 = 1.0 \text{ cm}; \qquad \theta_2 = 60°$$

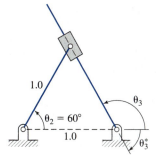

**Figure 4.11** Special geometry of an inverted slider crank mechanism.

the corresponding geometry of the mechanism is shown in Figure 4.11. In this instance, we should obtain

$$\theta_3 = 120°$$

However, from Equation (4.3-21)

$$\theta_3^* = -60°$$

Thus, the value of angular displacement obtained by using Equation (4.3-21) is incorrect, calling for an essential adjustment. Since

$$\tan \theta = \tan(\theta + 180°)$$

we adjust Equation (4.3-21) to obtain

$$\theta_3 = 180° + \theta_3^* = 180° + \tan^{-1}\left[\frac{\sin \theta_2}{\cos \theta_2 - \left(\dfrac{r_1}{r_2}\right)}\right] \qquad (4.3\text{-}22)$$

which will provide the correct value for the angular displacement.

In general, when evaluating an expression for a dependent variable that involves an inverse trigonometric function, it is necessary to check the equation against a specialized geometry and make adjustments as required.

We will now determine time derivative quantities of the dependent variables. We will employ both methods mentioned on page 125.

### Method #1 (Use the solved expressions for the dependent variables and differentiate them with respect to time.)

Taking the time derivative of Equation (4.3-19) gives

$$\dot{r}_3 = \tfrac{1}{2}\left(r_1^2 + r_2^2 - 2r_1r_2 \cos \theta_2\right)^{-1/2}(2r_1r_2 \sin \theta_2 \dot{\theta}_2) \qquad (4.3\text{-}23)$$

Substituting Equation (4.3-19) and simplifying,

$$\dot{r}_3 = \frac{r_1r_2}{r_3} \sin \theta_2 \dot{\theta}_2 \qquad (4.3\text{-}24)$$

Equation (4.3-24) is an expression for the time derivative of one of the dependent variables in terms of the independent variable as well as one of the two dependent variables. This is acceptable since we can solve for the dependent variable first, using Equation (4.3-19), and then determine the time derivative quantity.

Likewise, for the second time derivative, we can show that

$$\ddot{r}_3 = -\frac{\dot{r}_3^2}{r_3} + \frac{r_1 r_2}{r_3}\left(\dot{\theta}_2^2 \cos\theta_2 + \ddot{\theta}_2 \sin\theta_2\right) \tag{4.3-25}$$

To determine the first time derivative of $\theta_3$, we use Equation (4.3-22) and employ

$$\frac{d}{dt}\tan^{-1}u = \frac{1}{1+u^2}\frac{du}{dt}$$

After simplification

$$\dot{\theta}_3 = \dot{\theta}_2\left[\frac{1 - \left(\dfrac{r_1}{r_2}\right)\cos\theta_2}{1 + \left(\dfrac{r_1}{r_2}\right)^2 - 2\left(\dfrac{r_1}{r_2}\right)\cos\theta_2}\right] \tag{4.3-26}$$

Differentiating Equation (4.3-26) with respect to time,

$$\ddot{\theta}_3 = \dot{\theta}_2^2\left\{\frac{\left(\dfrac{r_1}{r_2}\right)\sin\theta_2\left[\left(\dfrac{r_1}{r_2}\right)^2 - 1\right]}{\left[1 + \left(\dfrac{r_1}{r_2}\right)^2 - 2\left(\dfrac{r_1}{r_2}\right)\cos\theta_2\right]^2}\right\} + \ddot{\theta}_2\left[\frac{1 - \left(\dfrac{r_1}{r_2}\right)\cos\theta_2}{1 + \left(\dfrac{r_1}{r_2}\right)^2 - 2\left(\dfrac{r_1}{r_2}\right)\cos\theta_2}\right] \tag{4.3-27}$$

For Equations (4.3-26) and (4.3-27), positive values of $\dot{\theta}_3$ and $\ddot{\theta}_3$ correspond to the counterclockwise direction.

A special case of Equations (4.3-25) and (4.3-27) is obtained if we consider a constant input rotational speed by setting $\ddot{\theta}_2 = 0$.

### Method #2 (Differentiate the real and imaginary components of the loop closure equation and solve for the time derivative quantities.)

We return to the real and imaginary components of the loop closure equation, differentiate with respect to time, and solve for the time derivatives of the dependent variables.

We take the time derivative of Equations (4.3-16) and (4.3-17):

$$-r_3 \sin\theta_3 \dot{\theta}_3 + \dot{r}_3 \cos\theta_3 = -r_2 \sin\theta_2 \dot{\theta}_2 \tag{4.3-28}$$

$$r_3 \cos\theta_3 \dot{\theta}_3 + \dot{r}_3 \sin\theta_3 = r_2 \cos\theta_2 \dot{\theta}_2 \tag{4.3-29}$$

To solve Equations (4.3-28) and (4.3-29) for $\dot{r}_3$, and to eliminate $\dot{\theta}_3$, we take

Equation (4.3-28) $\times \cos\theta_3 +$ Equation (4.3-29) $\times \sin\theta_3$

and obtain

$$\dot{r}_3(\cos^2\theta_3 + \sin^2\theta_3) = r_2\dot{\theta}_2(-\sin\theta_2\cos\theta_3 + \cos\theta_2\sin\theta_3) \tag{4.3-30}$$

Employing the identities

$$\cos^2\theta + \sin^2\theta = 1; \qquad \sin(\theta_3 - \theta_2) = \sin\theta_3 \cos\theta_2 - \sin\theta_2 \cos\theta_3$$

Equation (4.3-30) becomes

$$\dot{r}_3 = r_2\dot{\theta}_2 \sin(\theta_3 - \theta_2) \qquad (4.3\text{-}31)$$

Similarly

$$\dot{\theta}_3 = \frac{r_2\dot{\theta}_2}{r_3} \cos(\theta_3 - \theta_2) \qquad (4.3\text{-}32)$$

For the second time derivative quantities, we differentiate Equations (4.3-28) and (4.3-29) with respect to time and solve, to obtain

$$\ddot{r}_3 = -r_2\big[\dot{\theta}_2^2 \cos(\theta_3 - \theta_2) - \ddot{\theta}_2 \sin(\theta_3 - \theta_2)\big] + r_3\dot{\theta}_3^2 \qquad (4.3\text{-}33)$$

$$\ddot{\theta}_3 = \frac{r_2}{r_3}\big[\dot{\theta}_2^2 \sin(\theta_3 - \theta_2) + \ddot{\theta}_2 \cos(\theta_3 - \theta_2)\big] - \frac{2\dot{r}_3\dot{\theta}_3}{r_3} \qquad (4.3\text{-}34)$$

In comparing the two methods for determining time derivative quantities, we see that the first method, although more direct, has the disadvantage of generating complicated derivative quantities.

These two methods appear to provide very different expressions for time derivatives of the dependent variables. This is apparent in comparing Equations (4.3-24) and (4.3-26) with Equations (4.3-31) and (4.3-32). However, the calculated values using these equations will be identical.

It is possible to compare expressions for calculated quantities obtained by use of the graphical and the analytical techniques. For this example

$$|\dot{r}_3| = |\bar{v}_{B_3 B_2}|; \qquad |\ddot{r}_3| = \big|a^S_{B_3 B_2}\big|$$

In calculating dependent variables and their time derivatives, the Coriolis acceleration is not directly obtained. However, the Coriolis acceleration may be calculated from the dependent variables. Using Equation (2.4-8), we have

$$\big|a^C_{B_3 B_2}\big| = |2\dot{r}_{B_3 B_2}\dot{\theta}_3| = |2\dot{r}_3\dot{\theta}_3|$$

## EXAMPLE 4.1  Analysis of an Inverted Slider Crank Mechanism

For the mechanism shown in Figure 4.12(a), determine for the position shown

    (a) the rotational velocity of link 4
    (b) the Coriolis acceleration of point $B_3$ with respect to point $B_2$

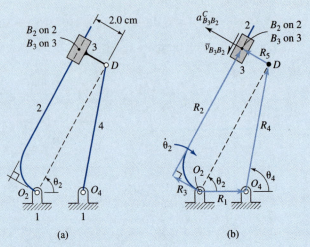

**Figure 4.12** Inverted slider crank.

Use a complex number approach.

$$r_{O_2O_4} = 3.0 \text{ cm}; \qquad r_{O_4D} = 8.0 \text{ cm}; \qquad \theta_2 = 60°; \qquad \dot\theta_2 = 30 \text{ rad/sec CW}$$

**SOLUTION**

The complex vector representation is shown in Figure 4.12(b). The loop closure equation is

$$R_3 + R_2 - R_5 - R_4 - R_1 = 0 \qquad (4.3\text{-}35)$$

However, since

$$R_3 = R_5 \qquad (4.3\text{-}36)$$

Equation (4.3-35) reduces to

$$R_2 - R_4 - R_1 = 0 \qquad (4.3\text{-}37)$$

The real component of Equation (4.3-37) is

$$r_2 \cos\theta_2 - r_4 \cos\theta_4 - r_1 = 0 \qquad (4.3\text{-}38)$$

The imaginary component of Equation (4.3-37) is

$$r_2 \sin\theta_2 - r_4 \sin\theta_4 = 0 \qquad (4.3\text{-}39)$$

The independent variable is $\theta_2$, and the dependent variables are $\theta_4$ and $r_2$.
Solve for dependent variable $\theta_4$ and eliminate $r_2$ by taking

$$\text{Equation (4.3-38)} \times \sin\theta_2 - \text{Equation (4.3-39)} \times \cos\theta_2$$

which gives

$$r_2 \cos\theta_2 \sin\theta_2 - r_4 \cos\theta_4 \sin\theta_2 - r_1 \sin\theta_2 - r_2 \sin\theta_2 \cos\theta_2 + r_4 \sin\theta_4 \cos\theta_2 = 0$$
$$(4.3\text{-}40)$$

Simplifying and rearranging Equation (4.3-40), and solving for $\theta_4$,

$$\theta_4 = \theta_2 + \sin^{-1}\left(\frac{r_1 \sin \theta_2}{r_4}\right) = 60° + \sin^{-1}\left(\frac{3.0 \sin 60°}{8.0}\right) = 78.9° \qquad (4.3\text{-}41)$$

From Equation (4.3-39), and solving for $r_2$,

$$r_2 = r_4 \frac{\sin \theta_4}{\sin \theta_2} = 8.0 \frac{\sin 78.9°}{\sin 60°} = 9.07 \text{ cm} \qquad (4.3\text{-}42)$$

The time derivative of Equation (4.3-38) is

$$-r_2\dot{\theta}_2 \sin \theta_2 + \dot{r}_2 \cos \theta_2 + r_4\dot{\theta}_4 \sin \theta_4 = 0 \qquad (4.3\text{-}43)$$

The time derivative of Equation (4.3-39) is

$$r_2\dot{\theta}_2 \cos \theta_2 + \dot{r}_2 \sin \theta_2 - r_4\dot{\theta}_4 \cos \theta_4 = 0 \qquad (4.3\text{-}44)$$

To solve for the rotational speed of link 4, we take

$$\text{Equation (4.3-43)} \times \sin \theta_2 - \text{Equation (4.3-44)} \times \cos \theta_2$$

which gives after simplification

$$\dot{\theta}_4 = \frac{r_2\dot{\theta}_2}{r_4 \cos(\theta_2 - \theta_4)} = \frac{9.07(-30)}{8.0 \cos(60° - 78.9°)} = -35.9 \text{ rad/sec} \qquad (4.3\text{-}45)$$

To solve for the time derivative of the second dependent variable, we take

$$\text{Equation (4.3-43)} \times \cos \theta_2 + \text{Equation (4.3-44)} \times \sin \theta_2$$

which gives after simplification

$$\dot{r}_2 = r_4\dot{\theta}_4 \sin(\theta_2 - \theta_4) = 8.0(-35.9) \sin(60° - 78.9°)$$
$$= -93.0 \text{ cm/sec} \qquad (4.3\text{-}46)$$

The solutions to the problem are as follows:

(a) The rotational velocity of link 4 is

$$\dot{\theta}_4 = -35.9 \text{ rad/sec} = 35.9 \text{ rad/sec CW}$$

(b) The magnitude of the Coriolis acceleration of point $B_3$ with respect to point $B_2$ is

$$\left|a^C_{B_3 B_2}\right| = 2|\dot{r}_{B_3 B_2}||\dot{\theta}_2| = 2|\dot{r}_2||\dot{\theta}_2| = 2 \times 93.0 \times 30.0 = 5{,}580 \text{ cm/sec}^2$$

The direction of this relative acceleration is illustrated in Figure 4.12(b).

### 4.3.3 Four-Bar Mechanism

For the four-bar mechanism shown in Figure 4.3(a), we consider the independent variable to be angle $\theta_2$. Then the two dependent variables are angular displacements $\theta_3$ and $\theta_4$.

The vector representation of this four-bar mechanism is shown in Figure 4.3(c). The loop closure equation may be written as

$$R_1 + R_4 = R_2 + R_3 \qquad (4.3\text{-}47)$$

Substituting Equation (4.3-2) in Equation (4.3-47) and extracting the real and imaginary components,

$$r_1 \cos \theta_1 + r_4 \cos \theta_4 = r_2 \cos \theta_2 + r_3 \cos \theta_3 \qquad (4.3\text{-}48)$$

$$r_1 \sin \theta_1 + r_4 \sin \theta_4 = r_2 \sin \theta_2 + r_3 \sin \theta_3 \qquad (4.3\text{-}49)$$

We now solve Equations (4.3-48) and (4.3-49) for $\theta_3$ and $\theta_4$. Rearranging Equations (4.3-48) and (4.3-49), and recognizing $\theta_1 = 0$,

$$r_3 \cos \theta_3 = r_1 - r_2 \cos \theta_2 + r_4 \cos \theta_4 \qquad (4.3\text{-}50)$$

$$r_3 \sin \theta_3 = -r_2 \sin \theta_2 + r_4 \sin \theta_4 \qquad (4.3\text{-}51)$$

Squaring and adding Equations (4.3-50) and (4.3-51) gives

$$r_3^2 = r_1^2 + r_2^2 + r_4^2 - 2r_2r_4 \cos \theta_2 \cos \theta_4 - 2r_1r_2 \cos \theta_2 + 2r_1r_4 \cos \theta_4 - 2r_2r_4 \sin \theta_2 \sin \theta_4 \qquad (4.3\text{-}52)$$

Rearranging gives

$$\cos \theta_4 (\cos \theta_2 - h_1) + \sin \theta_2 \sin \theta_4 = -h_3 \cos \theta_2 + h_5 \qquad (4.3\text{-}53)$$

where

$$h_1 = \frac{r_1}{r_2}; \qquad h_3 = \frac{r_1}{r_4}; \qquad h_5 = \frac{r_1^2 + r_2^2 - r_3^2 + r_4^2}{2r_2r_4} \qquad (4.3\text{-}54)$$

Equation (4.3-53) is in terms of one of the dependent variables, $\theta_4$. However, it appears in the argument of both the sine and cosine functions. To derive an explicit expression for the dependent variable, we employ the identities

$$\sin \theta = \frac{2 \tan \left( \dfrac{\theta}{2} \right)}{1 + \tan^2 \left( \dfrac{\theta}{2} \right)}; \qquad \cos \theta = \frac{1 - \tan^2 \left( \dfrac{\theta}{2} \right)}{1 + \tan^2 \left( \dfrac{\theta}{2} \right)}$$

in Equation (4.3-53). After simplification

$$d \tan^2 \left( \frac{\theta_4}{2} \right) + b \tan \left( \frac{\theta_4}{2} \right) + e = 0 \qquad (4.3\text{-}55)$$

where

$$d = -h_1 + (1 - h_3) \cos \theta_2 + h_5 \qquad (4.3\text{-}56)$$

$$b = -2 \sin \theta_2 \qquad (4.3\text{-}57)$$

$$e = h_1 - (1 + h_3) \cos \theta_2 + h_5 \qquad (4.3\text{-}58)$$

Equation (4.3-55) is a quadratic equation of $\tan(\theta_4/2)$. Solving for $\theta_4$ gives

$$\theta_4 = 2\tan^{-1}\left[\frac{-b \pm (b^2 - 4de)^{1/2}}{2d}\right] \tag{4.3-59}$$

Similarly, $\theta_3$ may be found by eliminating $\theta_4$ from Equations (4.3-48) and (4.3-49). The result is

$$\theta_3 = 2\tan^{-1}\left[\frac{-b \pm (b^2 - 4ac)^{1/2}}{2a}\right] \tag{4.3-60}$$

where

$$a = -h_1 + (1 + h_2)\cos\theta_2 + h_4 \tag{4.3-61}$$

$$c = h_1 - (1 - h_2)\cos\theta_2 + h_4 \tag{4.3-62}$$

$$h_2 = \frac{r_1}{r_3}; \qquad h_4 = \frac{-r_1^2 - r_2^2 - r_3^2 + r_4^2}{2\,r_2 r_3} \tag{4.3-63}$$

Equations (4.3-59) and (4.3-60) indicate that there are two values of $\theta_3$ and $\theta_4$ for a given set of link lengths and value of $\theta_2$. To illustrate this, Figure 4.13 shows two configurations of a mechanism having the same input angular displacement. The angular displacements of links 3 and 4 are

$$(\theta_3)_1 = 2\tan^{-1}\left[\frac{-b - (b^2 - 4ac)^{1/2}}{2a}\right]$$

$$(\theta_4)_1 = 2\tan^{-1}\left[\frac{-b - (b^2 - 4de)^{1/2}}{2d}\right] \tag{4.3-64}$$

**Figure 4.13** Two possible configurations of a four-bar mechanism for a given input angular displacement.

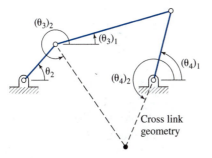

and

$$(\theta_3)_2 = 2\tan^{-1}\left[\frac{-b + (b^2 - 4ac)^{1/2}}{2a}\right]$$

$$(\theta_4)_2 = 2\tan^{-1}\left[\frac{-b + (b^2 - 4de)^{1/2}}{2d}\right]$$

(4.3-65)

In using Equations (4.3-59) and (4.3-60), one must select either the positive or negative sign. The proper selection is achieved by first calculating both angular displacements, and then selecting the one that matches a known geometry of the mechanism being analyzed. Equations (4.3-59) and (4.3-60) may be applied to all types of four-bar mechanisms.

We proceed to determine time derivative quantities of the dependent variables. In this example, we will restrict ourselves to the second method (i.e., differentiating the real and imaginary components of the loop closure equation, and then solving for the derivative quantities). Taking the time derivatives of Equations (4.3-50) and (4.3-51) gives the following two linear equations in terms of $\dot{\theta}_3$ and $\dot{\theta}_4$:

$$-r_2\dot{\theta}_2\sin\theta_2 - r_3\dot{\theta}_3\sin\theta_3 = -r_4\dot{\theta}_4\sin\theta_4 \qquad (4.3\text{-}66)$$

$$r_2\dot{\theta}_2\cos\theta_2 + r_3\dot{\theta}_3\cos\theta_3 = r_4\dot{\theta}_4\cos\theta_4 \qquad (4.3\text{-}67)$$

For Equations (4.3-66) and (4.3-67), values of $\theta_3$ and $\theta_4$ can be considered as known since they may be found using Equations (4.3-59) and (4.3-60). To solve for $\dot{\theta}_3$, we take

$$\text{Equation (4.3-66)} \times \cos\theta_4 + \text{Equation (4.3-67)} \times \sin\theta_4$$

to obtain

$$r_2\dot{\theta}_2(\sin\theta_2\cos\theta_4 - \sin\theta_4\cos\theta_2) + r_3\dot{\theta}_3(\sin\theta_3\cos\theta_4 - \sin\theta_4\cos\theta_3) = 0 \qquad (4.3\text{-}68)$$

Employing the identity

$$\sin(\theta_i - \theta_j) = \sin\theta_i\cos\theta_j - \sin\theta_j\cos\theta_i$$

and rearranging, we obtain

$$\dot{\theta}_3 = \frac{r_2\dot{\theta}_2}{r_3}\frac{\sin(\theta_2 - \theta_4)}{\sin(\theta_4 - \theta_3)} \qquad (4.3\text{-}69)$$

Similarly

$$\dot{\theta}_4 = \frac{r_2\dot{\theta}_2}{r_4}\frac{\sin(\theta_2 - \theta_3)}{\sin(\theta_4 - \theta_3)} \qquad (4.3\text{-}70)$$

To obtain expressions for $\ddot{\theta}_3$ and $\ddot{\theta}_4$, we differentiate Equations (4.3-66) and (4.3-67) with respect to time and solve, yielding

$$\ddot{\theta}_3 = \frac{-r_2\dot{\theta}_2^2 \cos(\theta_2 - \theta_4) - r_2\ddot{\theta}_2 \sin(\theta_2 - \theta_4) - r_3\dot{\theta}_3^2 \cos(\theta_3 - \theta_4) + r_4\dot{\theta}_4^2}{r_3 \sin(\theta_3 - \theta_4)} \quad (4.3\text{-}71)$$

$$\ddot{\theta}_4 = \frac{-r_2\dot{\theta}_2^2 \cos(\theta_2 - \theta_3) - r_2\ddot{\theta}_2 \sin(\theta_2 - \theta_3) - r_3\dot{\theta}_3^2 + r_4\dot{\theta}_4^2 \cos(\theta_3 - \theta_4)}{r_4 \sin(\theta_3 - \theta_4)} \quad (4.3\text{-}72)$$

## EXAMPLE 4.2  Analysis of a Four-Bar Mechanism

Figure 4.14(a) depicts a four-bar mechanism having

$$r_1 = 8.0\,\text{cm}; \qquad r_2 = 3.5\,\text{cm}; \qquad r_3 = 10.0\,\text{cm}; \qquad r_4 = 9.0\,\text{cm}$$

$$\theta_2 = 15°; \qquad \dot{\theta}_2 = 36.7\,\text{rad/sec (constant), and therefore } \ddot{\theta}_2 = 0$$

Determine the angular displacements, angular velocities, and angular accelerations of links 3 and 4.

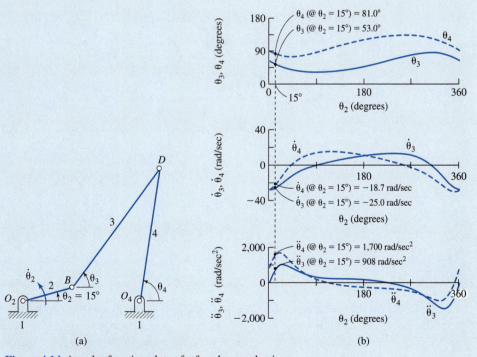

**Figure 4.14** Angular function plots of a four-bar mechanism.

## SOLUTION

Using Equations (4.3-54) and (4.3-63), we first calculate

$$h_1 = \frac{r_1}{r_2} = \frac{8.0}{3.5} = 2.286; \qquad h_3 = 0.889; \qquad h_4 = -1.361$$

Then, using Equations (4.3-57), (4.3-61), and (4.3-62),

$$a = -h_1 + (1 + h_2) \cos \theta_2 + h_4$$
$$= -2.286 + (1 + 0.800) \cos 15° - 1.361 = -1.908$$
$$b = -0.518; \qquad c = 0.732$$

Using Equations (4.3-64) and (4.3-65),

$$(\theta_3)_1 = 2 \tan^{-1} \left[ \frac{-b - (b^2 - 4ac)^{1/2}}{2a} \right]$$

$$= 2 \tan^{-1} \left[ \frac{-(-0.518) - ((-0.518)^2 - 4(-1.908) \times 0.732)^{1/2}}{2 \times (-1.908)} \right] = 53.0°$$

$$(\theta_3)_2 = -75.2°$$

from which we select by inspection the angular displacement of link 3 to be

$$\theta_3 = (\theta_3)_1 = 53.0°$$

Similarly, the angular displacement of link 4 is

$$\theta_4 = (\theta_4)_1 = 81.0°$$

Using Equation (4.3-69), the angular velocity of link 3 is

$$\dot{\theta}_3 = \frac{r_2 \dot{\theta}_2}{r_3} \frac{\sin(\theta_2 - \theta_4)}{\sin(\theta_4 - \theta_3)} = \frac{3.5 \times 36.7}{10.0} \frac{\sin(15.0° - 81.0°)}{\sin(81.0° - 53.0°)} = -25.0 \, \text{rad/sec}$$

Similarly, using Equation (4.3-70),

$$\dot{\theta}_4 = -18.7 \, \text{rad/sec}$$

Using Equations (4.3-71) and (4.3-72), the angular accelerations of links 3 and 4 are

$$\ddot{\theta}_3 = 908 \, \text{rad/sec}^2; \qquad \ddot{\theta}_4 = 1,700 \, \text{rad/sec}^2$$

A positive value of either rotational speed or angular acceleration is in the counterclockwise direction, whereas a negative value corresponds to the clockwise direction.

Figure 4.14(b) depicts plots of functions of angular displacement, angular velocity, and angular acceleration throughout one cycle of motion.

### 4.3.4 Slider Crank Mechanism

Consider the slider crank mechanism shown in Figure 4.4(a), for which the input is the rotation of link 2.

The complex vector representation is illustrated in Figure 4.4(b). The loop closure equation is

$$R_2 - R_3 - R_1 - R_4 = 0 \tag{4.3-73}$$

Substituting Equation (4.3-2) in (4.3-73), extracting the real and imaginary components, as well as recognizing that $\theta_1 = 90°$ and $\theta_4 = 0$,

$$r_2 \cos\theta_2 - r_3 \cos\theta_3 - r_4 = 0 \tag{4.3-74}$$

$$r_2 \sin\theta_2 - r_3 \sin\theta_3 - r_1 = 0 \tag{4.3-75}$$

The two dependent variables are $\theta_3$ and $r_4$. Equation (4.3-75) contains only one of the dependent variables, $\theta_3$, and therefore it is possible to solve for it explicitly as

$$\theta_3^* = \sin^{-1}\left(\frac{-r_1 + r_2 \sin\theta_2}{r_3}\right) \tag{4.3-76}$$

We consider the special case, where

$$r_1 = 0; \qquad r_2 = 1.0; \qquad r_3 = 1.0; \qquad \theta_2 = 60°$$

In this instance, we should obtain

$$\theta_3 = 120°$$

However, Equation (4.3-76) provides

$$\theta_3^* = 60°$$

The value of the angular displacement as provided by Equation (4.3-76) is thus incorrect, and an adjustment is necessary. Since

$$\sin\theta = \sin(180° - \theta)$$

we let

$$\theta_3 = 180° - \theta_3^* = 180° - \sin^{-1}\left(\frac{-r_1 + r_2 \sin\theta_2}{r_3}\right) \tag{4.3-77}$$

Dependent variable $\theta_3$ may now be calculated, and values may be substituted into the following expression, obtained by modifying Equation (4.3-74), to determine $r_4$:

$$r_4 = r_2 \cos\theta_2 - r_3 \cos\theta_3 \tag{4.3-78}$$

Note for this example it was not necessary to combine the real and imaginary components of the loop closure equation to generate equations that explicitly define the dependent variables.

Taking the time derivative of Equations (4.3-74) and (4.3-75) gives

$$-r_2\dot{\theta}_2 \sin\theta_2 + r_3\dot{\theta}_3 \sin\theta_3 - \dot{r}_4 = 0 \tag{4.3-79}$$

$$r_2\dot{\theta}_2 \cos\theta_2 - r_3\dot{\theta}_3 \cos\theta_3 = 0 \tag{4.3-80}$$

Solving Equations (4.3-79) and (4.3-80) for the first time derivative of the dependent variables gives

$$\dot{r}_4 = \frac{r_2 \dot{\theta}_2 \sin(\theta_3 - \theta_2)}{\cos \theta_3} \qquad (4.3\text{-}81)$$

$$\dot{\theta}_3 = \frac{r_2 \dot{\theta}_2 \cos \theta_2}{r_3 \cos \theta_3} \qquad (4.3\text{-}82)$$

Differentiating Equations (4.3-79) and (4.3-80) with respect to time, and solving for the second time derivative of the dependent variables, yields

$$\ddot{r}_4 = \frac{-r_2 \dot{\theta}_2^2 \cos(\theta_3 - \theta_2) + r_2 \ddot{\theta}_2 \sin(\theta_3 - \theta_2) + r_3 \dot{\theta}_3^2}{\cos \theta_3} \qquad (4.3\text{-}83)$$

$$\ddot{\theta}_3 = \frac{-r_2 \dot{\theta}_2^2 \sin \theta_2 + r_2 \ddot{\theta}_2 \cos \theta_2 + r_3 \dot{\theta}_3^2 \sin \theta_3}{r_3 \cos \theta_3} \qquad (4.3\text{-}84)$$

## EXAMPLE 4.3  Analysis of a Slider Crank Mechanism

Figure 4.15(a) depicts a slider crank mechanism. The link lengths and motion of the crank are

$$r_2 = 3.5 \text{ in}; \qquad r_3 = 9.0 \text{ in}$$

$$\theta_2 = 50°; \qquad \dot{\theta}_2 = 52.4 \text{ rad/sec (constant), and therefore } \ddot{\theta}_2 = 0$$

Determine the angular displacement, angular velocity, and angular acceleration of link 3 and the displacement, velocity, and acceleration of link 4.

### SOLUTION

Comparing the mechanism shown in Figure 4.15(a) with that illustrated in Figure 4.4

$$r_1 = -2.0 \text{ in}$$

Using Equation (4.3-77), the angular displacement of link 3 is

$$\theta_3 = 180° - \sin^{-1}\left( \frac{-r_1 + r_2 \sin \theta_2}{r_3} \right)$$

$$= 180° - \sin^{-1}\left( \frac{2.0 + 3.5 \sin 50°}{9.0} \right) = 148.7°$$

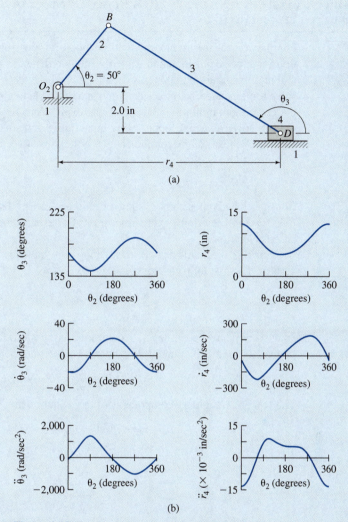

**Figure 4.15** Function plots of a slider crank mechanism.

Using Equation (4.3-78), the displacement of link 4 is

$$r_4 = r_2 \cos \theta_2 - r_3 \cos \theta_3 = 3.5 \cos 50° - 9.0 \cos 148.7° = 9.940 \text{ in}$$

Using Equations (4.3-81)–(4.3-84), the velocities and accelerations of links 3 and 4 are

$$\dot{\theta}_3 = 15.3 \text{ rad/sec}; \qquad \dot{r}_4 = -212 \text{ in/sec}$$

$$\ddot{\theta}_3 = 814 \text{ rad/sec}^2; \qquad \ddot{r}_4 = -4,170 \text{ in/sec}^2$$

Figure 4.15(b) depicts plots of functions of displacement, velocity, and acceleration throughout one cycle of motion.

## 4.4  COMPLEX VECTOR ANALYSIS OF A PLANAR MECHANISM WITH MULTIPLE LOOPS

For planar mechanisms having multiple loops, it is possible to generate a loop closure equation for each vector loop and solve for the two dependent variables of each loop from the real and imaginary components. In all cases where there are two or more loops, we must solve for dependent variables one loop at a time. We begin with the loop(s) associated with the given input motion(s). After deriving expressions for the dependent variables, we proceed to solve adjacent loops, using quantitative information obtained from the previous loop(s), by means of the common angles or lengths of vectors that are shared between loops. This procedure is illustrated by examples.

### 4.4.1  Inverted Slider Crank Mechanism with a Coupler Point

Consider the coupler point $C$ shown in Figure 4.16(a). For a prescribed angular motion of link 2, we will determine expressions for its Cartesian components of displacement, velocity, and acceleration.

For this mechanism, two vector loops are created as shown in Figures 4.16(b) and 4.16(c). The first loop closure equation is

$$R_2 - R_3 - R_1 = 0 \tag{4.4-1}$$

Equation (4.4-1) is the same as that presented in Section 4.3.2. Results from that previous analysis will now be employed to determine expressions for dependent variables of the second vector loop, which includes point $C$. The second loop closure equation is

$$R_1 + R_4 + R_5 - R_7 - R_6 = 0 \tag{4.4-2}$$

The real and imaginary components of this equation are

$$r_1 + r_4 \cos\theta_4 + r_5 \cos(\theta_4 - 90°) - r_6 = 0 \tag{4.4-3}$$

$$r_4 \sin\theta_4 + r_5 \sin(\theta_4 - 90°) - r_7 = 0 \tag{4.4-4}$$

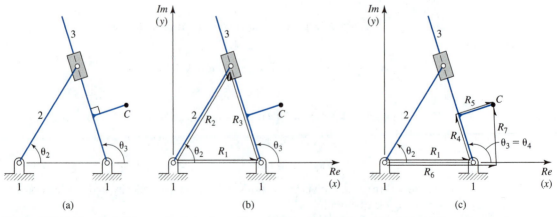

(a)                              (b)                              (c)

**Figure 4.16**  (a) Inverted slider crank mechanism with a coupler point. (b) First vector loop. (c) Second vector loop.

In the second vector loop, $r_6$ and $r_7$ are the dependent variables. Also

$$\theta_6 = 0; \qquad \theta_7 = 90°$$

Angle $\theta_4$ is a dependent variable of the first vector loop, and its value may be considered as known when analyzing the second vector loop. By inspection

$$\theta_4 = \theta_3 \tag{4.4-5}$$

Solving equations for the two dependent variables of the second loop, and simplifying,

$$r_6 = r_1 + r_4 \cos \theta_3 + r_5 \sin \theta_3 \tag{4.4-6}$$

$$r_7 = r_4 \sin \theta_3 - r_5 \cos \theta_3 \tag{4.4-7}$$

Equations (4.4-6) and (4.4-7) also correspond to the Cartesian components that locate the position of point $C$, that is,

$$x_C = r_6; \qquad y_C = r_7 \tag{4.4-8}$$

The position vector of point $C$ is therefore

$$\bar{r}_C = x_C \bar{i} + y_C \bar{j} \tag{4.4-9}$$

Differentiating Equations (4.4-6) and (4.4-7) with respect to time gives the Cartesian components of velocity as

$$\dot{r}_6 = \dot{x}_C = \dot{\theta}_3 \left( -r_4 \sin \theta_3 + r_5 \cos \theta_3 \right) \tag{4.4-10}$$

$$\dot{r}_7 = \dot{y}_C = \dot{\theta}_3 \left( r_4 \cos \theta_3 + r_5 \sin \theta_3 \right) \tag{4.4-11}$$

The velocity of point $C$ may be expressed as

$$\bar{v}_C = \dot{x}_C \bar{i} + \dot{y}_C \bar{j} \tag{4.4-12}$$

Further, the Cartesian components of acceleration are

$$\ddot{r}_6 = \ddot{x}_C = \dot{\theta}_3^2 \left( -r_4 \cos \theta_3 - r_5 \sin \theta_3 \right) + \ddot{\theta}_3 \left( -r_4 \sin \theta_3 + r_5 \cos \theta_3 \right) \tag{4.4-13}$$

$$\ddot{r}_7 = \ddot{y}_C = \dot{\theta}_3^2 \left( -r_4 \sin \theta_3 + r_5 \cos \theta_3 \right) + \ddot{\theta}_3 \left( r_4 \cos \theta_3 + r_5 \sin \theta_3 \right) \tag{4.4-14}$$

The acceleration of point $C$ is

$$\bar{a}_C = \ddot{x}_C \bar{i} + \ddot{y}_C \bar{j} \tag{4.4-15}$$

### 4.4.2 Four-Bar Mechanism with a Coupler Point

Consider the four-bar mechanism shown in Figure 4.7. We will determine the Cartesian components of the displacement, velocity, and acceleration of coupler point $C$ in terms of the angular motion of link 2.

For this mechanism, two vector loops are created as shown in Figures 4.7(b) and 4.7(c). The loop closure equations are

$$R_2 + R_3 - R_4 - R_1 = 0 \tag{4.4-16}$$

and

$$R_5 + R_6 + R_7 - R_9 - R_8 = 0 \tag{4.4-17}$$

The first vector loop is the same as that presented in Section 4.3.3. Results from that analysis may now be employed to determine expressions for dependent variables of the second vector loop.

For the second vector loop

$$\theta_8 = 0; \qquad \theta_9 = 90° \tag{4.4-18}$$

and

$$r_5 = r_2; \qquad \theta_5 = \theta_2; \qquad \theta_6 = \theta_3; \qquad \theta_7 = \theta_3 + 90° \tag{4.4-19}$$

Therefore, $\theta_5$, $\theta_6$, and $\theta_7$ are now known. The dependent variables of the second loop are the magnitudes of vectors $R_8$ and $R_9$.

Taking the real and imaginary components of Equation (4.4-17), isolating the dependent variables of the loop onto the left-hand side of the equations, and employing Equations (4.4-18) and (4.4-19),

$$r_8 = r_2 \cos \theta_2 + r_6 \cos \theta_3 - r_7 \sin \theta_3 \tag{4.4-20}$$

$$r_9 = r_2 \sin \theta_2 + r_6 \sin \theta_3 + r_7 \cos \theta_3 \tag{4.4-21}$$

Equations (4.4-20) and (4.4-21) are also related to the Cartesian components of the position of point $C$, that is,

$$x_C = r_8; \qquad y_C = r_9 \tag{4.4-22}$$

Differentiating Equations (4.4-20) and (4.4-21) with respect to time gives the horizontal and vertical components of velocity as

$$\dot{r}_8 = \dot{x}_C = -r_2\dot{\theta}_2 \sin \theta_2 - \dot{\theta}_3 (r_6 \sin \theta_3 + r_7 \cos \theta_3) \tag{4.4-23}$$

$$\dot{r}_9 = \dot{y}_C = r_2\dot{\theta}_2 \cos \theta_2 + \dot{\theta}_3 (r_6 \cos \theta_3 - r_7 \sin \theta_3) \tag{4.4-24}$$

Differentiating Equations (4.4-23) and (4.4-24) with respect to time yields the following expressions for the horizontal and vertical components of acceleration:

$$\ddot{r}_8 = \ddot{x}_C = -r_2\dot{\theta}_2^2 \cos \theta_2 - r_2\ddot{\theta}_2 \sin \theta_2 - \dot{\theta}_3^2(r_6 \cos \theta_3 - r_7 \sin \theta_3) \\ - \ddot{\theta}_3(r_6 \sin \theta_3 + r_7 \cos \theta_3) \tag{4.4-25}$$

$$\ddot{r}_9 = \ddot{y}_C = -r_2\dot{\theta}_2^2 \sin \theta_2 + r_2\ddot{\theta}_2 \cos \theta_2 - \dot{\theta}_3^2(r_6 \sin \theta_3 + r_7 \cos \theta_3) \\ + \ddot{\theta}_3(r_6 \cos \theta_3 - r_7 \sin \theta_3) \tag{4.4-26}$$

## EXAMPLE 4.4  Analysis of a Four-Bar Mechanism with a Coupler Point

Figure 4.17(a) depicts a four-bar mechanism with a coupler point having

$$r_1 = 8.0 \text{ cm}; \qquad r_2 = 3.5 \text{ cm}; \qquad r_3 = 10.0 \text{ cm}$$

$$r_4 = 9.0 \text{ cm}; \qquad r_6 = 3.0 \text{ cm}; \qquad r_7 = 4.0 \text{ cm}$$

$$\theta_2 = 15°; \qquad \dot{\theta}_2 = 36.7 \text{ rad/sec (constant)}$$

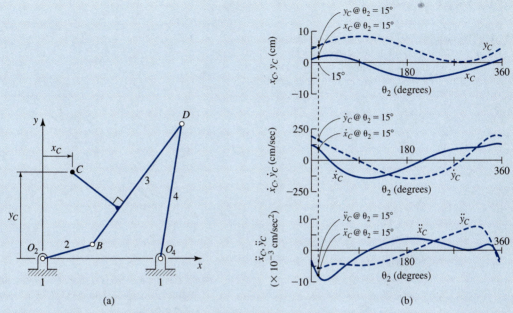

**Figure 4.17** Coupler point path plots of a four-bar mechanism.

Determine the displacement, velocity, and acceleration of coupler point $C$ with respect to the given coordinate system.

### SOLUTION

Some of the required quantities were determined in Example 4.2.
Using Equations (4.4-20) and (4.4-22),

$$x_C = r_2 \cos \theta_2 + r_6 \cos \theta_3 - r_7 \sin \theta_3$$
$$= 3.5 \times \cos 15° + 3.0 \times \cos 53° - 4.0 \times \sin 53° = 1.99 \text{ cm}$$

Similarly

$$y_C = 5.71 \text{ cm}; \qquad \dot{x}_C = 86.6 \text{ cm/sec}; \qquad \dot{y}_C = 159 \text{ cm/sec}$$
$$\ddot{x}_C = -8{,}050 \text{ cm/sec}^2; \qquad \ddot{y}_C = -5{,}470 \text{ cm/sec}^2$$

From Equations (4.4-9), (4.4-12), and (4.4-15)

$$\bar{r}_C = x_C \bar{i} + y_C \bar{j} = 1.99\bar{i} + 5.71\bar{j} \text{ cm}$$
$$\bar{v}_C = 86.6\bar{i} + 159\bar{j} \text{ cm/sec}$$
$$\bar{a}_C = -8{,}050\bar{i} - 5{,}470\bar{j} \text{ cm/sec}^2$$

Figure 4.17(b) shows plots of functions of components of displacement, velocity, and acceleration of the coupler point throughout one cycle of motion.

# PROBLEMS

**P4.1**   For the mechanism shown in Figure P4.1, determine the rotational speed of link 2 and the sliding velocity of link 3 with respect to link 2 if $\bar{v}_A = 2.0$ in/sec in the direction shown. Use a complex number approach.

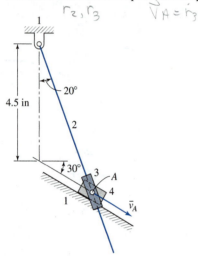

**Figure P4.1**

**P4.2**   For the mechanism shown in Figure P4.2, determine the angular acceleration of link 3. Use a complex number approach.

$$r_{O_2A} = 6.0 \text{ cm}; \qquad \theta_2 = 60°$$

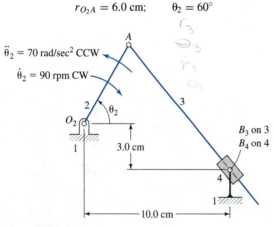

**Figure P4.2**

**P4.3**   Figure P4.3 depicts a skeleton diagram representation of an Oldham coupling. Determine $r_2, r_3, \dot{r}_2,$ $\dot{r}_3, \ddot{r}_2,$ and $\ddot{r}_3$ in terms of $r_1, \theta_2, \dot{\theta}_2,$ and $\ddot{\theta}_2$. Use a complex number approach.

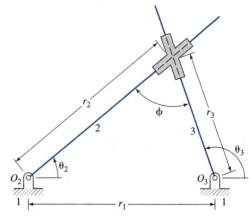

**Figure P4.3**

**P4.4**   For the mechanism shown in Figure P4.4, determine the velocity of point $C$. Use a complex number approach.

$$r_{O_2A} = 4.0 \text{ cm}; \qquad r_{AB} = 3.0 \text{ cm}$$
$$r_{BC} = 2.0 \text{ cm}; \qquad r_{AD} = 8.0 \text{ cm}$$
$$\theta_2 = 225°; \qquad \dot{\theta}_2 = 150 \text{ rpm CW}$$

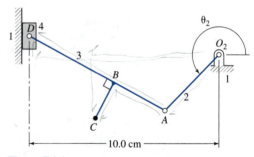

**Figure P4.4**

**P4.5**   For the mechanism shown in Figure P4.5, determine
   (a) the angular velocity of link 3
   (b) the velocity of point $A_2$ with respect to point $A_3$
   (c) the Coriolis acceleration of point $A_2$ with respect to point $A_3$

Use a complex number approach.

$$r_{O_2A_2} = 5.0 \text{ cm}; \qquad \theta_2 = 135°$$
$$\dot{\theta}_2 = 80 \text{ rpm CW}$$

**P4.6**   For the mechanism shown in Figure P4.6, determine, for the position shown,
   (a) the rotational velocity of link 3

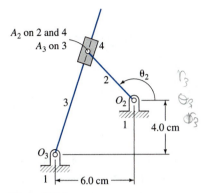

**Figure P4.5**

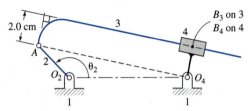

**Figure P4.6**

(b) the Coriolis acceleration of point $B_3$ with respect to point $B_4$

Use a complex number approach.

$$r_{O_2A} = 3.0 \text{ cm}; \qquad r_{O_2O_4} = 8.0 \text{ cm}$$
$$\theta_2 = 135°; \qquad \dot{\theta}_2 = 30 \text{ rad/sec CW}$$

**P4.7** For the scaled mechanism shown in Figure P4.7, determine

(a) the acceleration of point $B$
(b) the sliding acceleration of $A_2$ with respect to $A_4$

Use a complex number approach.

$$r_{O_2A_2} = 5.0 \text{ cm}; \qquad \theta_2 = 45°$$

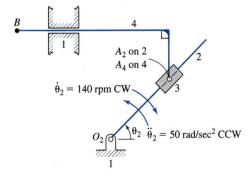

**Figure P4.7**

**P4.8** Determine the rotational speed of link 3 of the mechanism given in Figure P4.8 for the position shown. Use a complex number approach.

$$r_{O_2A} = 4.0 \text{ cm}; \qquad r_{AB} = 10.0 \text{ cm}; \qquad \theta_2 = 45°$$

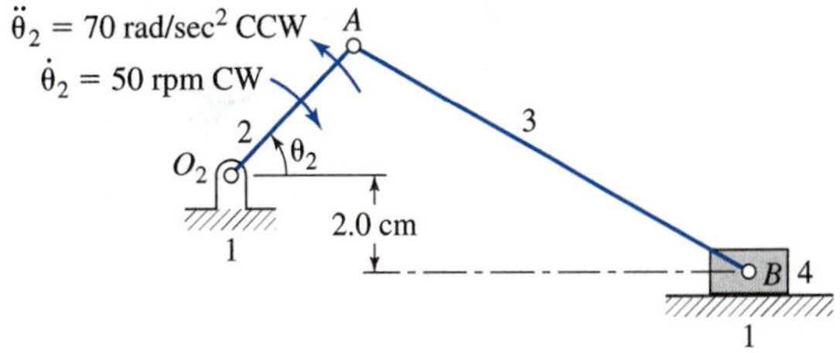

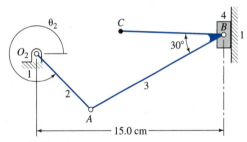

**Figure P4.8**

**P4.9** For the mechanism shown in Figure P4.9, determine

(a) the velocity of point $C$
(b) the sliding velocity of point $B$ with respect to the base link

Use a complex number approach.

$$r_{O_2A} = 6.0 \text{ cm}; \qquad r_{AB} = 12.0 \text{ cm}$$
$$r_{BC} = 9.0 \text{ cm}$$
$$\theta_2 = 315°; \qquad \dot{\theta}_2 = 180 \text{ rpm CW}$$

**Figure P4.9**

**P4.10** For the mechanism shown in Figure P4.10, determine

(a) the angular velocity of link 3
(b) the velocity of point $B_3$ with respect to point $B_1$
(c) the Coriolis acceleration of point $B_3$ with respect to point $B_1$
(d) the equations for the components of velocity of point $C$ in terms of given parameters and quantities determined in parts (a), (b), and (c)

Use a complex number approach.

$$r_{O_2A} = 4.0 \text{ cm}; \qquad r_{AD} = 3.0 \text{ cm}$$
$$r_{DC} = 2.0 \text{ cm}$$
$$\theta_2 = 315°; \qquad \dot{\theta}_2 = 150 \text{ rpm CCW (constant)}$$

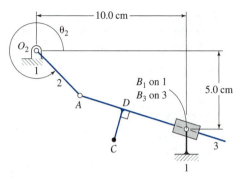

**Figure P4.10**

**P4.11** Figure P4.11 shows an intermittent-motion mechanism known as the Geneva mechanism. The circular peg $P$ is attached to link 2 and slides in the slots of link 3. Use a complex number approach, when

$$\theta_2 = 160°$$

Determine

(a) the rotational velocity of link 3
(b) the sliding velocity of peg $P$ with respect to the slot in link 3

$$r_{O_2P} = 4.0\sqrt{2} \text{ cm;} \qquad r_{O_2O_3} = 8.0 \text{ cm}$$
$$\dot{\theta}_2 = 30 \text{ rad/sec CCW (constant)}$$

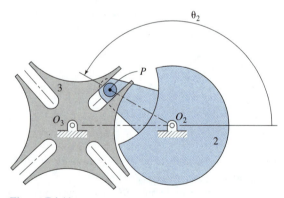

**Figure P4.11**

**P4.12** For the mechanism shown in Figure P4.12, determine the velocity of point $C$ for the given configuration. Use a complex number approach.

$$r_{O_2A} = 1.5 \text{ in;} \qquad r_{AB} = 1.5 \text{ in}$$
$$r_{BC} = 1.0 \text{ in;} \qquad r_{AD} = 4.0 \text{ in}$$
$$\theta_2 = 45°; \qquad \dot{\theta}_2 = 150 \text{ rpm CCW (constant)}$$

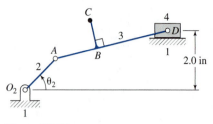

**Figure P4.12**

**P4.13** For the mechanism shown in Figure P4.13, determine
(a) the speed of point $C$ as a function of $\theta_2$
(b) the magnitude of the relative velocity of $B_4$ with respect to $B_2$ as a function of $\theta_2$
(c) the magnitude of the sliding acceleration of $B_4$ with respect to $B_2$ as a function of $\theta_2$
(d) the maximum magnitude of the sliding velocity of $B_4$ with respect to $B_2$ during a complete rotation of link 2, and the corresponding value(s) of $\theta_2$ where this takes place (Hint: Use the expression from part (c) to determine value(s) of $\theta_2$ where the sliding acceleration is zero.)

Use a complex number approach.

$$r_{O_2B_2} = 1.5 \text{ in;} \qquad \dot{\theta}_2 = 75 \text{ rpm CW (constant)}$$

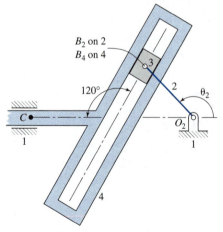

**Figure P4.13**

**P4.14** For the mechanism shown in Figure P4.14, determine
(a) the angular velocity of link 3
(b) the velocity of point $C$.

Use a complex number approach.

$$r_{O_2B} = 4.0 \text{ cm;} \qquad r_{BC} = 15.0 \text{ cm}$$
$$\theta_2 = 225°; \qquad \dot{\theta}_2 = 50 \text{ rad/sec CCW}$$

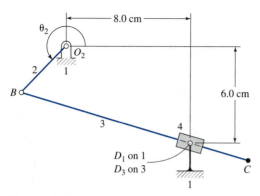

**Figure P4.14**

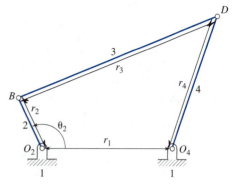

**Figure P4.16**

**P4.15** For the mechanism shown in Figure P4.15, determine
(a) the angular velocity of link 3
(b) the velocity of point $D_3$ with respect to point $D_1$
Use a complex number approach.

$$r_{O_2B} = 4.0 \text{ cm}; \qquad \theta_2 = 135°$$

$$\dot{\theta}_2 = 120 \text{ rpm CW (constant)}$$

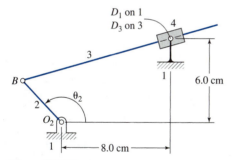

**Figure P4.15**

**P4.16** The nomenclature for this group of problems is given in Figure P4.16, and the dimensions and data are given in Table P4.16. For each, determine the angular displacements, angular velocities, and angular accelerations of links 3 and 4. Use a complex number approach. (**Mathcad program: fourbarkin**)

**P4.17** The nomenclature for this group of problems is given in Figure P4.17, and the dimensions and data are given in Table P4.17. For each, determine the angular displacement, angular velocity, and angular acceleration of link 3, and the displacement, velocity, and acceleration of link 4. Use a complex number approach. (**Mathcad program: slidercrankkin**)

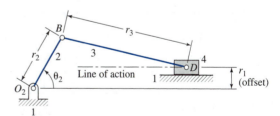

**Figure P4.17**

**TABLE P4.16**

| | $r_1$ (cm) | $r_2$ (cm) | $r_3$ (cm) | $r_4$ (cm) | $\theta_2$ (degrees) | $\dot{\theta}_2$ (rad/sec) | $\ddot{\theta}_2$ (rad/sec$^2$) |
|---|---|---|---|---|---|---|---|
| (a) | 5.0 | 1.0 | 4.0 | 3.0 | 30 | 30 | 0 |
| (b) | 8.0 | 2.0 | 5.0 | 4.0 | 60 | 45 | 0 |
| (c) | 10.0 | 32.0 | 20.0 | 20.0 | 120 | −50 | 40 |
| (d) | 5.0 | 1.0 | 6.0 | 4.0 | 210 | 100 | −40 |

**P4.18** The nomenclature for this group of problems is given in Figure P4.18, and the dimensions and data are given in Table P4.18. For each, determine the linear displacement, linear velocity, and linear acceleration of point $C$. Use a complex number approach. (**Mathcad program: fourbarcpkin**)

**P4.19** The nomenclature for this group of problems is given in Figure P4.19, and the dimensions and data are given in Table P4.19. For each, determine the linear displacement, linear velocity, and linear acceleration of point $C$. Use a complex number approach. (**Mathcad program: slidercrankcpkin**)

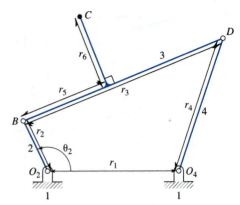

**Figure P4.18**

**TABLE P4.17**

|     | $r_1$ (cm) | $r_2$ (cm) | $r_3$ (cm) | $\theta_2$ (degrees) | $\dot{\theta}_2$ (rad/sec) | $\ddot{\theta}_2$ (rad/sec²) |
|-----|-----|-----|------|-----|-----|-----|
| (a) | 0   | 5.0 | 10.0 | 30  | 50  | 0   |
| (b) | 0   | 3.0 | 4.0  | 60  | 45  | 0   |
| (c) | 2.0 | 3.0 | 6.0  | 120 | −50 | 40  |
| (d) | −2.0| 4.0 | 10.0 | 150 | 100 | −40 |

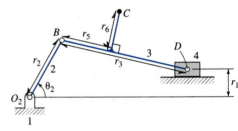

**Figure P4.19**

**TABLE P4.18**

|     | $r_1$ (cm) | $r_2$ (cm) | $r_3$ (cm) | $r_4$ (cm) | $r_5$ (cm) | $r_6$ (cm) | $\theta_2$ (degrees) | $\dot{\theta}_2$ (rad/sec) | $\ddot{\theta}_2$ (rad/sec²) |
|-----|-----|-----|------|-----|------|------|-----|-----|-----|
| (a) | 1.0 | 0.2 | 0.4  | 1.0 | 0.2  | 0.5  | 30  | 30  | 0   |
| (b) | 5.0 | 1.0 | 4.0  | 3.0 | −2.0 | 0    | 60  | 45  | 0   |
| (c) | 8.0 | 2.0 | 10.0 | 6.0 | 12.0 | −1.0 | 120 | −50 | 40  |
| (d) | 0.2 | 1.0 | 0.4  | 0.9 | 0.2  | 0.4  | 210 | 100 | −40 |

**TABLE P4.19**

|     | $r_1$ (cm) | $r_2$ (cm) | $r_3$ (cm) | $r_5$ (cm) | $r_6$ (cm) | $\theta_2$ (degrees) | $\dot{\theta}_2$ (rad/sec) | $\ddot{\theta}_2$ (rad/sec²) |
|-----|------|-----|------|-----|------|-----|-----|-----|
| (a) | 0    | 5.0 | 10.0 | −4  | 0    | 30  | 30  | 0   |
| (b) | 0    | 2.0 | 6.0  | 2.0 | 4.0  | 60  | 45  | 0   |
| (c) | 2.0  | 4.0 | 8.0  | 12.0| 0    | 120 | −50 | 40  |
| (d) | −2.0 | 3.0 | 7.0  | 5.0 | −3.0 | 315 | 100 | −40 |

# 5 Gears

## 5.1 INTRODUCTION

This chapter covers both friction and toothed gear systems. The following sections present a mathematical model for analyzing friction gears, descriptions of various types of toothed gears, and common manufacturing methods for gears.

## 5.2 FRICTION GEARING

A simple means of transmitting rotational motion and power from one shaft to another is by use of a pair of friction gears, as shown in Figure 5.1(a). Here, transmission relies on the friction force between the cylinders in contact.

A free body diagram of each cylinder is shown in Figure 5.1(b). Tangent to both cylinders is a force of magnitude

$$F = |\overline{F}| \tag{5.2-1}$$

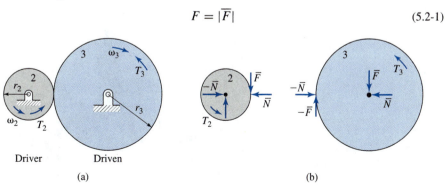

Driver     Driven

(a)           (b)

**Figure 5.1** Friction gears: (a) friction gear set, (b) free body diagrams.

In the arrangement shown, gear 2 is the *driver* gear, and gear 3 is the *driven* gear. Power is supplied to the system through the shaft connected to gear 2, and power is removed from the system through the shaft attached to gear 3. The torque driving gear 2 is in the same direction as its rotational speed. The magnitude of that torque is

$$|T_2| = Fr_2 \tag{5.2-2}$$

Torque applied to gear 3 is in the opposite direction to its rotation. The magnitude of that torque is

$$|T_3| = Fr_3 \tag{5.2-3}$$

Combining Equations (5.2-2) and (5.2-3),

$$\left| \frac{T_2}{T_3} \right| = \frac{r_2}{r_3} \tag{5.2-4}$$

If no slippage occurs,

$$r_2 |\omega_2| = r_3 |\omega_3| \tag{5.2-5}$$

The magnitude of the speed ratio is

$$\left| \frac{\omega_3}{\omega_2} \right| = \frac{r_2}{r_3} \tag{5.2-6}$$

Combining Equations (5.2-4) and (5.2-6),

$$|T_2\omega_2| = |T_3\omega_3| \tag{5.2-7}$$

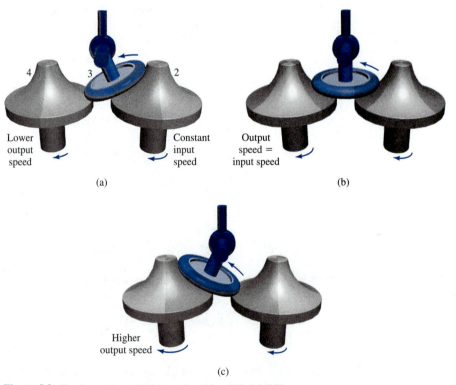

(a)                                                          (b)

(c)

**Figure 5.2** Continuously variable traction drive [Model 5.2].

However, multiplication of a torque by a rotational speed yields *power* (Appendix D, Section D.10.3). Therefore, Equation (5.2-7) states that assuming no slippage between the friction gears, no power is lost in the transmission.

Frictional forces tangent to the cylinders are obtained by imposing a normal force of magnitude

$$N = |\overline{N}| \tag{5.2-8}$$

Since for the friction model

$$F \leq \mu N \tag{5.2-9}$$

in which the *coefficient of friction*, $\mu$, is generally less than unity, $N$ must therefore be greater than $F$. These relatively large normal forces must be supported by the bearings, as indicated in Figure 5.1(b), and must be taken into account in a design.

One definite advantage of using friction gearing is the ease of generating a continuous range of speed ratios. Consider the system shown in Figure 5.2. Changes in speed ratio are accomplished by altering the orientation of the idler wheel, link 3, and thereby varying the radii of contact between links 2 and 4. In the illustration, three positions of the idler wheel are shown that produce three distinct speed ratios. If the input cone 2 rotates at a constant rate, then the output of cone 4 may have a varying rotational speed.

Figure 5.3 shows an alternative system that can produce a continuous range of speed ratios. This system employs a variation of the open loop friction drive (Figure 1.17). In this

**Model 5.2**

Continuously
Variable Traction
Drive

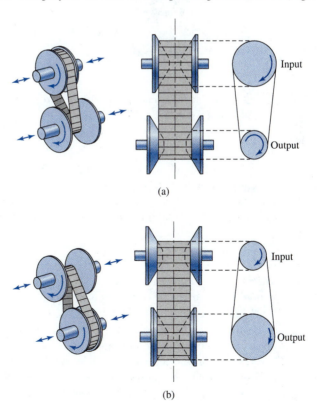

(a)

(b)

**Figure 5.3** Continuously variable belt drive: (a) high speed ratio, (b) low speed ratio.

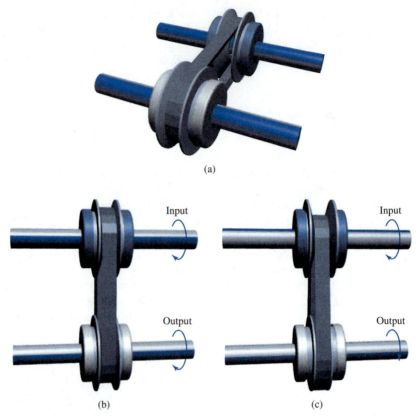

**Figure 5.4** (a) Continuously variable belt drive [Model 5.4]. (b) Unit speed ratio.
(c) Low speed ratio.

| Model 5.4 |
|---|
| Continuously Variable Belt Drive |

case, a range of speed ratios is obtained by allowing the belt to contact the pulleys at different radii, accomplished by adjusting the positions of the two halves of the pulleys. Two positions are shown in Figures 5.3(a) and 5.3(b), resulting in two distinct speed ratios. Figure 5.4 shows additional illustrations of a similar system.

## 5.3  COMMON TYPES OF TOOTHED GEARS

| Model 5.5A |
|---|
| Straight Spur Gears |

*Straight spur gears* (Figure 5.5(a)): These gears have straight teeth, parallel to the axis of rotation. They are relatively easy to manufacture.

| Model 5.5B |
|---|
| Helical Spur Gears |

   *Helical spur gears* (Figure 5.5(b)): The teeth of a helical spur gear are at an oblique angle with respect to the axis of rotation. Compared with straight spur gears, there is a longer line of contact with a meshing gear, resulting in increased strength and durability. Helical spur gears are commonly used in automotive transmissions and machine tools. Figure B.46 (Appendix B) shows helical spur gears employed in a power drill.

   The *helix angle*, $\psi$, for helical spur gears is illustrated in Figure 5.6. *Right-hand* and *left-hand* varieties are shown. A pair of meshing helical spur gears with parallel axes of

(a)                                              (b)

**Figure 5.5** Spur gears: (a) straight spur gears [Model 5.5A], (b) helical spur gears [Model 5.5B].

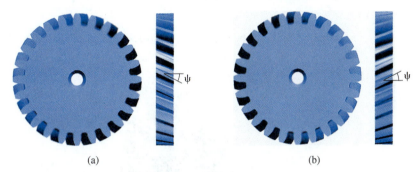

(a)                                              (b)

**Figure 5.6** Helical gears: (a) right-hand, (b) left-hand.

**Figure 5.7** Miter gears [Model 5.7].

rotation must include a gear with left-hand teeth and another with right-hand teeth. The helix angles must be equal.

A specific type of a helical spur gear, called a *miter gear*; it employs a helix angle of 45°. For miter gears, the centerlines of the axes of rotation can be perpendicular to one another, as shown in Figure 5.7. In this instance, two left-hand gears have been employed.

Despite the considerable advantages of helical spur gears relative to straight spur gears, because there is oblique contact between pairs of teeth of mating gears, thrust forces

| Model 5.7 |
| Miter Gears |

(a)                                                    (b)

**Figure 5.8** Herringbone gears: (a) [Model 5.8A], (b) [Model 5.8B].

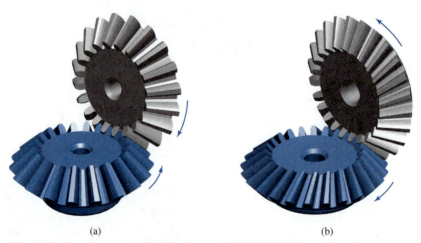

(a)                                                    (b)

**Figure 5.9** Bevel gears: (a) plain bevel gears [Model 5.9A], (b) spiral bevel gears [Model 5.9B].

<table>
<tr><td>

**Model 5.8A**
Narrow
Herringbone
Gears

**Model 5.8B**
Wide
Herringbone
Gears

**Model 5.9A**
Plain Bevel
Gears

**Model 5.9B**
Spiral Bevel
Gears

</td></tr>
</table>

occur along the axes of rotation. These axial loads may be supported by adding thrust bearings.

*Herringbone gears* (Figures 5.8(a) and 5.8(b)): Herringbone gears are equivalent to two helical gears with left- and right-hand teeth. This causes the axial load produced on one side to be counterbalanced by that produced on the other. Herringbone gears are well suited for heavy load applications such as large turbines and generators.

*Plain bevel gears* (Figure 5.9(a)): These gears permit transmission of motion between two shafts angled relative to each other.

*Spiral bevel gears* (Figure 5.9(b)): The teeth of spiral bevel gears are cut obliquely so that the length of the line of contact between meshing gears is longer compared to plain bevel gears, thus providing greater tooth strength and durability.

*Hypoid gears* (Figure 5.10): A hypoid gear set resembles spiral bevel gears. However, the teeth are cut such that the axes of rotation of the two gears in mesh do not intersect. A

**Figure 5.10** Hypoid gear set [Video 5.10].

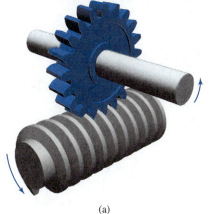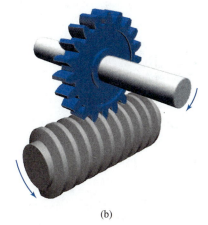

(a)      (b)

**Figure 5.11** Worm and wheel gears: (a) right-hand worm [Model 5.11A], (b) left-hand worm [Model 5.11B].

common use of hypoid gears is in the differential assembly of rear wheel drive vehicles. They allow reduction in height of body styles by lowering the drive shaft to the rear wheels.

*Worm and wheel gears* (Figures 5.11 and 5.12): A *worm* gear is similar in shape to a screw thread. The mating gear, called the *wheel*, often has teeth that are curved at their tips to permit greater contact area. Power is always supplied to the worm. In fact, worm and wheel gear sets are typically self-locking (i.e., motion cannot be transmitted by driving the wheel gear). Worm and wheel gear sets provide high reductions of speed in a compact space. For each worm and wheel gear set shown in Figure 5.11, one rotation of the worm gear will cause the wheel to advance one tooth. Therefore, the magnitude of the speed ratio, $e$, is

$$|e| = \frac{1}{N_w} \tag{5.3-1}$$

where $N_w$ is the number of teeth on the wheel.

The gear set shown in Figure 5.11(a) incorporates a *right-hand worm*, and in Figure 5.11(b) a *left-hand worm* is employed. Although the magnitudes of the speed ratios of

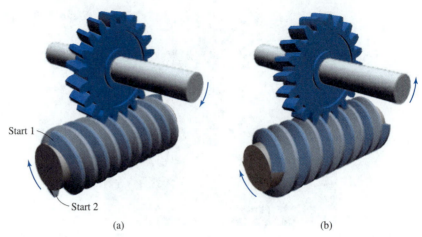

Start 1

Start 2

(a)                (b)

**Figure 5.12** Worm and wheel, double-start worm: (a) right-hand worm [Model 5.12A], (b) left-hand worm [Model 5.12B].

| |
|---|
| **Model 5.12A** |
| Double-Start |
| Right-Hand |
| Worm |

both gear sets are the same, for the same input motion of the worms, the direction of the rotation of the wheels is opposite.

Figure 5.12 shows two other versions of worm and wheel sets. Each incorporates a *double-start worm*. Both left- and right-hand worm gears are illustrated. For both cases shown, for each rotation of the worm gear, the wheel advances two teeth. Therefore, the magnitude of the speed ratio is

| |
|---|
| **Model 5.12B** |
| Double-Start |
| Left-Hand |
| Worm |

$$|e| = \frac{2}{N_w} \tag{5.3-2}$$

Worm gears can be designed to have as many as four starts.

## 5.4 FUNDAMENTAL LAW OF TOOTHED GEARING

For meshing gears, it is usually critical to maintain a constant ratio of speeds. Slight variations in the ratio would lead to unwanted vibrations and noise, caused by fluctuating loads and speeds. This section presents the requirement for the shape of gear teeth necessary to maintain constant speed ratio.

Figure 5.13 shows two members, links 2 and 3, in contact at point $Q$. The links are pivoted about points $O_2$ and $O_3$. The *line of centers* contains $O_2$ and $O_3$. Rotational speeds of the links are designated as $\omega_2$ and $\omega_3$. By inspection, $\omega_2$ and $\omega_3$ must be in opposite directions if contact is to be maintained as the links rotate. The point of contact, $Q$, is actually composed of two points: $Q_2$ on link 2, and $Q_3$ on link 3.

When point $Q$ is on the line of centers (Figure 5.13(a)), there is a rolling action between the links. In this case, there is no relative sliding at $Q$ since

$$\bar{v}_{Q_2} = \bar{v}_{Q_3}$$

Figure 5.13(b) shows an alternate configuration of the links where point $Q$ is not on the line of centers. In this instance, there is both relative sliding and turning, and point $Q$ is a two

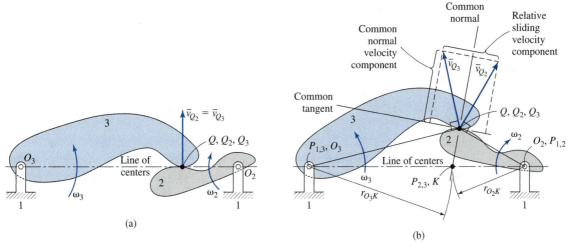

**Figure 5.13** Two members in contact.

degree of freedom kinematic pair. The *common tangent* and *common normal* to the surfaces at $Q$ are indicated. The common normal intersects the line of centers at $K$.

    The ratio of rotational speeds between links may be readily found using the concept of
instantaneous centers (Section 2.8). As indicated in Figure 5.13(b), two of the instantaneous
centers, $P_{1,2}$ and $P_{1,3}$, are at the points $O_2$ and $O_3$, respectively. Based on Kennedy's Theorem
(Section 2.9), the third instantaneous center coincides with $K$. For this system we may write

$$|\bar{v}_{P_{2,3}}| = r_{P_{1,2}P_{2,3}}|\omega_2| = r_{P_{1,3}P_{2,3}}|\omega_3| \tag{5.4-1}$$

which is equivalent to

$$r_{O_2K}|\omega_2| = r_{O_3K}|\omega_3| \tag{5.4-2}$$

and

$$\left|\frac{\omega_3}{\omega_2}\right| = \frac{r_{O_2K}}{r_{O_3K}} \tag{5.4-3}$$

Equation (5.4-3) indicates that the magnitude of the speed ratio for two links in contact is
equal to the ratio of the lengths of the line segments that are along the line of centers. The
lengths of the line segments involve the location of $K$ and the axes of rotation of the links.

    The above may be applied for the specific case where the links are meshing gears.
Also, we may deduce the condition to maintain a constant speed ratio between a pair of
meshing gears, known as the *fundamental law of toothed gearing*. The law may be stated
as follows: To maintain a constant speed ratio between a pair of gears, the common normal
at the point of contact between meshing teeth must always intersect the line of centers at a
fixed point.

    This fixed point is also called the *pitch point*. Even though the location of the point of
contact may change, the fundamental law of toothed gearing can still be satisfied as long as
the location of the pitch point remains fixed.

## 5.5 INVOLUTE TOOTH GEARING

Having established the fundamental law of toothed gearing, it is now possible to search for those shapes of teeth which will provide a constant speed ratio between meshing gears. This section presents gear teeth that have involute form, which is by far the most common shape of gear teeth.

Figure 5.14(a) shows two counterrotating cylinders, also referred to as *base circles*, of radii $r_{b2}$ and $r_{b3}$. Their rotational speeds are $\omega_2$ and $\omega_3$ in the directions shown. A piece of string is wrapped around both base circles, similar to that of the cross belt friction drive shown in Figure 1.19. A bead is attached to the string between the base circles. In Figure 5.14(a), the bead is located on the line of centers of the base circles, and Figure 5.14(b) shows the bead in an alternate position. As the base circles rotate, the string unwinds from base circle 2 and is taken up by base circle 3. In order for the string to remain taut, the unwinding and take-up speeds must be equal, that is,

$$r_{b2}|\omega_2| = r_{b3}|\omega_3| \tag{5.5-1}$$

As the base circles rotate, the bead moves on a straight path with respect to the stationary base link. However, motion of the bead relative to a base circle may be observed by holding one base circle stationary, and then either winding or unwinding the string around the base circle, as shown in Figure 5.15(a). Under these conditions, the curve traced out by the bead is called an *involute*. At any position, the direction of the unwound string is normal to the involute.

**Model 5.15**
Involute
Generation

The model shown in Figure 5.15(b) consists of two base circles. In this instance, the bead traces out two involutes simultaneously, one with respect to each of the base circles. At all positions, the bead corresponds to the point of contact of the involutes, and the orientation of the string between the base circles is normal to both involutes.

Let us now remove the string from the demonstration model shown in Figure 5.15(b). Instead, as illustrated in Figure 5.16, motion is transmitted by meshing surfaces having involute profiles. These profiles are in contact at point $A$. A common normal to the involute

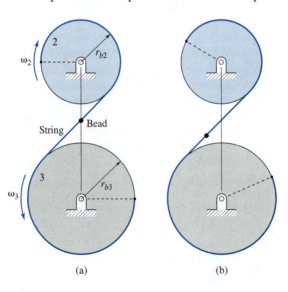

**Figure 5.14** Two counterrotating cylinders connected by a string.

(a)                                    (b)

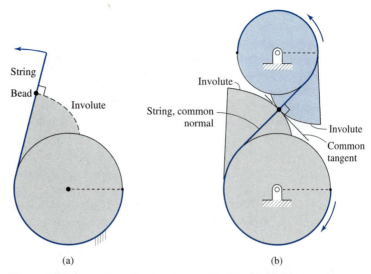

**Figure 5.15** Generation of an involute profile [Model 5.15]: (a) one-cylinder model, (b) two-cylinder model.

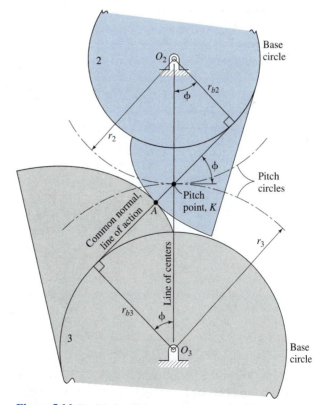

**Figure 5.16** Double involute.

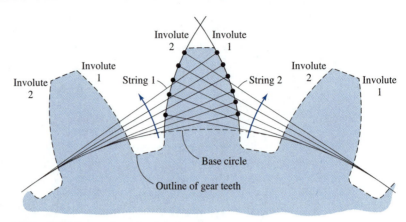

**Figure 5.17** Generation of an involute gear.

profiles is drawn through the point of contact. It has the same orientation as that which the string had between the base circles. The common normal intersects the line of centers at the pitch point $K$. As the base circles rotate, the location of the pitch point remains fixed, even though the point of contact changes. Since the pitch point does not move, the meshing involute profiles satisfy the fundamental law of toothed gearing, that is, they generate a constant speed ratio.

Involute profiles are incorporated in gear teeth to generate a constant speed ratio between a pair of meshing gears. Each gear employs multiple teeth positioned around its periphery. The shape of both sides of every tooth is an involute. Each tooth, referred to as an *involute gear tooth*, makes use of two involutes. As illustrated in Figure 5.17, the involute on each side of a gear tooth is equivalent to unwrapping a string from the base circle. The beads on string 1 and string 2 trace out involute 1 and involute 2. Figure 5.18 shows a pair of gears with involute gear teeth. All teeth on a gear employ the same involute profile and have uniform height.

For the animation provided through **[Model 5.18]**, pairs of teeth in turn come into and out of mesh and maintain a constant speed ratio as the gears rotate. Figure 5.19 shows an external gear meshing with an internal gear. Figure 5.20 shows a rack and pinion gear set.

The common normal at the point of contact for a pair of gears is also called the *line of action*. Gears incorporating involute profiles have a fixed line of action. Assuming no friction, the line of action also represents the orientation of interactive forces between meshing teeth. The angle between the line of action and a direction perpendicular to the line of centers is called the *pressure angle*, $\phi$, as illustrated in Figure 5.16.

Figure 5.16 shows portions of two *pitch circles* associated with the gears. These circles are centered at $O_2$ and $O_3$, the same as for the base circles, and both pass through the pitch point. Their radii are $r_2$ and $r_3$. Applying Equation (5.4-2) in this instance gives

$$r_2|\omega_2| = r_3|\omega_3| \tag{5.5-2}$$

Equation (5.5-2) is identical to Equation (5.2-5), which was obtained by considering a pair of friction gears of radii $r_2$ and $r_3$, for which there was no slippage. Thus, a pair of toothed gears is kinematically equivalent to friction gears with radii equal to the pitch circles. Either side of Equation (5.5-2) represents the magnitude of the *pitch line velocity*, $v_p$.

**Model 5.18**
External-
External
Involute Gears
Meshing

**Model 5.19**
External-
Internal
Involute Gears
Meshing

**Model 5.20**
Rack and Pinion

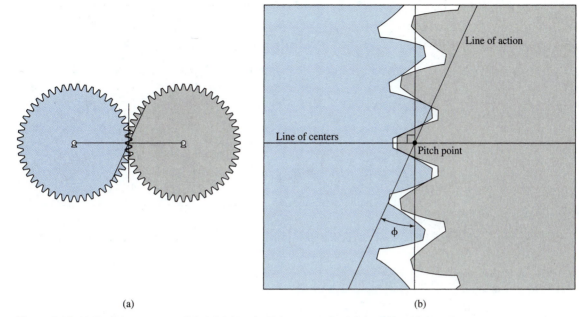

(a)                                    (b)

**Figure 5.18** (a) Straight spur gears [Model 5.18]. (b) Enlargement in vicinity of meshing teeth.

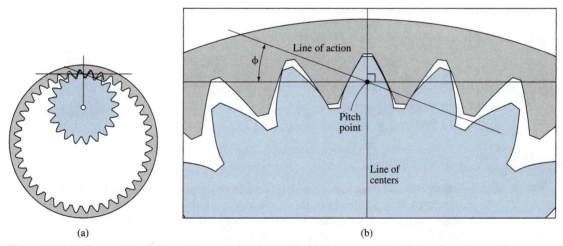

(a)                                    (b)

**Figure 5.19** (a) External-internal meshing gears [Model 5.19]. (b) Enlargement in vicinity of meshing teeth.

Using Figure 5.16, the relation between radii of the base and pitch circles is

$$ r_{bj} = r_j \cos \phi \tag{5.5-3} $$

The shape of an involute for a gear tooth is a function of its base circle diameter and is independent of the *center-to-center distance* between a pair of meshing gears. As an illustration, Figure 5.21 shows two base circles with a string wrapped around both of them. Two

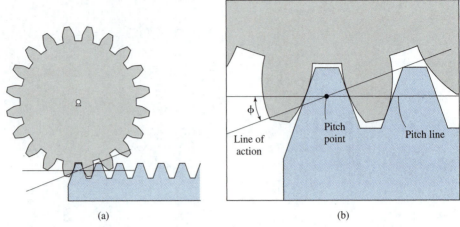

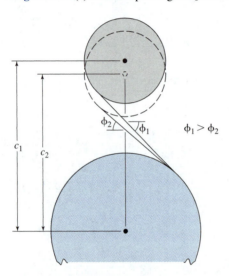

**Figure 5.20** (a) Rack and pinion gears [Model 5.20]. (b) Enlargement in vicinity of meshing teeth.

**Figure 5.21** Variation of pressure angle.

center-to-center distances, $c_1$ and $c_2$, are shown. For each, the line of action and pressure angle are different. However, the involute profiles generated for each base circle are identical. Thus, as long as contact is kept between gear teeth, even if the center-to-center distance is changed, a constant speed ratio will be maintained.

## 5.6  CYCLOIDAL TOOTH GEARING

Gear teeth having the shape of a cycloid, or *cycloidal gear teeth*, also satisfy the fundamental law of toothed gearing. As illustrated in Figure 5.22, a *cycloid* is formed by rolling circles on the inside and outside of a pitch circle, and drawing the path of a point on the periphery of the rolling circles. The curves inside and outside of the pitch circle are called the *hypocycloid* and *epicycloid*, respectively. Sizes of pitch circles for two gears in mesh may differ. However, sizes of the rolling circles for the epicycloid and hypocycloid of one

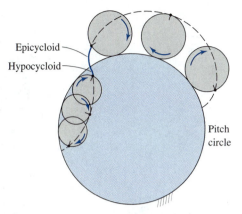

**Figure 5.22** Generation of a cycloid [Model 5.22].

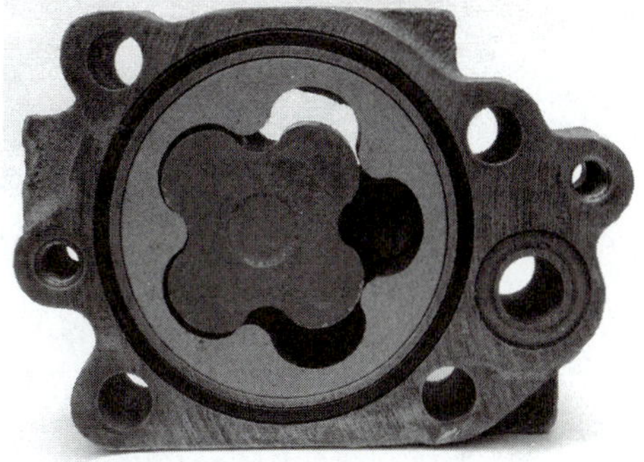

**Figure 5.23** Gerotor pump [Video 5.23].

**Figure 5.24** Cycloidal gerotor pump.

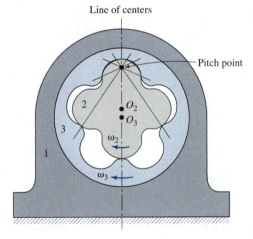

gear must respectively match those generated by the hypocycloid and epicycloid of the meshing gear.

**Video 5.23**
Gerotor Pump

Gears with cycloidal teeth are often employed in positive-displacement pumps. Figure 5.23 shows an oil pump from an automobile engine. A pump of this form is also called a *cycloidal gerotor pump*. Another such pump is shown in Figure 5.24. It consists of two gears with cycloidal teeth, indicated as links 2 and 3. Link 3 is an internal gear, and both gears have a fixed axis of rotation. There are multiple points of contact between the gears. For the configuration shown, we draw the common normal through all points of contact. Since there must be only one constant speed ratio of the gears, then as required, all common normals intersect the line of centers at the same fixed pitch point.

The directions of the common normals at the points of contact are not fixed as the gears rotate. This is distinct from gears having involute gear teeth, for which the direction of the common normal remains fixed.

**Model 5.25**
Gerotor Pump

Figure 5.25 shows a series of positions of the gears. The angle indicated for each configuration denotes the rotation of link 3 from the selected starting point. For demonstration

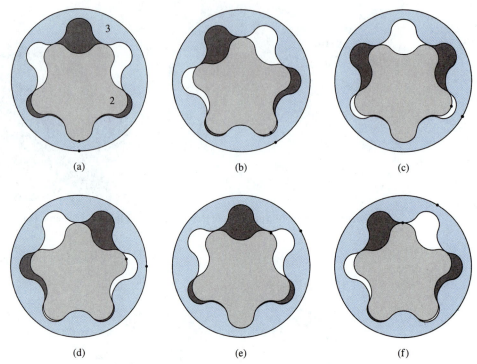

**Figure 5.25**  Cycloidal gerotor pump [Model 5.25]: (a) $\theta_3 = 0°$, (b) $\theta_3 = 30°$, (c) $\theta_3 = 60°$, (d) $\theta_3 = 90°$, (e) $\theta_3 = 120°$, (f) $\theta_3 = 150°$.

purposes, pockets between points of contact are alternately coloured black and white to distinguish one from the next. In actual use, all pockets carry the same fluid. During rotation, each pocket increases in size, drawing fluid in through an inlet port (not shown) that is located on the sides of the gears. Once a pocket reaches its maximum size, the source from the inlet port is cut off. Upon further rotation, the size of the pocket decreases, and fluid is discharged through an outlet port.

Another application of cycloidal gear teeth is in a blower employed in superchargers. It consists of two counterrotating lobes as shown in Figure 5.26. The angle shown for each configuration indicates the rotation of the right lobe from the selected starting configuration. Each lobe is in fact a gear having two cycloidal teeth. For the starting position illustrated, there cannot be any driving torque between the gears. This is because the point of contact lies on the line of centers, and the common normal at the point of contact coincides with the line of centers. Therefore, in addition to the gears shown, it is necessary to include two other gears of equal size in mesh with their same axes of rotation. The additional gears ensure a continuous torque transmission between the driver and driven shafts.

The above two examples employing cycloidal gears make use of the entire epicycloid and hypocycloid. However, it is also possible to use only a portion of the cycloid, similar to that of a pair of gears using involute profiles (Figure 5.18). Such a pair of gears is shown in Figure 5.27.

**Model 5.26**
Supercharger

**Model 5.27**
External-
External
Cycloidal Gears
Meshing

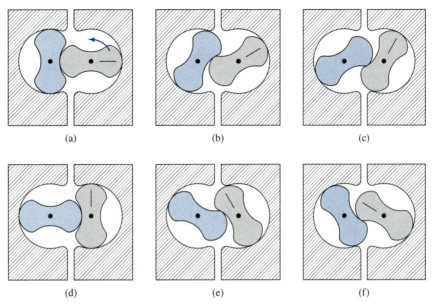

**Figure 5.26** Supercharger blower [Model 5.26]: (a) $\theta = 0°$, (b) $\theta = 30°$, (c) $\theta = 60°$, (d) $\theta = 90°$, (e) $\theta = 120°$, (f) $\theta = 150°$.

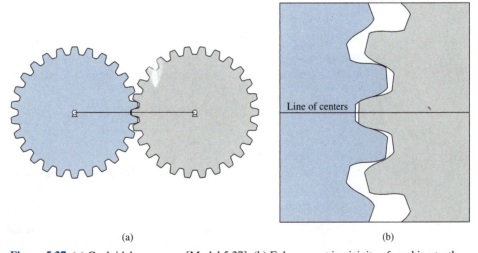

**Figure 5.27** (a) Cycloidal spur gears [Model 5.27]. (b) Enlargement in vicinity of meshing teeth.

## 5.7 SIZING OF INVOLUTE GEAR TEETH

*Circular pitch* and *diametral pitch* are two means used to designate the size of straight spur gears incorporating an involute profile, using the U.S. Customary system of units.

*Circular pitch*, $p_c$, is the curvilinear distance in inches, measured along the pitch circle, from a point on one tooth to the corresponding point on an adjacent tooth. It is illustrated in Figure 5.28 and is expressed as

$= \text{if in mesh}.$

$$p_c = \frac{\pi d}{N} = \frac{2\pi r}{N} \qquad (5.7\text{-}1)$$

where $d$ is the pitch circle diameter, and $N$ is the number of teeth on the gear.

*Diametral pitch*, $P$, is the number of gear teeth per inch of pitch circle diameter, and has units of in$^{-1}$. It is expressed as

$$P = \frac{N}{d} = \frac{N}{2r} \qquad (5.7\text{-}2)$$

Combining Equations (5.7-1) and (5.7-2),

$$P p_c = \pi \qquad (5.7\text{-}3)$$

*Module* is the term used to indicate the size of involute gear teeth in the SI system. It is defined as the pitch circle diameter, in millimeters, divided by the number of teeth on the gear. Module has units of mm, and is expressed as

$$m = \frac{d}{N} = \frac{2r}{N} \qquad (5.7\text{-}4)$$

Two gears in mesh must have the same diametral pitch, circular pitch, and module. This requirement is often simply stated that "gears in mesh must have the same pitch." Meshing gears must also have the same pressure angle. The standardized values of pressure angles are 14.5°, 20°, and 25°. Today, 20° is the most common.

From Equations (5.7-1), (5.7-2), and (5.7-4)

$$d = 2r = \frac{N p_c}{\pi} = \frac{N}{P} = mN \qquad (5.7\text{-}5)$$

Thus, for a given pitch, the number of teeth on a gear is proportional to its pitch circle diameter.

Proportions of a gear tooth are shown in Figure 5.28(a). Included in this figure are

- *addendum*, $a$: the radial distance between the pitch circle and a circle (also called the *addendum circle*) drawn through the tops of the gear teeth
- *dedendum*, $b$: the radial distance between the pitch circle and a circle (also called the *dedendum circle*) defining the bottom land between teeth
- *clearance*, $b - a$: the difference between the dedendum and the addendum

Figures 5.28(b) and 5.28(c) show the addendum and dedendum associated with a rack and an internal gear. A rack has a *pitch line*, rather than a pitch circle.

Values of addendum and dedendum have been standardized in order to facilitate interchangeability of gears. They may be expressed in terms of either circular pitch, diametral pitch, or module. Table 5.1 provides a summary of standardized values in terms of diametral

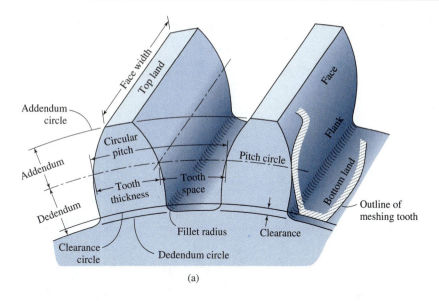

(a)

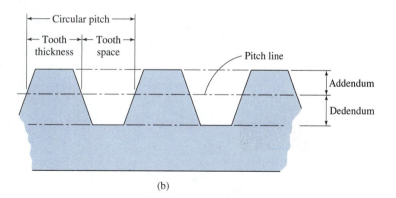

(b)

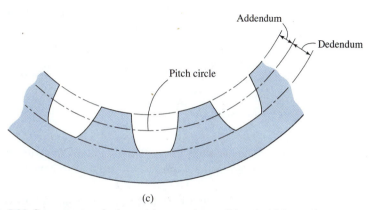

(c)

**Figure 5.28** Spur gear terminology: (a) external gear, (b) rack, (c) internal gear.

**TABLE 5.1** Standard Proportions of Gear Teeth as a Function of Diametral Pitch

|  | Addendum $a$ | Dedendum $b$ | Whole Depth $a + b$ | Clearance $b - a$ |
|---|---|---|---|---|
| **Full-Depth Tooth ($\phi = 14.5°$)** | $\dfrac{1.0}{P}$ | $\dfrac{1.157}{P}$ | $\dfrac{2.157}{P}$ | $\dfrac{0.157}{P}$ |
| **Stub-Tooth ($\phi = 20°$)** | $\dfrac{0.8}{P}$ | $\dfrac{1.0}{P}$ | $\dfrac{1.8}{P}$ | $\dfrac{0.2}{P}$ |
| **Full-Depth Tooth ($\phi = 20°$)** | $\dfrac{1.0}{P}$ | $\dfrac{1.25}{P}$ | $\dfrac{2.25}{P}$ | $\dfrac{0.25}{P}$ |
| **Full-Depth Tooth ($\phi = 25°$)** | $\dfrac{1.0}{P}$ | $\dfrac{1.25}{P}$ | $\dfrac{2.25}{P}$ | $\dfrac{0.25}{P}$ |

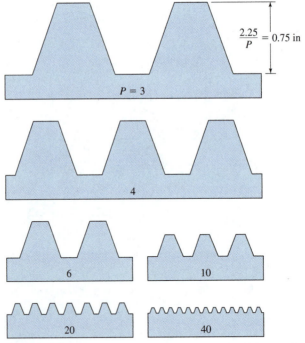

**Figure 5.29**  Sizes of gear teeth for typical values of diametral pitch (full depth, $\phi = 20°$).

pitch. Straight spur gears have two standard forms when $\phi = 20°$: *stub* and *full-depth*. Values are provided for both forms. Full-depth teeth have a larger value of addendum than stub teeth.

Figure 5.29 illustrates teeth on racks with various values of diametral pitch, using the proportions listed in Table 5.1 for full-depth teeth and $\phi = 20°$.

The proportions in Table 5.1 show that the value of the dedendum is greater than the addendum. The clearance, defined as the difference between the dedendum and the addendum, also corresponds to the distance between the top land of one tooth and the bottom land of a meshing tooth.

As indicated in Figure 5.28(a), the *tooth thickness* and *tooth space* are measured along the pitch circle. The tooth thickness is nominally one-half of the circular pitch, and thus the tooth thickness and tooth space between two teeth of a gear are equal. In this instance, if the pitch circles of two meshing gears touch at the pitch point, then contact will simultaneously occur on both sides of a gear tooth with the meshing gear. Obviously, the thickness cannot be greater than the tooth space or the gears will not mesh.

## EXAMPLE 5.1

A spur gear, having 30 teeth and a diametral pitch of 6 in$^{-1}$, is rotating at 200 rpm. Determine the circular pitch and the magnitude of the pitch line velocity.

### SOLUTION

From Equation (5.7-3), the circular pitch is

$$p_c = \frac{\pi}{P} = \frac{\pi}{6} = 0.524 \text{ in}$$

To find the magnitude of the pitch line velocity, $v_p = r\omega$, we first have to determine the pitch circle diameter of the gear. From Equation (5.7-2)

$$d = \frac{N}{P} = \frac{30}{6} = 5.0 \text{ in}$$

The angular velocity of the gear in terms of radians per second is

$$\omega = 200 \text{ rev/min} \times 2\pi \text{ rad/rev} \times \frac{1}{60} \text{ min/sec} = 20.94 \text{ rad/sec}$$

and finally the magnitude of the pitch line velocity is

$$v_p = r\omega = \frac{d}{2}\omega = 2.5 \text{ in} \times 20.94 \text{ rad/sec} = 52.4 \text{ in/sec}$$

## EXAMPLE 5.2

Two spur gears are in mesh. Driven gear 3 has a magnitude of rotational speed that is one-half that of driver gear 2. Gear 2 rotates at 500 rpm, has a module of 3 mm, and has 48 teeth. Determine

(a) the number of teeth of gear 3
(b) the magnitude of the pitch line velocity

## SOLUTION

(a) From Equation (5.7-4)

$$d_2 = mN_2 = 3 \times 48 = 144 \text{ mm}; \qquad r_2 = 72 \text{ mm}$$

From Equation (5.5-2)

$$r_3 = \frac{r_2|\omega_2|}{|\omega_3|} = \frac{r_2\omega_2}{\left(\frac{1}{2}\omega_2\right)} = 2r_2 = 2 \times 72 \text{ mm} = 144 \text{ mm}$$

$$d_3 = 288 \text{ mm}$$

The number of teeth on gear 3 is then

$$N_3 = \frac{d_3}{m} = \frac{288}{3} = 96$$

where, as required, the same value of module has been used for both gears.

(b) The magnitude of the pitch line velocity is

$$v_P = r_2|\omega_2| = 72 \text{ mm} \times 500 \times \frac{2\pi}{60} \text{ rad/sec} = 3,770 \text{ mm/sec} = r_3|\omega_3|$$

## 5.8 BACKLASH AND ANTIBACKLASH GEARS

The limiting condition that permits proper meshing between two gears is when the tooth thicknesses and tooth spaces are equal. For this reason, the tooth thickness of a gear is often manufactured to be slightly less than the tooth space. *Backlash* is the difference between the tooth space and tooth thickness as measured along a pitch circle. Figure 5.30 illustrates the backlash for a pair of meshing gears. Although an insignificant amount of backlash is often

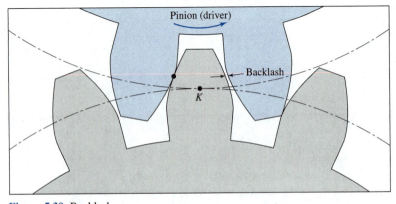

**Figure 5.30** Backlash.

 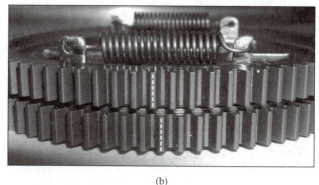

(a)                                                     (b)

**Figure 5.31**  (a) Antibacklash gear. (b) Close-up of antibacklash gear.

incorporated into a design, excessive backlash would cause impacts between teeth, especially if a reversal of rotation occurs. This can give rise to increased noise and tooth loading.

An *antibacklash gear* is shown in Figure 5.31(a). A close-up of the gear teeth is shown in Figure 5.31(b). The gear is composed of two halves that can move rotationally relative to one another about the same central axis against the action of spring forces. The teeth of the two halves are staggered when the gear is not in mesh. Dashed lines on the top of a tooth on the two sides indicate the stagger. Through meshing, the stagger is either reduced or eliminated. The relative rotation between the two halves is accomplished through forces applied by the meshing teeth.

> **Model 5.32**
> Antibacklash Gear

Figure 5.32 shows a pinion that meshes with two gears. One is a regular gear, and the other is an antibacklash gear. In Figure 5.32(a), the gears are separated. As the animation of **[Model 5.32]** proceeds, the two gears are brought into mesh with the pinion (Figure 5.32(b)). The teeth on the two meshing regular gears do not have simultaneous contact on both sides of a gear tooth. However, for meshing of the pinion and antibacklash gear, contact is maintained on both sides, even when the directions of rotation are reversed.

## 5.9 GEOMETRICAL CONSIDERATIONS IN THE DESIGN OF REVERTED GEAR TRAINS

The topic of gear trains will be covered in detail in Chapter 8. For the time being, we will study geometrical considerations that relate to the design of gear trains. This will be illustrated by one particular form of gear train, called a *reverted gear train*.

Reverted gear trains are characterized by having their input and output axes of rotation collinear. A typical reverted gear train is illustrated in Figure 5.33. Gear 2 is mounted on the input shaft, and gear 5 is mounted on the output shaft. Gears 3 and 4 are rigidly connected through a shaft and have a common axis of rotation. They mesh with gears 2 and 5, respectively. The center-to-center distance, $c$, between gears 2 and 3 is equal to that between gears 4 and 5, that is,

$$c = \frac{d_2}{2} + \frac{d_3}{2} = \frac{d_4}{2} + \frac{d_5}{2}$$

(5.9-1)

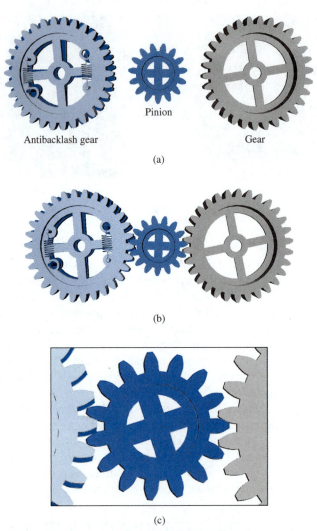

Pinion

Antibacklash gear                                          Gear

(a)

(b)

(c)

**Figure 5.32**  Antibacklash gear [Model 5.32]: (a) separated gears, (b) assembled gears, (c) close-up of meshing gears.

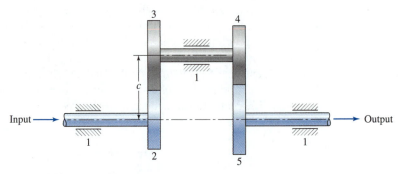

**Figure 5.33**  Reverted gear train.

where $d_2$, $d_3$, $d_4$, and $d_5$ denote pitch circle diameters. If the teeth are characterized by their module, then substituting Equation (5.7-4) in Equation (5.9-1) and simplifying,

$$m_2 N_2 + m_3 N_3 = m_4 N_4 + m_5 N_5 \qquad (5.9\text{-}2)$$

Gears in mesh must have teeth of the same module. For the gear train shown in Figure 5.33, we require

$$m_2 = m_3 \qquad \text{and} \qquad m_4 = m_5 \qquad (5.9\text{-}3)$$

and Equation (5.9-2) becomes

$$m_2(N_2 + N_3) = m_4(N_4 + N_5) \qquad (5.9\text{-}4)$$

Similar expressions to Equations (5.9-2)–(5.9-4) may be generated in cases where the teeth are defined by their circular pitch and diametral pitch.

In the instance when the sizes of all gear teeth are equal (e.g., $m_2 = m_4$), Equation (5.9-4) simplifies to

$$N_2 + N_3 = N_4 + N_5 \qquad (5.9\text{-}5)$$

## EXAMPLE 5.3

Table 5.2 provides information that pertains to the reverted gear train shown in Figure 5.33.

**TABLE 5.2** Parameters of the Reverted Gear Train Shown in Figure 5.33

| Gear No. | No. of Teeth | Diametral Pitch (in$^{-1}$) |
|----------|--------------|------------------------------|
| 2 | 30 | 6 |
| 3 | 48 | — |
| 4 | — | 8 |
| 5 | 72 | — |

All gears have teeth with 20° pressure angle and are full depth. Determine

(a) the circular pitch of gear 2
(b) the base circle radius of gear 3
(c) the pitch circle diameter of gear 4
(d) the addendum circle diameter of gear 5
(e) the center-to-center distance, $c$, between gears 2 and 3

### SOLUTION

We begin by determining the missing information in Table 5.2. Since gears 2 and 3 are in mesh, and gears 4 and 5 are in mesh,

$$P_3 = P_2 = 6 \text{ in}^{-1} \qquad \text{and} \qquad P_5 = P_4 = 8 \text{ in}^{-1}$$

Also, substituting Equation (5.7-2) in Equation (5.9-1),

$$\frac{N_2}{2P_2} + \frac{N_3}{2P_3} = \frac{N_4}{2P_4} + \frac{N_5}{2P_5}$$

$$\frac{30}{2 \times 6} + \frac{48}{2 \times 6} = \frac{N_4}{2 \times 8} + \frac{72}{2 \times 8}$$

from which

$$N_4 = 32$$

Based on the above, the solutions are as follows:

(a)  From Equation (5.7-3)

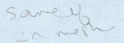

$$p_{c2} = \frac{\pi}{P_2} = \frac{\pi}{6} = 0.524 \text{ in}$$

(b)  From Equations (5.5-3) and (5.7-2)

$$r_{b3} = r_3 \cos\phi = \frac{d_3}{2}\cos\phi = \frac{N_3}{2P_3}\cos\phi = \frac{48}{2 \times 6}\cos 20° = 3.76 \text{ in}$$

(c)  From Equation (5.7-2)

$$d_4 = \frac{N_4}{P_4} = \frac{32}{8} = 4.00 \text{ in}$$

(d)  From Equation (5.7-2) and Table 5.1, the addendum circle diameter of gear 5 is

$$d_5 + 2a_5 = \frac{N_5}{P_5} + 2\frac{1}{P_5} = \frac{72}{8} + \frac{2}{8} = 9.25 \text{ in}$$

(e)  From Equations (5.9-1) and (5.7-2)

$$c = \frac{d_2}{2} + \frac{d_3}{2} = \frac{N_2}{2P_2} + \frac{N_3}{2P_3} = \frac{30}{2 \times 6} + \frac{48}{2 \times 6} = 6.50 \text{ in}$$

## 5.10  CONTACT RATIO

In the transmission of rotational motion through two meshing gears, it is essential that at any time one or more pairs of teeth be in contact. Otherwise, there would be instances when there is no smooth transfer of motion.

For a pair of meshing gears turning at a constant rate, *contact ratio* is the average number of pairs of gear teeth in contact over time. Theoretically, the value of contact ratio must be greater than 1.00. However, 1.40 is generally accepted as the practical minimum value.

To illustrate, Figure 5.34 shows a pair of meshing gears for which the teeth have involute form. Gear 2 is the driver gear, and gear 3 is the driven gear. Initial contact, labelled as point A in Figure 5.34(a), occurs when the outer tip of the driven gear tooth touches the driver gear. Final contact, labelled as point B in Figure 5.34(c), occurs when the outer tip

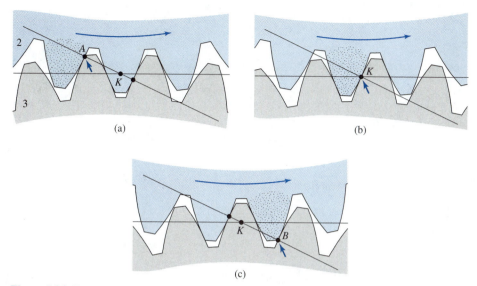

**Figure 5.34** Two spur gears in mesh: (a) initial contact, (b) contact at pitch point, (c) final contact.

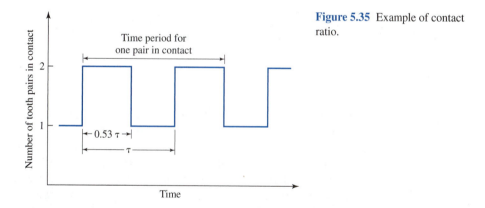

**Figure 5.35** Example of contact ratio.

of the driver gear tooth contacts the driven gear. For both of these configurations there is a second pair of teeth in contact. Thus, prior to and after the noted teeth come into and out of mesh, there is another pair of teeth in mesh, which ensures continuous and uniform transmission of motion. Figure 5.34(b) illustrates another configuration where the point of contact is at the pitch point. At this instant, this is the sole point of contact between the gears.

At any instant in time, there is an integer number of pairs of gear teeth in mesh. For the gears shown in Figure 5.34, during a portion of the cycle, one pair of gear teeth is in mesh, whereas at other times there are two pairs. Figure 5.35 shows the function of the number of pairs of gear teeth in mesh over time for the gear set illustrated in Figure 5.34. The time average of the curve, which is the contact ratio, is

$$m_c = \frac{1 \times 0.47\tau + 2 \times 0.53\tau}{\tau} = 1.53$$

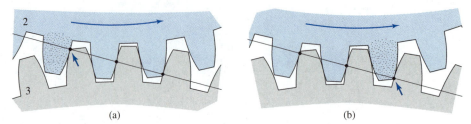

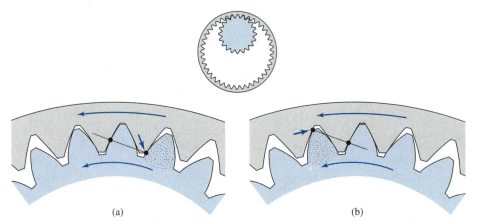

**Figure 5.36** Two spur gears in mesh: (a) initial contact, (b) final contact.

**Figure 5.37** External-internal meshing gear set: (a) initial contact, (b) final contact.

Figure 5.36 shows a different pair of gears in mesh. In this instance, for the initial and final points of contact for the noted pair of teeth, there are three points of contact between the gears. For other configurations (not shown), there are two points of contact. This gear set has a higher value of contact ratio than that shown in Figure 5.34.

Figure 5.37 shows an external-internal meshing gear set. The external gear is the driver gear. Configurations for the initial and final points of contact are illustrated.

Figure 5.38 combines the initial and final configurations of a meshing pair of gear teeth in the same drawing. The initial and final points of contact are labelled $A$ and $B$, respectively. Length $r_{AB}$ is referred to as the *length of action*, and given symbol $Z$. During meshing action, the point of contact moves along the straight line from $A$ to $B$. Points $S$ and $S'$ lie at the intersection of the pitch circle and side of a tooth of gear 2, when this gear makes initial and final contact with meshing gear 3. Points $T$ and $T'$ have similar definitions pertaining to gear 3. Arcs $SS'$ and $TT'$ are the *arcs of action* and must be equal for the pair of meshing gears. The corresponding angular displacements of the gears between the configurations of initial and final contact may be broken into two parts as shown in Figure 5.38. Quantities $\alpha_2$ and $\alpha_3$ are the *angles of approach*, and $\beta_2$ and $\beta_3$ are the *angles of recess*. The contact ratio may be expressed in terms of the arc of action and the circular pitch as

$$m_c = \frac{\text{arc of action}}{p_c} = \frac{(\alpha_2 + \beta_2)r_2}{p_c} = \frac{(\alpha_3 + \beta_3)r_3}{p_c} \qquad (5.10\text{-}1)$$

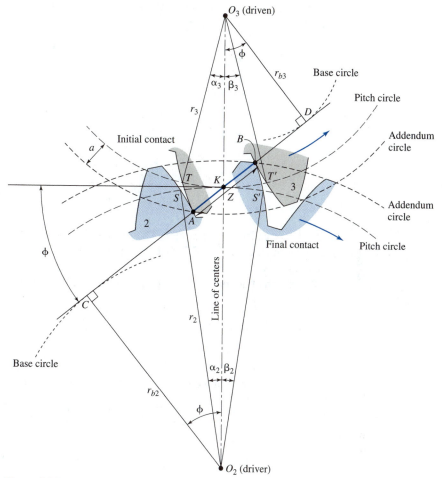

**Figure 5.38** Two spur gears in mesh.

An alternative expression for contact ratio can be found by recognizing that the transmission of rotation between involute gears in mesh is equivalent to a cross belt friction drive. Taking this into consideration and referring to Figure 5.38, the angle turned by either base circle while tooth contact moves from $A$ to $B$ can be obtained by dividing the length of action by the corresponding base circle radius. For either gear, this angle may be expressed as

$$\frac{r_{AB}}{r_b} = \frac{Z}{r_b} \tag{5.10-2}$$

The angle subtended by one gear tooth is

$$\frac{2\pi}{N} \tag{5.10-3}$$

An expression for contact ratio is then found by taking the ratio of the angles given in (5.10-2) and (5.10-3), yielding

$$m_c = \frac{\left(\dfrac{Z}{r_b}\right)}{\left(\dfrac{2\pi}{N}\right)} = \frac{Z}{p_b} \tag{5.10-4}$$

where

$$p_b = p_c \cos\phi$$

$$p_b = \frac{2\pi r_b}{N} \tag{5.10-5}$$

Quantity $p_b$ is called the *base pitch*. The form of the equation for base pitch is the same as for circular pitch, Equation (5.7-1), except that the base circle radius has replaced the pitch circle radius.

The length of action, $Z$, is found by examining Figure 5.38, where

$$Z = r_{AB} = r_{AK} + r_{KB} \tag{5.10-6}$$

and

$$r_{AK} = r_{AD} - r_{KD}; \qquad r_{KB} = r_{CB} - r_{CK} \tag{5.10-7}$$

To obtain $r_{AK}$, we determine lengths $r_{AD}$ and $r_{KD}$. In right triangle $O_3DA$

$$r_{O_3A} = r_3 + a; \qquad r_{O_3D} = r_{b3} = r_3 \cos\phi \tag{5.10-8}$$

and therefore

$$r_{AD} = \left[(r_{O_3A})^2 - (r_{O_3D})^2\right]^{1/2} = \left[(r_3 + a)^2 - r_3^2\cos^2\phi\right]^{1/2} \tag{5.10-9}$$

Also

$$r_{KD} = r_3 \sin\phi \tag{5.10-10}$$

Combining Equations (5.10-7), (5.10-9), and (5.10-10),

$$r_{AK} = r_{AD} - r_{KD} = \left[(r_3 + a)^2 - r_{b3}^2\right]^{1/2} - r_3 \sin\phi \tag{5.10-11}$$

Similarly

$$r_{KB} = \left[(r_2 + a)^2 - r_{b2}^2\right]^{1/2} - r_2 \sin\phi \tag{5.10-12}$$

Combining Equations (5.10-6), (5.10-11), and (5.10-12),

$$Z = \left[(r_2 + a)^2 - r_{b2}^2\right]^{1/2} + \left[(r_3 + a)^2 - r_{b3}^2\right]^{1/2} - (r_2 + r_3) \sin\phi \tag{5.10-13}$$

For a rack and pinion set, the rack is equivalent to a sector of a gear of infinite radius, and therefore Equation (5.10-13) cannot be used. In this instance, if the pinion is link 2, it can be shown that the length of action is

$$Z = \left[(r_2 + a)^2 - r_{b2}^2\right]^{1/2} - r_2 \sin\phi + \frac{a}{\sin\phi}$$

(5.10-14)

For an external-internal pair of meshing gears, for which gear 3 is the internal gear, the length of action is

$$Z = \left[(r_2 + a)^2 - r_{b2}^2\right]^{1/2} - \left[(r_3 - a)^2 - r_{b3}^2\right]^{1/2} + (r_3 - r_2)\sin\phi$$

(5.10-15)

## EXAMPLE 5.4  Contact Ratio for Meshing Gears

Figure 5.39 shows two pitch and addendum circles for two full-depth gears in mesh. From the information provided, determine the contact ratio.

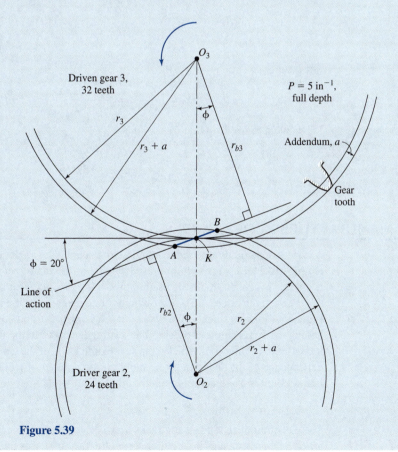

**Figure 5.39**

## SOLUTION

The pitch circle radii (Equation (5.7-2)) are

$$r_2 = \frac{d_2}{2} = \frac{N_2}{2P} = \frac{24}{2 \times 5} = 2.400 \text{ in}; \qquad r_3 = \frac{N_3}{2P} = 3.200 \text{ in}$$

The gear tooth addendum for both gears (Table 5.1) is

$$a = \frac{1.0}{P} = \frac{1}{5} = 0.200 \text{ in}$$

The base circle radii (Equation (5.5-3)) are

$$r_{b2} = r_2 \cos \phi = 2.4 \cos 20° = 2.255 \text{ in}; \qquad r_{b3} = r_3 \cos \phi = 3.007 \text{ in}$$

The addendum circle radii are

$$r_2 + a = 2.400 + 0.200 = 2.600 \text{ in}; \qquad r_3 + a = 3.400 \text{ in}$$

The length of contact (Equation (5.10-13)) is

$$Z = \left[ (r_2 + a)^2 - r_{b2}^2 \right]^{1/2} + \left[ (r_3 + a)^2 - r_{b3}^2 \right]^{1/2} - (r_2 + r_3) \sin \phi$$
$$= [2.600^2 - 2.255^2]^{1/2} + [3.400^2 - 3.007^2]^{1/2} - (2.400 + 3.200) \sin 20°$$
$$= 0.966 \text{ in}$$

The base pitch (Equation (5.10-5)) is

$$p_b = p_{b2} = \frac{2\pi r_{b2}}{N_2} = \frac{2\pi \times 2.255}{24} = 0.590 \text{ in} = p_{b3}$$

The contact ratio (Equation (5.10-4)) is

$$m_c = \frac{Z}{p_b} = \frac{0.966}{0.590} = 1.637$$

## 5.11  MANUFACTURING OF GEARS

A variety of methods are used to manufacture gears. They include molding, milling, and generating processes. This section presents some of the main methods used to produce gears. Emphasis is given to methods used in producing involute gear teeth.

For a given pitch, the shape of an involute depends on the number of teeth on the gear. Figure 5.40 shows shapes of teeth on gears having 6, 20, and 45 teeth. The tooth on the gear having the larger number of teeth has a straighter face. In the limit, as the number of teeth increases to infinity (i.e., a rack), the shape of the side of the gear teeth is a straight line. The dependence of the shape of an involute on the number of teeth on the gear must be taken into account in the machining operation.

### 5.11.1  Form Milling

Teeth of straight spur gears and racks may be cut using *form milling*. Teeth are produced by repeatedly passing a milling cutter across the face of a gear blank (i.e., workpiece). Such a

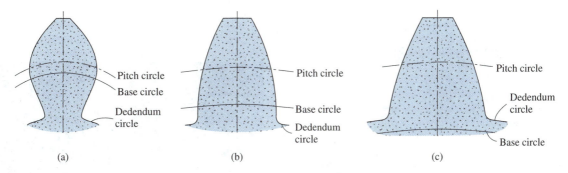

Figure 5.40  Typical shapes of gear teeth: (a) 6 teeth, (b) 20 teeth, (c) 45 teeth.

(a)

(b)

**Figure 5.41** (a) Milling cutter. (b) Cross section of cutter tooth.

**Video 5.43**
Form Milling

milling cutter is shown in Figure 5.41. During each cut, the workpiece is held stationary. Figure 5.42 shows a schematic of a form milling operation. The setup for the form milling of a straight spur gear is shown in Figure 5.43. An indexing head rotates the workpiece between cuts by the angle subtended by a gear tooth. For a rack, the workpiece is translated between cuts.

This procedure is relatively slow compared to other methods presented below. Also, accuracy of the machining process depends on the correctness of the form cutter. Since for a given pitch the shape of teeth is a function of the number of teeth on a gear, there must be a series of cutters for each tooth pitch and pressure angle. In fact, to be totally accurate, a different cutter would be needed for each number of teeth. However, generally a set of eight cutters is employed for all gears having the same pitch and pressure angle, which leads to approximations of the shape of teeth.

### 5.11.2 Generating Processes

#### Hobbing

*Hobbing* is one of the fastest, most versatile methods of making rough and/or finish cuts of gear teeth. A typical cutter, also called a *hob*, is illustrated in Figure 5.44(a). The cross section of a row of cutter teeth of a hob is illustrated in Figure 5.44(b). The *hob pitch*, $p_h$, is the

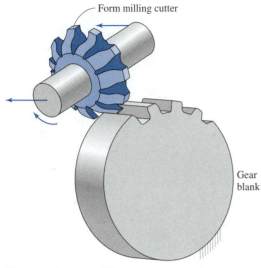

Form milling cutter

Gear blank

**Figure 5.42** Form milling.

Cutter

Indexer

Gear

**Figure 5.43** Form milling of a straight spur gear [Video 5.43].

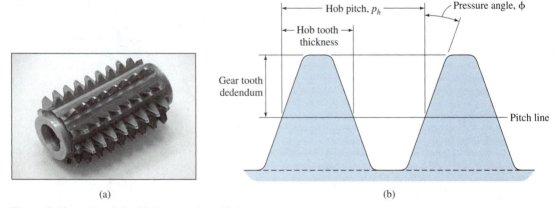

(a)

(b)

Hob pitch, $p_h$

Hob tooth thickness

Pressure angle, $\phi$

Gear tooth dedendum

Pitch line

**Figure 5.44** (a) Gear hob. (b) Cross section of hob teeth.

**Model 5.45**
Straight Spur
Gear Hobbing

spacing between the cutter teeth. Figure 5.45 illustrates a hob and the side view of a gear. As the hob rotates, the workpiece (i.e., gear) simultaneously turns in a carefully coordinated manner. For each revolution of the hob, the gear turns about its axis the angle to be subtended by one tooth. Relative motion of the hob and workpiece generates the involute shape. Figure 5.46 shows two series of accumulated images of the positions of the cutter teeth with respect to the workpiece; the first series shows the cutter teeth advancing and generating an involute profile on one side of a gear tooth, and the second series shows the cutter teeth receding while producing an involute profile on an adjacent tooth.

In addition to the relative rotational motion described above, the hob is slowly driven across the face of the workpiece, creating involute profiles over the entire face of the gear.

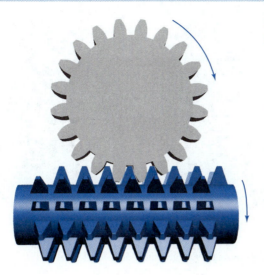

**Figure 5.45** Straight spur gear and hob [Model 5.45].

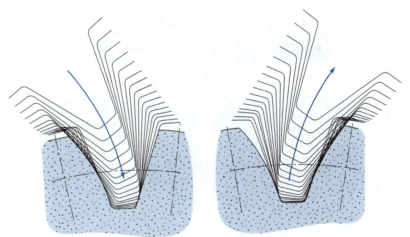

**Figure 5.46** Hobbing operation—Relative Motion between hob and gear blank.

Figures 5.47(a) and 5.47(b) respectively show the relative positions of the hob and gear before and after a cutting operation. A corresponding animation, which includes both relative rotational motions and translation of the hob across the face, is shown through **[Model 5.45]**. In actual operation, prior to the cut, the gear starts as a circular disc without any teeth. A picture of the hobbing of a straight spur gear is shown in Figure 5.48.

**Video 5.48**
Straight Spur
Gear Hobbing

In the hobbing of a gear, spacing between gear teeth is dictated by the hob. For a straight spur gear, the hob will generate a gear with

$$p_c = p_h \qquad (5.11\text{-}1)$$

A single hob may be used to produce any size of straight spur gear having a given circular pitch and pressure angle.

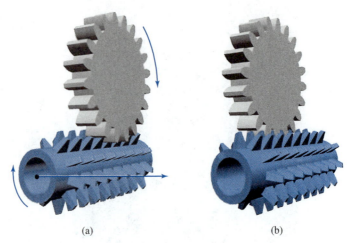

**Figure 5.47**  Straight spur gear and hob [Model 5.45].

(a)                                          (b)

**Figure 5.48**  Hobbing of a straight spur gear [Video 5.48].

**Model 5.50**
Helical Spur
Gear Hobbing

**Video 5.51**
Helical Spur
Gear Hobbing

A hob can also be used to produce a helical spur gear. In this instance, the hob is rotated by the *helix angle*, $\psi$, from the position used to cut a straight spur gear (see Figure 5.49). A typical setup for cutting a helical gear is shown in Figure 5.50. A picture of the hobbing of a helical spur gear is shown in Figure 5.51.

For a helical spur gear, the hob generates teeth by having its spacing defined by the *normal pitch*, $p_n$, as indicated in Figure 5.49. The relationships between the normal pitch, hob pitch, $p_h$, and circular pitch, $p_c$, of a helical spur gear are

$$p_n = p_h = p_c \cos \psi \qquad (5.11\text{-}2)$$

As an example, the hob pitch of the hob shown in Figure 5.44(a) is 0.375 inches. Therefore, using Equation (5.7-1), the pitch circle diameter of a straight spur gear (i.e., $p_c = p_h$) having $N = 62$ teeth is

$$d = \frac{N p_c}{\pi} = \frac{62 \times 0.375}{\pi} = 7.400 \text{ in} \qquad (5.11\text{-}3)$$

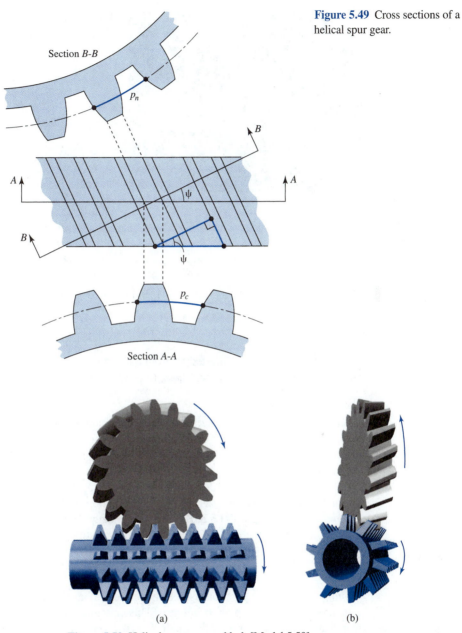

**Figure 5.49** Cross sections of a helical spur gear.

Section B-B

$p_n$

$\psi$

$\psi$

$p_c$

Section A-A

(a)                                        (b)

**Figure 5.50** Helical spur gear and hob [Model 5.50].

If we now wish to employ the same hob to cut a helical spur gear having the same pitch circle diameter, spacing between teeth cut by the hob is the normal pitch, that is,

$$p_n = 0.375 \text{ in} \tag{5.11-4}$$

Using Equation (5.11-2), the circular pitch is greater than the normal pitch, and therefore for the same size of gear, we will have fewer teeth. If $N = 60$ on the helical spur gear, and

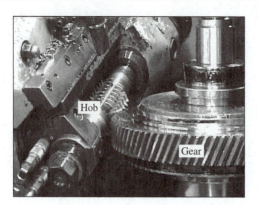

**Figure 5.51** Hobbing of a helical spur gear [Video 5.51].

it is to be the same size as the straight spur gear, then the circular pitch is

$$p_c = \frac{\pi d}{N} = \frac{\pi \times 7.400}{60} = 0.3875 \text{ in} \qquad (5.11\text{-}5)$$

Combining Equations (5.11-2), (5.11-4), and (5.11-5),

$$\psi = \cos^{-1}\left(\frac{p_n}{p_c}\right) = \cos^{-1}\left(\frac{0.375}{0.3875}\right) = 14.57°$$

Hobbing is generally limited to producing external gears. However, hobbing may be completed with special spherical hobs to produce internal gears having large diameters.

### Shaping

*Shaping* is another type of generating process. A shaping operation is shown in Figure 5.52 where the cutter reciprocates with respect to the workpiece. The cutter and workpiece are indexed between each cutting stroke. Relative motion between the teeth of the cutter and workpiece ensures that the required involute profile is generated.

Two other illustrations are shown in Figure 5.53 for the cutting of an external and internal gear. The same cutter is used for both. Motions of the animations provided through **[Model 5.53A]** and **[Model 5.53B]** have been exaggerated to illustrate rotation of the gear between strokes of the cutter. Figure 5.54 shows pictures of the cutting of a straight spur gear and an internal gear.

### Planing

*Planing* is a cutting operation used to manufacture bevel gears. Figure 5.55 illustrates a schematic for cutting a plain bevel gear. This operation involves two reciprocating cutters that shape both sides of a tooth simultaneously. The two cutters are mounted on a cradle (not shown in the figure). For machining each tooth, the cradle gives the cutters the same relative motion with respect to the workpiece as a meshing gear. Figure 5.56 shows a setup for cutting of a plain bevel gear.

### Rotary Broaching

*Rotary broaching* of plain bevel gears has some similarities to form milling of straight spur gears. Figure 5.57(a) shows a setup for a rotary broaching operation. It includes a rotating

**Model 5.53A**
Shaping of a Straight Spur Gear

**Model 5.53B**
Shaping of an Internal Gear

**Video 5.54A**
Shaping of a Straight Spur Gear

**Video 5.54B**
Shaping of an Internal Gear

**Video 5.56**
Planing of a Plain Bevel Gear

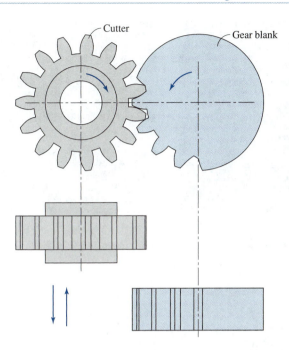

**Figure 5.52** Shaping of a straight spur gear.

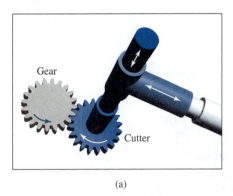
(a)

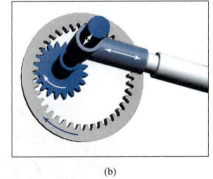
(b)

**Figure 5.53** Shaping of straight spur gears: (a) external gear [Model 5.53A], (b) internal gear [Model 5.53B].

cutter. A separate picture of the cutter is shown in Figure 5.57(b). There are two distinct portions of cutter teeth on the periphery: a portion where the cutter teeth progressively increase in depth, and a portion in which the cutter teeth have constant depth.

Another illustration of rotary broaching is shown in Figure 5.58. Starting from the configuration shown in Figure 5.58(a), the cutter is swung into position (Figure 5.58(b)). While holding the gear blank stationary, a slot is cut in the gear blank. This is accomplished with the teeth of increasing depth (Figure 5.58(c)). Then, the gear blank is translated vertically upward (Figure 5.58(d)). As the cutter continues to rotate, the gear blank moves downward while at the same time the cutter teeth, with constant depth, cut the slot

**Model 5.58**
Rotary
Broaching

(a)                                           (b)

**Figure 5.54**  Shaping of gears: (a) straight spur gear [Video 5.54A], (b) internal gear [Video 5.54B].

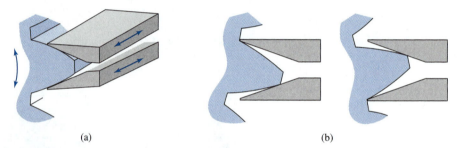

(a)                                           (b)

**Figure 5.55**  Schematic of a planing operation.

**Figure 5.56**  Planing of a plain bevel gear [Video 5.56].

(Figure 5.58(e)). This relative motion permits the bottom of the slot to be cut in a straight line. A portion of the periphery of the cutting wheel has no teeth. This permits the gear blank to be indexed in preparation for cutting of the next slot. While the gear blank is being indexed, the cutter is rotated to the position for the start of another cutting operation. The operation is repeated until all slots are cut and the machined gear is then swung back into its starting position (Figure 5.58(f)).

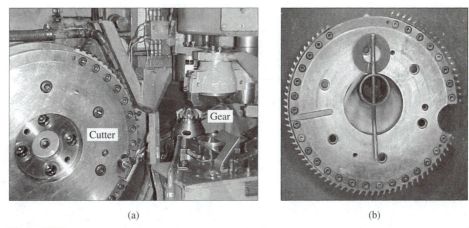

(a)                                                                    (b)

**Figure 5.57** (a) Rotary broaching of plain bevel gear. (b) Cutter.

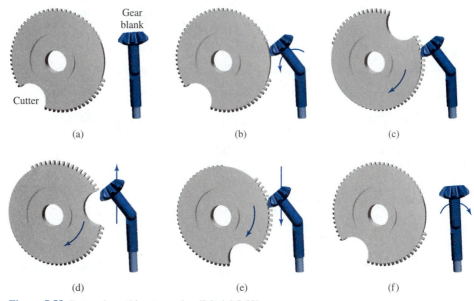

**Figure 5.58** Rotary broaching operation [Model 5.58].

## EXAMPLE 5.5

Table 5.3 provides information that pertains to the reverted gear train shown in Figure 5.33.
All gears were manufactured using a hob cutter having 20° pressure angle, 0.375-inch hob pitch, and full depth. Determine

(a) the base circle radius of gear 3
(b) the center-to-center distance between gears 2 and 3

(c) the circular pitch of gear 4

(d) the helix angle of gear 5

(e) the pitch circle diameter of gear 4

(f) the addendum circle diameter of gear 4 (i.e., the diameter of the gear blank)

**TABLE 5.3** Parameters of the Reverted Gear Train Shown in Figure 5.33

| Gear No. | No. of Teeth | Type of Gear |
|----------|--------------|--------------|
| 2 | 30 | Straight spur |
| 3 | 48 | — |
| 4 | 29 | — |
| 5 | 47 | Helical spur |

## SOLUTION

By inspection, gear 3 is a straight spur gear, and gear 4 is a helical spur gear.

(a) For gears 2 and 3

$$p_{c2} = p_{c3} = 0.375 \text{ in} = p_h$$

From Equation (5.7-3)

$$P_2 = P_3 = \frac{\pi}{p_{c3}} = \frac{\pi}{0.375} = 8.38 \text{ in}^{-1}$$

From Equations (5.5-3) and (5.7-2)

$$r_{b3} = r_3 \cos\phi = \frac{d_3}{2} \cos\phi = \frac{N_3}{2P_3} \cos\phi = \frac{48}{2 \times 8.38} \cos 20° = 2.69 \text{ in}$$

(b) From Equations (5.9-1) and (5.7-2)

$$c = \frac{d_2}{2} + \frac{d_3}{2} = \frac{N_2}{2P_2} + \frac{N_3}{2P_3} = \frac{30}{2 \times 8.38} + \frac{48}{2 \times 8.38} = 4.65 \text{ in}$$

(c) Employing the result from part (b), and combining Equations (5.9-1) and (5.7-1),

$$c = 4.65 \text{ in} = \frac{d_4}{2} + \frac{d_5}{2} = \frac{N_4 p_{c4}}{2\pi} + \frac{N_5 p_{c5}}{2\pi} = \frac{29 p_{c4}}{2\pi} + \frac{47 p_{c5}}{2\pi}$$

Noting that

$$p_{c4} = p_{c5}$$

we obtain

$$p_{c4} = 0.384 \text{ in}$$

(d) For gears 4 and 5, we recognize that

$$p_{n4} = p_{n5}$$

From Equation (5.11-2)

$$p_{n4} = p_{c4} \cos \psi_4 = p_h$$

$$0.375 = 0.384 \cos \psi_4$$

and therefore

$$\psi_4 = 12.7° = \psi_5$$

(e) From Equation (5.7-1)

$$d_4 = \frac{N_4 p_{c4}}{\pi} = \frac{29 \times 0.384}{\pi} = 3.54 \text{ in}$$

(f) For a given hob, the addendum obtained for a helical spur gear is the same as that for a straight spur gear. Therefore

$$a_4 = a_2 = \frac{1}{P_2} = \frac{1}{8.38} \text{ in}$$

Using the results of part (e), the addendum circle diameter of gear 4 is

$$d_4 + 2a_4 = 3.54 + \frac{2}{8.38} = 3.78 \text{ in}$$

## 5.12 INTERFERENCE AND UNDERCUTTING OF GEAR TEETH

Based on Equation (5.7-5), the number of teeth on a gear is proportional to its pitch circle radius. Also, according to Equation (5.5-3), an involute gear with a given pressure angle has a base circle radius that is proportional to the pitch circle radius. Furthermore, the dedendum and whole depth of gear teeth of standard size (see Table 5.1) are functions of diametral pitch. By combining these relationships, for a given pressure angle and depending on the number of gear teeth, the base circle radius may be either greater than or less than the dedendum circle radius. This is illustrated below.

Figure 5.40 shows three gear teeth. All have the same pitch, pressure angle, and whole depth. However, because each tooth is from a gear having a different number of teeth, the position of the base circle relative to the tooth profile is distinct from one to the next. Figure 5.40(a) illustrates a tooth from a gear having only six teeth. In this instance, the base circle is larger than the dedendum circle. Since an involute can only be generated outward from a base circle (see Figure 5.15), the portion of the tooth profile inside of the base circle is not an involute. In Figure 5.40(a), this portion of the tooth profile is drawn simply from the circle as a radial line to the center of the gear, with a fillet at the base of the tooth. Figure 5.40(b) illustrates a tooth on a gear having 20 teeth. Here, there is a reduced portion of the tooth profile inside of the base circle, which is not an involute. Figure 5.40(c) shows a tooth on a gear having 45 teeth. Now, the base circle is smaller than the dedendum circle, and therefore the entire profile of the tooth may be manufactured with an involute profile.

The meshing of a pair of involute gears was shown to be equivalent to a string wound between two base circles (Section 5.5). The bead on the string traces out two involutes, each with respect to its base circle. These involutes match the profiles of the involute gear teeth.

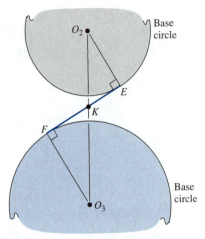

Figure 5.59  Meshing limits for two gears.

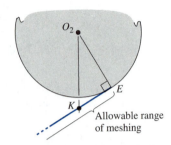

Figure 5.60  Meshing limits for rack and pinion.

Figure 5.59 shows two base circles and a string (i.e., line of action) wrapped between them. Points $E$ and $F$ are at the intersections of the base circles and the line of action. They represent the limits between which two involutes can be generated. For a corresponding pair of gears, these are the limits within which meshing between two involute profiles can take place. These points are called the *meshing limits*. Outside of these limits, it is impossible to have meshing between two involute profiles. Also shown in Figure 5.59 is the pitch point, $K$, located on the line of action between the two meshing limits.

A rack is equivalent to a gear of infinite radius. For a rack and pinion set, there is only one meshing limit to be considered. Figure 5.60 indicates the range over which meshing can take place between an involute on the pinion and the rack.

Figure 5.61 illustrates a pinion having eight teeth meshing with a rack. The teeth are full-depth, with a 20° pressure angle (see Table 5.1). The base circle of the pinion is greater than its dedendum circle. If the portion of tooth profile inside of the base circle is a radial line to the center of the pinion, as shown, it would be impossible for the gears to mesh properly. Two regions are depicted in this figure where there is overlapping (i.e., *interference*) of the teeth on the pinion with those on the rack.

Figure 5.62 shows an enlargement of Figure 5.61 in the vicinity of meshing between a pair of teeth. Starting from the configuration shown in Figure 5.62(a), examine the path of point $A$ on the rack with respect to the pinion. As shown in Figure 5.62(b), point $A$ initially contacts point $H$ on the pinion. Figure 5.62(c) shows the path of $A$ with respect to the pinion as a dashed line during meshing. Since it is inside the boundary of the pinion, it therefore reveals the amount of interference that would take place. The dashed line shows the permissible outline of one side of a pinion tooth to avoid interference. *Undercutting* is the required removal of material on the pinion that will prevent interference. Figure 5.62(d) shows one side of a pinion tooth that has been undercut to eliminate its interference with one corner (point $A$) of the rack tooth. However, point $D$ on the rack would also cause

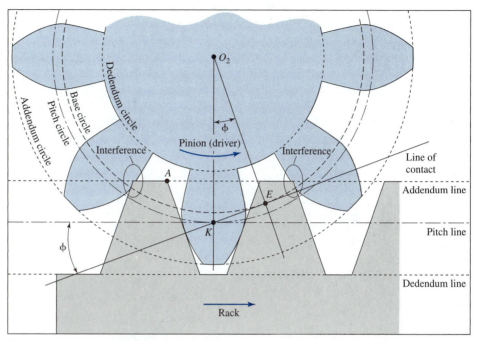

**Figure 5.61** Rack and pinion with interference.

similar interference with the pinion teeth. Therefore, both sides of the pinion teeth must be undercut. Figure 5.63 shows a pinion that has been undercut to prevent interference.

For the undercut pinion shown in Figure 5.63, the thickness of the teeth at the flank is less than that at the base circle. Furthermore, a small portion of the involute outside of the base circle has to be removed. Each gear tooth is essentially a cantilever beam, and motion is transmitted by means of the interactive forces between gear teeth along the line of action. It follows that decreasing thicknesses of the gear teeth near their bases has the detrimental effect of reducing the strength of the gear. Obviously, undercutting in design is to be avoided.

The interference shown in Figure 5.61 may be eliminated by having more teeth on the pinion by reducing the pitch of the teeth, and/or increasing the pitch circle diameter of the pinion. Figure 5.64 shows an alternative means. Here, the pinion is identical to that shown in Figure 5.61 for which there is no undercutting. However, the addendum of the rack teeth in Figure 5.64 is smaller than that in Figure 5.61, that is, the rack shown in Figure 5.64 does not possess the standard proportions of teeth as listed in Table 5.1. In Figure 5.64, the addendum of the rack reaches only to the meshing limit, $E$, which is on the base circle. In this instance, points of contact are never inside of the base circle, and it is not necessary to remove a portion of the involute to eliminate interference. If the addendum of the rack were greater than that shown in Figure 5.64, it would be necessary to remove a portion of the involute on the pinion to prevent interference.

When a portion of an involute is cut off in order to prevent interference, expressions for the length of action (Equations (5.10-13)–(5.10-15)) do not apply. Also, reducing the addendum results in a decrease of the length of action as well as the contact ratio.

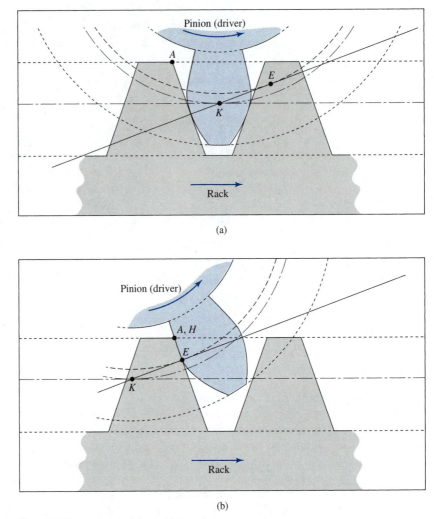

**Figure 5.62** Rack and pinion with interference.

Figure 5.65 illustrates the same rack and pinion set as shown in Figure 5.64. The outline of a meshing gear is added as a dashed line. The addendum of the gear is identical to that of the rack. The addendum circle of the gear is tangent to the addendum line of the rack at point X. Point J at the corner of the gear tooth is slightly lower than point E. Points of contact between the pinion and the gear never reach the interference limit and always remain in the permissible range for meshing without requiring undercutting. It is therefore concluded that if a pinion can mesh with a rack without interference, it will also properly mesh with another gear having teeth possessing the same specifications as the rack, and containing an identical or larger number of teeth than the pinion.

For a given form of gear tooth, it is possible to determine the minimum number of teeth on a pinion that will mesh with a rack without requiring removal of a portion of the

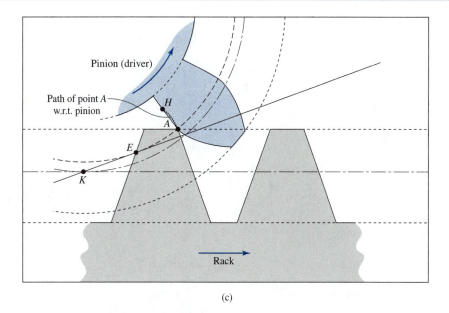

(c)

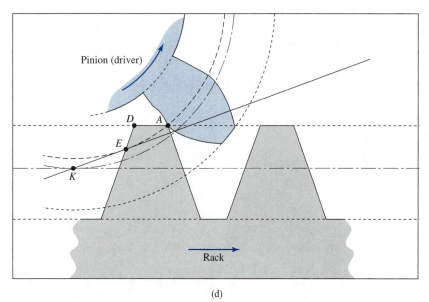

(d)

**Figure 5.62** (*Continued*)

involute profile. The limiting case has the addendum line of the rack passing through the meshing limit. Figure 5.66 includes the essential dimensions of such a rack and pinion set. The pitch point and meshing limit are denoted by $K$ and $E$, respectively. Therefore

$$\sin \phi = \frac{r_{KE}}{r} \qquad (5.12\text{-}1)$$

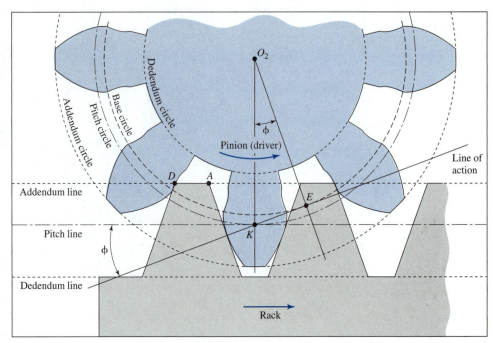

**Figure 5.63** Undercut pinion.

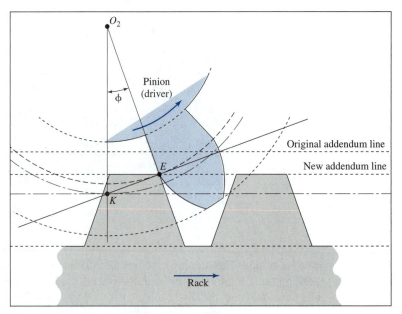

**Figure 5.64** Rack and pinion—reduced rack.

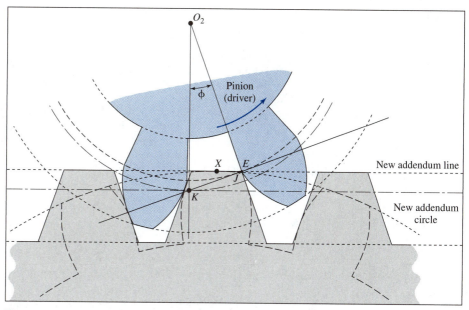

**Figure 5.65** Pinion meshing with a rack and gear.

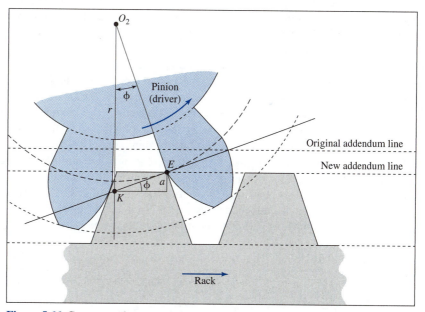

**Figure 5.66** Geometry of prevent undercutting.

Also

$$\sin\phi = \frac{a}{r_{KE}} = \frac{(\alpha/P)}{r_{KE}} \tag{5.12-2}$$

where $\alpha$ is a constant that, when divided by the diametral pitch, $P$, gives the addendum. Applying this to the standardized tooth dimensions listed in Table 5.1, for full-depth teeth, $\alpha = 1.0$, and for stub teeth, $\alpha = 0.8$.

Multiplying Equations (5.12-1) and (5.12-2) together gives

$$\sin^2\phi = \frac{\alpha}{rP} \tag{5.12-3}$$

Combining Equations (5.12-3) and (5.7-2) gives

$$\sin^2\phi = \frac{2\alpha}{N} \tag{5.12-4}$$

or

$$\boxed{N = \frac{2\alpha}{\sin^2\phi}} \tag{5.12-5}$$

Equation (5.12-5) can be used to calculate the smallest number of teeth on a pinion that will mesh with a rack without requiring the elimination of a portion of the involute. Values are listed in Table 5.4 for common gear tooth systems. Because values in the table were calculated for the pinion meshing with a rack, they can also be conservatively used as minimums for a pinion meshing with a gear of equal or larger size.

Figure 5.67 shows a pinion having five teeth meshing with a larger gear. The base circle of the pinion is illustrated. The pinion teeth have been undercut so that there is no interference. However, the contact ratio is less than unity, and there is actually no contact for the configuration illustrated. As a result, for this pair of gears it is impossible to maintain a constant speed ratio.

The relative motions of hob cutter teeth with respect to a gear blank (Section 5.11.2) has some similarities to the action of a pinion meshing with a rack. In this instance, however, material is being removed from the gear during the relative motions. If a hob is set up to manufacture a gear where the base circle will be greater than the dedendum circle, the required undercutting will be made during the machining process. Also, Equation (5.12-5) may be employed to determine the minimum number of teeth on a gear that may be cut by a hob without removing a portion of the involute profile. By inspection of Figure 5.44, the "addendum" of a hob cutter tooth is equal to the dedendum of a gear tooth being cut. Therefore, $\alpha$ is obtained by examining the values of dedendum listed in Table 5.1. Considering full-depth teeth, the values are 1.157 for a 14.5° pressure angle, and 1.25 for a 20° and 25° pressure angle. Calculated minimums are listed in Table 5.4.

**TABLE 5.4**  Minimum Number of Teeth on a Gear to Avoid Undercutting

|  | Full-Depth Tooth ($\phi = 14.5°$) | Stub Tooth ($\phi = 20°$) | Full-Depth Tooth ($\phi = 20°$) | Full-Depth Tooth ($\phi = 25°$) |
|---|---|---|---|---|
| Meshing with a Rack | 32 | 14 | 18 | 12 |
| Manufactured Using a Hob | 37 | 18 | 22 | 14 |

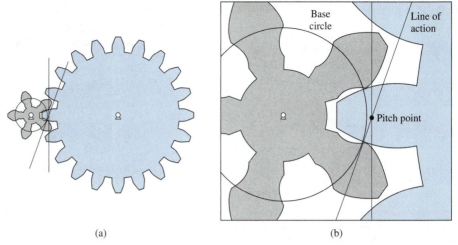

**Figure 5.67** (a) Meshing of gears requiring undercutting [Model 5.67]. (b) Enlargement in vicinity of meshing teeth.

# PROBLEMS

**P5.1** What is the pitch circle diameter of a 50-tooth spur gear having a circular pitch of 0.375 in?

**P5.2** How many revolutions per minute is a spur gear turning at if it has 32 teeth, a circular pitch of 0.75 in, and a magnitude of the pitch line velocity of 15 ft/sec?

**P5.3** How many revolutions per minute is a spur gear turning at if it has 48 teeth, a module of 3 mm, and a magnitude of the pitch line velocity of 500 mm/sec?

**P5.4** A spur gear having 40 teeth is rotating at 450 rpm and is to drive another spur gear at 600 rpm.

(a) What is the value of the speed ratio?
(b) How many teeth must the second gear have?

**P5.5** Two spur gears have a diametral pitch of 5 in$^{-1}$, a magnitude of speed ratio of 0.20, and a center-to-center distance of 12 in. How many teeth do the gears have?

**P5.6** Two spur gears in mesh have a module of 2.5 mm, a center-to-center distance of 50 mm, and a magnitude of speed ratio of 0.6. How many teeth do the gears have? If the pinion is rotating at 1,200 rpm, what is the magnitude of the pitch line velocity?

**P5.7** Two spur gears in mesh are designated as $A$ and $B$. Gear $A$ has 40 teeth of 4 in$^{-1}$ diametral pitch, and

gear $B$ has a pitch circle diameter of 16 in. The teeth have a standard pressure angle of $14.5°$ and involute form. Both gears have external teeth. Determine

(a) the number of teeth on gear $B$
(b) the center-to-center distance between the gears
(c) the circular pitch
(d) the outside diameter of gear $B$
(e) the base circle diameter of gear $B$

**P5.8** A gear with a diametral pitch of 3 in$^{-1}$ and a 20° full-depth involute form has 35 teeth and meshes with a 70-tooth gear. The larger gear rotates at 300 rpm CW. Both gears have external teeth. Determine

(a) the pitch circle diameters
(b) the base circle radius of the smaller gear
(c) the circular pitch
(d) the center-to-center distance between the gears
(e) the rotational speed of the smaller gear
(f) the contact ratio

**P5.9** Derive the expression for the length of contact for a rack and pinion gear set (Equation 5.10-14).

**P5.10** Derive the expression for the length of contact for an external-internal gear set (Equation 5.10-15).

**P5.11** Table P5.11 provides information that pertains to the reverted gear train shown in Figure 5.33.

**TABLE P5.11** Parameters of the Reverted Gear Train Shown in Figure 5.33

| Gear No. | No. of Teeth | Module (mm) |
|----------|--------------|-------------|
| 2 | 24 | 4 |
| 3 | — | — |
| 4 | 32 | — |
| 5 | 62 | 6 |

All gears are straight spur with a 20° pressure angle. Determine

(a) the circular pitch of gear 2
(b) the base circle radius of gear 3
(c) the pitch circle diameter of gear 4
(d) the center-to-center distance between gears 2 and 3

**P5.12** Table P5.12 provides information that pertains to the reverted gear train shown in Figure 5.33.

**TABLE P5.12** Parameters of the Reverted Gear Train Shown in Figure 5.33

| Gear No. | No. of Teeth | Type of Gear |
|----------|--------------|--------------|
| 2 | 30 | — |
| 3 | 48 | Straight spur |
| 4 | 28 | Helical spur |
| 5 | 46 | — |

All gears were manufactured using a hob cutter having a 20° pressure angle, 0.375-in hob pitch, and full depth. Determine

(a) the base circle radius of gear 3
(b) the center-to-center distance between gears 2 and 3
(c) the circular pitch of gear 4
(d) the helix angle of gear 5
(e) the pitch circle diameter of gear 4
(f) the addendum circle diameter of gear 4 (i.e., the diameter of the gear blank)

**P5.13** A straight spur gear $A$ has 24 teeth. It meshes with an internal gear $B$. Both gears have a 25° pressure angle, 12 in$^{-1}$ diametral pitch, and full-depth teeth. The center-to-center distance between the gears is

1.50 in. Determine

(a) the addendum circle diameter of gear $A$
(b) the number of teeth on internal gear $B$
(c) the base circle radius of internal gear $B$
(d) the thickness of a gear tooth on internal gear $B$, measured along its pitch circle

**P5.14** A pinion has 32 teeth and has been manufactured using a hob having a 20° pressure angle, 8 in$^{-1}$ diametral pitch, and stub teeth. It meshes with a rack. Determine

(a) the length of action
(b) the contact ratio

**P5.15** The information in Table P5.15 pertains to the gear train shown in Figure P5.15.

**TABLE P5.15**

| Gear No. | No. of Teeth | Type of Gear |
|----------|--------------|--------------|
| 1 | 18 | Helical spur |
| 2 | — | — |
| 3 | 62 | — |
| 4 | 20 | — |
| 5 | 64 | Straight spur |

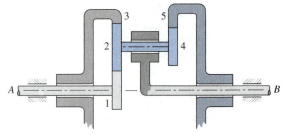

**Figure P5.15**

External gears 1, 2, and 4 were manufactured using a hob cutter having a 20° pressure angle and 0.500-in hob pitch. Gears 3 and 5 are internal. All gears have full-depth teeth. Shafts $A$ and $B$ are collinear. Determine

(a) the base circle radius of gear 4
(b) the center-to-center distance between gears 1 and 2
(c) the contact ratio between gears 4 and 5
(d) the circular pitch of gear 1
(e) the helix angle of gear 3
(f) the pitch circle diameter of gear 3
(g) the addendum circle diameter of gear 3

# 6    Gear Trains

## 6.1 INTRODUCTION

A combination of gears arranged for the purpose of transmitting torque and rotational motion from an input shaft to an output shaft is called a *gear train*. Gear trains are used to transmit torque and rotary motion to an alternate location and/or change rotational speed.

The number of gears employed in a gear train can range from two to several dozen. Normally gears are assembled in a sturdy housing that supports the shafts, using ball or roller bearings. Gears are often keyed to their shafts, and the housing is usually enclosed and provided with ample lubrication. In heavy-duty gear applications, the lubricant is circulated, filtered, and sometimes cooled.

Figure 6.1 shows a gear train employed in a lathe, where an electric motor provides a single input rotational speed. Multiple output turning speeds of the spindle and workpiece may be obtained by changing the meshed gears.

A gear train has at least one input shaft and one output shaft. However, a gear train may be designed to accommodate multiple inputs, deliver multiple outputs, or both. A gear train is characterized by its speed ratio(s) between the input(s) and output(s). The size of a gear train depends on its power rating, which can range from a fraction of a watt to megawatts.

In this chapter, the types of gear trains are described and classified. Methods of determining speeds of the components are also covered.

### 6.1.1 Speed Ratio

The *speed ratio* of component $j$ with respect to component $i$ in a gear train is defined as

$$e_{j/i} = \frac{\text{rotational speed of component } j}{\text{rotational speed of component } i} \qquad (6.1\text{-}1)$$

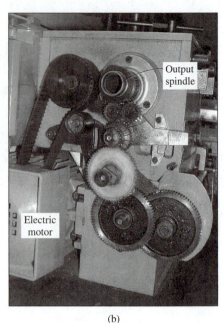

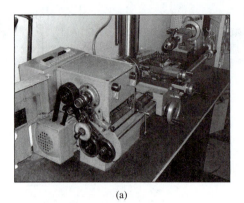

(a)                                      (b)

**Figure 6.1** Gear train in a lathe.

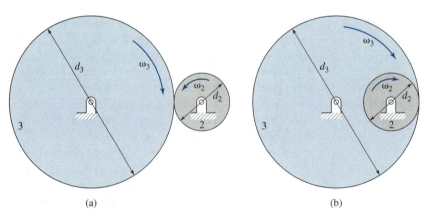

(a)                                      (b)

**Figure 6.2** Meshing gears represented by their pitch circles: (a) external-external pair, (b) external-internal pair.

Positive $e_{j/i}$ values indicate that components $j$ and $i$ both turn in the same direction. Negative $e_{j/i}$ values signify an opposite direction of rotation. Equation (6.1-1) may be applied to the overall speed ratio of the gear train. In that instance, $j$ corresponds to the output component of the gear train, and $i$ corresponds to the input component.

Figure 6.2 illustrates pitch circles of two meshing pairs of gears. For the external-external meshing pair shown in Figure 6.2(a), the gears rotate in opposite directions, and the speed ratio is negative. For the external-internal pair, Figure 6.2(b), the gears rotate in the same direction, and the speed ratio is positive.

The sign convention defined above for speed ratio can only be applied to components that have parallel axes of rotation. For cases where the axes of rotation are not parallel, such as with worm and wheel, miter, and bevel gears, only the magnitude of the speed ratio is defined by Equation (6.1-1), and the direction of rotation may be determined using a suitably prepared sketch.

From Equation (5.7-5), the number of teeth on a gear is proportional to its pitch circle diameter. Gears in mesh must be of the same pitch. From Equation (5.5-2), the magnitude of the speed ratio of a pair of meshing gears is inversely proportional to the ratio of the radii. Thus, we conclude that the magnitude of the speed ratio is inversely proportional to the ratio the number of gear teeth. For the external-external meshing pair shown in Figure 6.2(a), having $N_2$ and $N_3$ teeth on gears 2 and 3, respectively, the directions of rotation are opposite. The speed ratio of gear 3 with respect to gear 2 is

$$e_{3/2} = \frac{\omega_3}{\omega_2} = -\frac{N_2}{N_3} \qquad (6.1\text{-}2)$$

For the external-internal meshing pair shown in Figure 6.2(b), the speed ratio is

$$e_{3/2} = \frac{\omega_3}{\omega_2} = +\frac{N_2}{N_3} \qquad (6.1\text{-}3)$$

It is usually desired to design a gear train to provide a specific speed ratio. In instances where the magnitude of the specified speed ratio is a fraction involving small whole numbers, for example 2/3, there is a wide selection of the arrangement of gears and numbers of teeth that may be employed. Such combinations would be 20 and 30, 40 and 60, etc. Since gears must have integer numbers of teeth, some speed ratios are either impossible or impractical to obtain exactly. For instance, a magnitude of speed ratio involving a fraction of two large prime numbers, such as 123/177, would require meshing gears with 123 and 177 teeth to provide the exact speed ratio. Alternative gear trains, however, can be developed if a small deviation in the speed ratio is allowed from the desired value. For example, two gears with 36 and 25 teeth provide a magnitude of speed ratio of 0.6944, which is very close to the specified value of $123/177 = 0.6949$.

### 6.1.2 Classification of Gear Trains

Gear trains may be classified into two groups: *ordinary* and *planetary* (or *epicyclic*), (Figure 6.3). In an *ordinary gear train*, axes of all gears are stationary relative to the base link, which is normally the housing (i.e., the base link) of the gear train. In a *planetary gear*

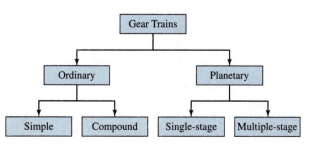

**Figure 6.3** Classification of gear trains.

*train*, the axis of at least one gear moves along a circular path relative to the base link. These two groups of gear trains are described further in the following sections.

## 6.2  ORDINARY GEAR TRAINS

There are two types of ordinary gear trains: *simple gear trains* and *compound gear trains*.

### 6.2.1  Simple Gear Trains

In a simple gear train, two or more gears mesh in consecutive sequence, and there is only one gear mounted on each axis of rotation. Two examples of simple gear trains are shown in Figure 6.4.

> **Model 6.4A**
> Simple Gear
> Train

The speed ratio for any two gears in mesh of a simple gear train can be obtained using either Equation (6.1-2) or (6.1-3). Thus, for the gear train shown in Figure 6.4(a), the speed

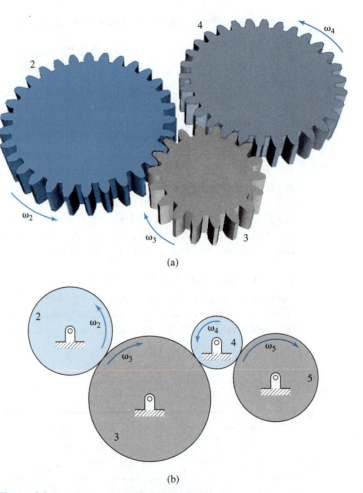

(a)

(b)

**Figure 6.4** Simple gear trains [Model 6.4A].

ratios of gears 2 and 3, and of gears 3 and 4, are

$$e_{3/2} = \frac{\omega_3}{\omega_2} = -\frac{N_2}{N_3} \quad \text{and} \quad e_{4/3} = \frac{\omega_4}{\omega_3} = -\frac{N_3}{N_4} \qquad (6.2\text{-}1)$$

respectively. If gear 2 is the input, and gear 4 is the output, then employing Equations (6.2-1), the speed ratio of the gear train is

$$e_{4/2} = \frac{\omega_4}{\omega_2} = e_{4/3} \times e_{3/2} = \left(-\frac{N_3}{N_4}\right)\left(-\frac{N_2}{N_3}\right) = \frac{N_2}{N_4} \qquad (6.2\text{-}2)$$

For the gear train shown in Figure 6.4(a)

$$N_2 = 31; \qquad N_3 = 18; \qquad N_4 = 31$$

and using Equation (6.2-2),

$$e_{4/2} = +1.0$$

This gear train would have the same magnitude of speed ratio as when gears 2 and 4 are directly in mesh. However, the speed ratio is changed from negative to positive, due to the presence of gear 3, referred to as an *idler gear*.

Even when the number of idler gears in a simple gear train is increased, the magnitude of speed ratio will still depend only on the ratio of the number of teeth in the gears mounted on the input and output shafts. Whether the speed ratio is positive or negative depends on the number of idler gears. Employing external gears only in a simple gear train, an odd number of idler gears produces a positive speed ratio, and with an even number of idler gears, as shown in Figure 6.4(b), the speed ratio is negative, that is,

$$e_{5/2} = -\frac{N_2}{N_5} \qquad (6.2\text{-}3)$$

In this example, gears 3 and 4 are idler gears.

Idler gears are added to provide the required output direction of rotation and/or transmit rotary motion to an output shaft located at a specified distance from the input shaft. If the distance between the input and output shafts is large, it may be more economical to use one or more idler gears instead of using large input and output gears to bridge the distance. Figure 6.5 illustrates two alternative simple gear trains that have the same speed ratio and center-to-center distance between the input and output shafts.

## 6.2.2  Compound Gear Trains

A compound gear train has four or more gears of which at least two gears are mounted on the same shaft. Those gears mounted on the same axis of rotation are constrained to turn at the same rate and in the same direction. Figure 6.6 shows three examples of compound gear trains. In each example, gears 3 and 4 are mounted on the same axis of rotation and both are idler gears. For the gear train shown in Figure 6.6(a), which is also called a *reverted gear train*, the input and output axes of rotation are collinear.

The speed ratio for a compound gear train is found in a similar manner to that used for a simple gear train. Consider the gear train in Figure 6.6(a), where gear 2 is attached to the input shaft, gear 5 to the output shaft, and idler gears 3 and 4 are mounted on the idler shaft

**Model 6.6A**
Reverted Gear
Train

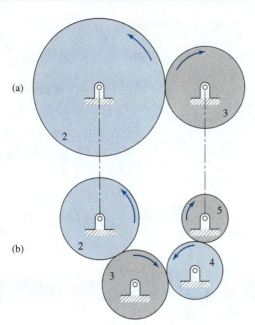

**Figure 6.5** Two simple gear trains with the same speed ratio.

(a)

(b)

and turn at the same speed. Speed ratios for meshing gears 2 and 3 and gears 4 and 5 are

$$e_{3/2} = \frac{\omega_3}{\omega_2} = -\frac{N_2}{N_3} \qquad \text{and} \qquad e_{5/4} = \frac{\omega_5}{\omega_4} = -\frac{N_4}{N_5}$$

The speed ratio for the gear train is

$$e_{5/2} = \frac{\omega_5}{\omega_2} = \frac{\omega_5 \, \omega_3}{\omega_4 \, \omega_2} = \left(-\frac{N_4}{N_5}\right)\left(-\frac{N_2}{N_3}\right) = \frac{N_2 N_4}{N_3 N_5} \tag{6.2-4}$$

In this instance, unlike the case of simple gear trains, the numbers of teeth of the idler gears influence the magnitude of the speed ratio. Using a similar analysis, the speed ratio for the compound gear train in Figure 6.6(b) is

$$e_{6/2} = -\frac{N_2 N_4}{N_3 N_6} \tag{6.2-5}$$

Idler gear 5 has no effect on the magnitude of the speed ratio; although it affects the direction of output speed and the distance between the input and output shafts.

The compound gear train illustrated in Figure 6.6(c) incorporates an external-internal meshing pair. The numbers of teeth are

$$N_2 = 9; \qquad N_3 = 62; \qquad N_4 = 20$$
$$N_5 = 51; \qquad N_6 = 15; \qquad N_7 = 54$$

The speed ratio is

$$e_{7/2} = \frac{\omega_7}{\omega_2} = \left(+\frac{N_6}{N_7}\right)\left(-\frac{N_4}{N_5}\right)\left(-\frac{N_2}{N_3}\right) = \frac{N_2 N_4 N_6}{N_3 N_5 N_7} = \frac{9 \times 20 \times 15}{62 \times 51 \times 54} = 0.0158$$

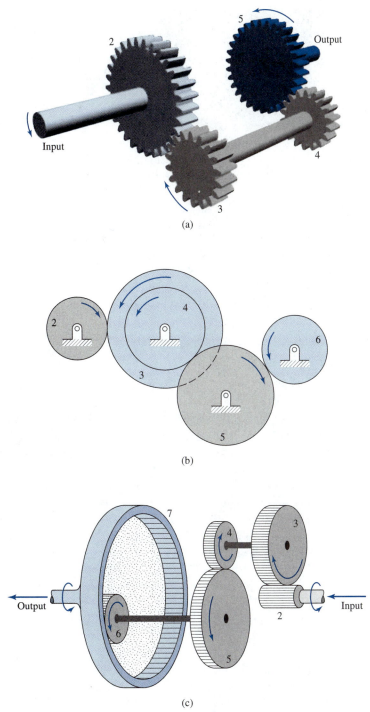

(a)

(b)

(c)

**Figure 6.6** Compound gear trains [Model 6.6A].

**Figure 6.7** Gear train used in a winch [Video 6.7].

This arrangement of gears is employed in the design of a winch. Figure 6.7 shows a corresponding photograph of the system.

A compound gear train may be designed to provide multiple speed ratios through one output shaft by transmitting motion through various pairs of gears. This is illustrated by the following two examples.

## EXAMPLE 6.1 Speed Ratios of a Manual Transmission

Figure 6.8 shows a schematic of a manual transmission, which employs a compound gear train. The shifter allows pairs of gears to be disengaged and reengaged to obtain three different speed ratios. Determine the speed ratios of the gear train.

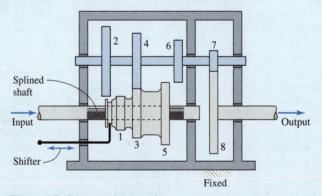

**Figure 6.8** Compound gear train.

### SOLUTION

Each configuration is equivalent to the reverted compound gear train shown in Figure 6.6(a). Table 6.1 indicates those gears which are in mesh for the three different transmission configurations and the speed ratio for each.

TABLE 6.1 Speed Ratios of the Manual Transmission Gearbox Shown in Figure 6.8

| Output Speed No. | Gears in Mesh | Speed Ratio |
|---|---|---|
| 1 | 1, 2, 7, 8 | $e_{8/1} = \dfrac{N_1 N_7}{N_2 N_8}$ |
| 2 | 3, 4, 7, 8 | $e_{8/3} = \dfrac{N_3 N_7}{N_4 N_8}$ |
| 3 | 5, 6, 7, 8 | $e_{8/5} = \dfrac{N_5 N_7}{N_6 N_8}$ |

## EXAMPLE 6.2 Designing for the Numbers of Teeth of a Reverted Gear Train

Find suitable tooth numbers for the gear train shown in Figure 6.8, to generate the speed ratios 1/4.00, 1/2.45, and 1/1.55. All gears are to have at least 12 teeth. The center-to-center distance between the idler and input/output shafts is 72 mm, and all gears have a module of 4 mm.

### SOLUTION

Because all gears are to have the same module, and meshing gears must share a common center-to-center distance, then employing Equation (5.7-4) for a pair of meshing gears designated as $i$ and $j$,

$$c = 72 \text{ mm} = \frac{d_i}{2} + \frac{d_j}{2} = \frac{m(N_i + N_j)}{2} = \frac{4(N_i + N_j)}{2} \qquad (6.2\text{-}6)$$

from which

$$N_i + N_j = 36$$

Therefore

$$N_1 + N_2 = N_3 + N_4 = N_5 + N_6 = N_7 + N_8 = 36 \qquad (6.2\text{-}7)$$

Expressions for the speed ratios are given in Table 6.1. Subject to the constraint given by Equation (6.2-7), the first desired speed ratio, $e_{8/1}$, can match exactly the desired value by selecting

$$N_7 = 12; \qquad N_8 = 24 \qquad (6.2\text{-}8)$$

$$N_1 = 12; \qquad N_2 = 24 \qquad (6.2\text{-}9)$$

It is not possible to obtain exactly the desired values of the second speed ratio, $e_{8/3}$, and third speed ratio, $e_{8/5}$. The numbers of teeth on gears 7 and 8 have already been specified (Equation (6.2-8)), and there are constraints on the numbers of teeth on gears that determine the second and third speed ratios (Equation (6.2-7)). Through a trial and error procedure of testing for various permissible combinations of numbers of teeth, we determine that the closest approximations are obtained using

$$N_3 = 16; \qquad N_4 = 20$$
$$N_5 = 20; \qquad N_6 = 16$$

which yields

$$e_{8/3} = \frac{1}{2.50}; \qquad e_{8/5} = \frac{1}{1.60}$$

An alternate set of solutions can be found by selecting

$$N_7 = 13; \qquad N_8 = 23$$

Values closest to the desired speed ratios, while satisfying the constraint for the minimum number of teeth on a gear, are obtained using

$$N_1 = 12; \qquad N_2 = 24$$
$$N_3 = 15; \qquad N_4 = 21$$
$$N_5 = 19; \qquad N_6 = 17$$

In this case

$$e_{8/1} = \frac{1}{3.54}; \qquad e_{8/3} = \frac{1}{2.48}; \qquad e_{8/5} = \frac{1}{1.58}$$

and now we only obtain an approximation of the desired first speed ratio. Actually, for this alternate set of solutions, employing

$$N_1 = 11; \qquad N_2 = 25$$

would give a better approximation of the desired first speed ratio. However, this violates the constraint of the minimum number of teeth on a gear.

---

**Model 6.9**
Manual
Transmission

A manual automobile transmission incorporates a similar arrangement of gears to that presented in Examples 6.1 and 6.2. Figure 6.9 illustrates a typical transmission that has three forward speed ratios and a reverse. Gear 1 is rigidly connected to the *input shaft*. Gears 2 through 5, inclusive, are rigidly connected to the *countershaft*. Gear 6 has its own axis of rotation and is employed to generate the speed ratio for reverse motion. Splines on the *output shaft* and gear 8 ensure both components share the same rotational motion, but do permit axial movement of gear 8 along the output shaft. By employing the *shifter*, gear 8 can be moved from the position shown in Figure 6.9 to mesh with either gear 4 or gear 6. Gear 7, being disengaged, is free to rotate relative to the output shaft. When required, one of the *synchronizers* (see Appendix B, Section B.13) rigidly connects gear 7 to the output shaft.

Figure 6.10 illustrates the configurations for all speed ratios of the manual transmission. While shifting between each speed ratio, the *clutch* (see Appendix B, Section B.14) is

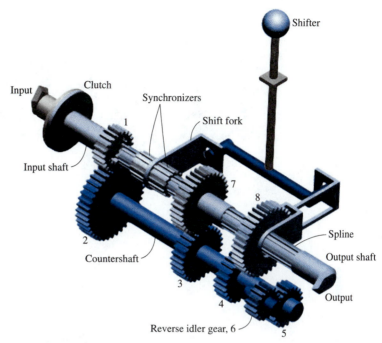

**Figure 6.9** Manual transmission [Model 6.9].

disengaged. For reverse (Figure 6.10(a)), motion is transmitted through gears 1, 2, 5, 6, and 8. Gear 6 is an idler gear and is used solely to change the direction of rotation. For neutral (Figure 6.10(b)), no motion is transmitted to the output shaft. This is because gear 7 is free to rotate with respect to the output shaft, although gears 1, 2, 3, and 7 are in mesh. For the first forward speed ratio (Figure 6.10(c)), gear 8 has been translated along the splined portion of the output shaft and meshes with gear 4. Motion is transmitted through gears 1, 2, 4, and 8. For the second forward speed ratio (Figure 6.10(d)), gear 8 no longer meshes with any other gear, and motion is transmitted through gears 1, 2, 3, and 7. A synchronizer (see Appendix B, Section B.13) is used to connect gear 7 to the output shaft. For the third forward speed ratio (Figure 6.10(e)), another synchronizer is employed to provide a direct coupling between the input and output shafts, and the speed ratio is unity. Under these conditions, the countershaft has no influence on the speed ratio.

Through use of a synchronizer, it is not necessary to engage and disengage gears 3 and 7 when shifting into and out of the second forward speed ratio. A synchronizer therefore eliminates the possibility of clashing the teeth of these gears. It also allows the possibility of employing helical spur gears, which are particularly difficult to engage and disengage through relative axial sliding.

## 6.2.3 Gear Trains with Bevel Gears

For the gear trains presented in Sections 6.2.1 and 6.2.2, all axes of rotation are parallel. However, other ordinary gear trains can be designed by employing meshing gears having

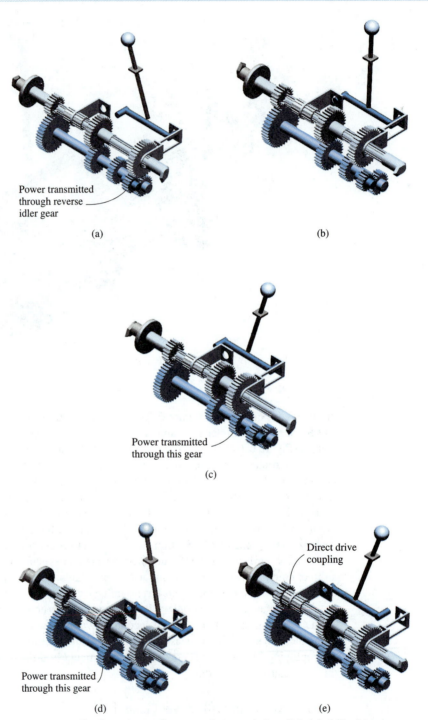

Power transmitted
through reverse
idler gear

(a)

(b)

Power transmitted
through this gear

(c)

Power transmitted
through this gear

(d)

Direct drive
coupling

(e)

**Figure 6.10** Configurations of a manual transmission [Model 6.9]: (a) reverse, (b) neutral, (c) first, (d) second, (e) third.

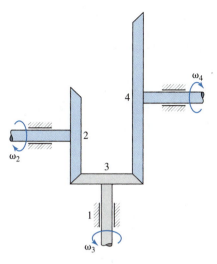

Figure 6.11 Gear train incorporating bevel gears.

nonparallel axes of rotation. Such gear trains can incorporate bevel gears, hypoid gears, and worm and wheel sets.

The magnitude of the speed ratio between any two bevel gears in mesh is inversely proportional to their number of gear teeth. For example, Figure 6.11 shows an ordinary gear train consisting of three bevel gears. Gear 2 is the input, gear 4 is the output, and gear 3 is an idler gear. The magnitudes of the speed ratios between gears 2 and 3, and between gears 3 and 4, are

$$|e_{3/2}| = \left|\frac{\omega_3}{\omega_2}\right| = \frac{N_2}{N_3}; \qquad |e_{4/3}| = \left|\frac{\omega_4}{\omega_3}\right| = \frac{N_3}{N_4} \tag{6.2-10}$$

The magnitude of the speed ratio of the gear train is

$$|e_{4/2}| = |e_{4/3}| \times |e_{3/2}| = \left|\frac{\omega_4}{\omega_2}\right| \tag{6.2-11}$$

Combining Equations (6.2-10) and (6.2-11),

$$e_{4/2} = -\frac{N_3}{N_4}\frac{N_2}{N_3} = -\frac{N_2}{N_4} \tag{6.2-12}$$

A minus sign in Equation (6.2-12) indicates that the input and output turn in opposite directions. The sign convention can be applied in this instance since the input and output shafts of the gear train are parallel to each other.

## 6.3 PLANETARY GEAR TRAINS

In a *planetary gear train*, the axis of at least one gear, called a *planet gear*, moves on a circular path relative to the base link.

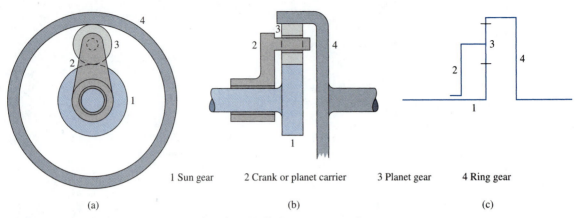

|   |   |   |   |
|---|---|---|---|
| 1 Sun gear | 2 Crank or planet carrier | 3 Planet gear | 4 Ring gear |

(a)                                (b)                                (c)

**Figure 6.12** (a) Planetary gear train. (b) Side view. (c) Skeleton representation.

An example of a planetary gear train is shown in Figure 6.12. It consists of a *sun gear*, having $N_1$ teeth, a concentric *ring gear*, having $N_4$ teeth, and planet gears, each with $N_3$ teeth. Planet gears mesh with the external teeth of the sun gear and with the internal teeth of the ring gear. Shafts of the planet gears are attached to the *crank* or *planet carrier*. The pitch of the sun, planet, and ring gears must be identical for this gear train.

It is often convenient to represent a planetary gear train using a highly simplified skeleton drawing. For instance, Figure 6.12(b) shows the side view of the planetary train illustrated in Figure 6.12(a), and the corresponding skeleton form is shown in Figure 6.12(c).

A planetary gear train has the following advantages over ordinary gear trains:

- It has compact space requirements, particularly when the input and output axes are collinear.
- Both static and dynamic forces are balanced when multiple planets are equally spaced about the central axis of the gear train.
- High torque capacity is possible, by using multiple planets.
- It can provide a wide range of speed ratios.

### 6.3.1 Number of Teeth on Gears in Planetary Gear Trains

The numbers of teeth on the gears of a planetary gear train cannot be selected arbitrarily. Consider the planetary gear train shown in Figure 6.12; the following geometrical relationship exists between the pitch circle diameters:

$$d_1 + 2d_3 = d_4 \tag{6.3-1}$$

Since in this instance the pitch of all meshing gears must be equal, it follows from Equations (5.7-5) and (6.3-1) that

$$N_1 + 2N_3 = N_4 \tag{6.3-2}$$

or

$$N_3 = \frac{N_4 - N_1}{2} \qquad (6.3\text{-}3)$$

## 6.3.2  Number of Planet Gears

For the planetary gear train shown in Figure 6.12, only one planet gear is needed to transmit motion. However, multiple planet gears are usually employed as in the example shown in Figure 6.13. The use of multiple planets allows more gears to share the transmission of loads through alternate paths simultaneously, which as a result increases the load capacity of the gear train without increasing its overall size. In practice, to improve the dynamic characteristics of the gear train, planet gears are usually equally spaced, so loads are balanced about the central axis.

The number of equally spaced planet gears, $n$, that can be placed between a sun and ring gear depends on the space available and the numbers of teeth of the gears. The following two criteria must be satisfied [3]:

1.  Prevention of addendum circle overlap of planet gears:

$$n < \frac{180°}{\sin^{-1}\left(\dfrac{N_3 + 2\alpha}{N_1 + N_3}\right)} \qquad (6.3\text{-}4)$$

   where $\alpha$ takes on values of 0.8 and 1.0 for stub teeth and full-depth teeth, respectively.

2.  Meshing of gear teeth:

$$\frac{N_1 + N_4}{n} = \text{integer} \qquad (6.3\text{-}5)$$

Figure 6.13  Planetary gear train [Model 6.13].

## EXAMPLE 6.3  Number of Equally Spaced Planet Gears

For the planetary gear train shown in Figure 6.13, determine the number of equally spaced planet gears that may be employed. The teeth are to have full depth. The numbers of teeth are

$$N_1 = 27; \qquad N_3 = 18; \qquad N_4 = 63$$

### SOLUTION

The number of equally spaced planet gears may be found by applying Equations (6.3-4) and (6.3-5). Substituting the numbers of teeth in Equation (6.3-4) and employing $\alpha = 1.0$ for full-depth teeth,

$$n < \frac{180°}{\sin^{-1}\left(\dfrac{N_3 + 2\alpha}{N_1 + N_3}\right)} = \frac{180°}{\sin^{-1}\left(\dfrac{18 + 2.0}{27 + 18}\right)} = 6.82 \qquad (6.3\text{-}6)$$

Since the number of planet gears must be an integer, according to this constraint, up to six equally spaced planets may be employed. From Equation (6.3-5)

$$\frac{N_1 + N_4}{n} = \frac{27 + 63}{n} = \frac{90}{n} = \text{integer} \qquad (6.3\text{-}7)$$

Substituting the values $n = 1, 2, 3, 4, 5$, and 6 in Equation (6.3-7), we find that the values $n = 1, 2, 3, 5$, and 6 satisfy the equation. Therefore, either 1, 2, 3, 5, or 6 equally spaced planet gears may be employed in this gear train. It is not possible to use four equally placed planet gears.

## 6.3.3  Classification of Planetary Gear Trains

Lévai [6] identified twelve basic types of *single-stage planetary gear trains* incorporating spur gears, where all axes of rotation are parallel. Their skeleton drawings are shown in Figure 6.14. For example, the planetary gear train shown in Figure 6.12 is Type A. Additional variations of planetary gear trains can be obtained by employing gears having non-parallel axes of rotation. Such examples are presented later in this chapter.

Multiple-stage planetary gear trains employ a combination of single-stage gear trains. Each stage may be one of the types illustrated in Figure 6.14 or may employ other types of gears, such as bevel gears or worm and wheel.

The mobility of each of the planetary gear trains shown in Figure 6.14 equals two. Therefore, the rotational speeds of all components may be determined if two of the rotational speeds are specified. The two input speeds may be equal or unequal, in the same direction or opposite. One of the two input speeds may be zero. For the gear train shown in Figure 6.12, if the crank is not allowed to rotate, all gears have fixed axes of rotation, and the planetary gear train is treated as an ordinary gear train.

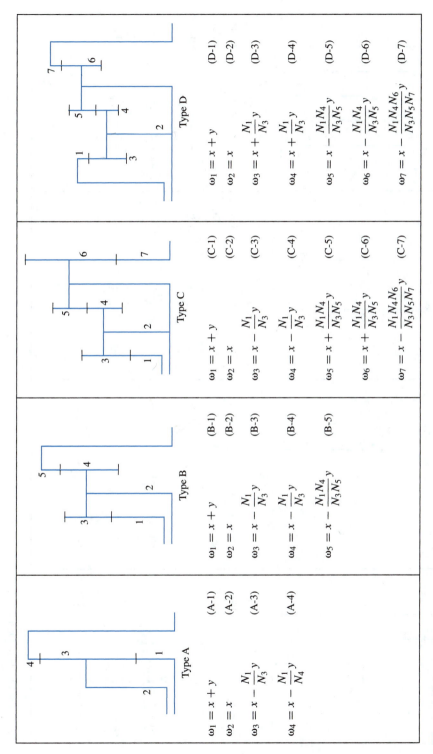

**Figure 6.14** Basic planetary gear train types [7].

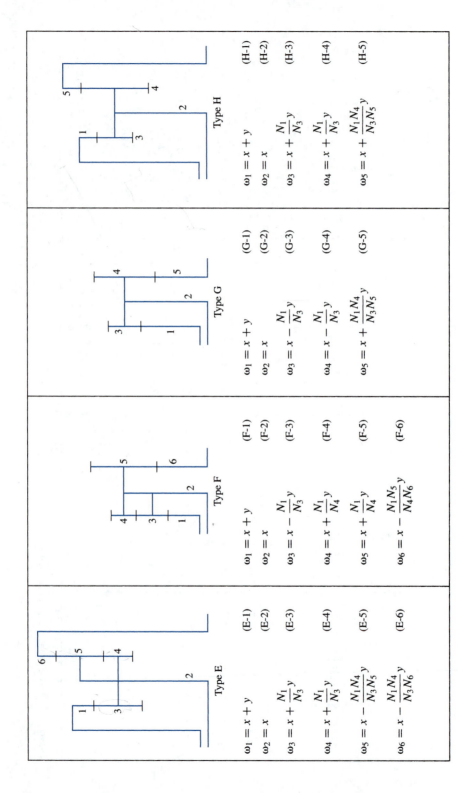

**Type E**

$\omega_1 = x + y$    (E-1)

$\omega_2 = x$    (E-2)

$\omega_3 = x + \dfrac{N_1}{N_3} y$    (E-3)

$\omega_4 = x + \dfrac{N_1}{N_3} y$    (E-4)

$\omega_5 = x - \dfrac{N_1 N_4}{N_3 N_5} y$    (E-5)

$\omega_6 = x - \dfrac{N_1 N_4}{N_3 N_6} y$    (E-6)

**Type F**

$\omega_1 = x + y$    (F-1)

$\omega_2 = x$    (F-2)

$\omega_3 = x - \dfrac{N_1}{N_3} y$    (F-3)

$\omega_4 = x + \dfrac{N_1}{N_4} y$    (F-4)

$\omega_5 = x + \dfrac{N_1}{N_4} y$    (F-5)

$\omega_6 = x - \dfrac{N_1 N_5}{N_4 N_6} y$    (F-6)

**Type G**

$\omega_1 = x + y$    (G-1)

$\omega_2 = x$    (G-2)

$\omega_3 = x - \dfrac{N_1}{N_3} y$    (G-3)

$\omega_4 = x - \dfrac{N_1}{N_3} y$    (G-4)

$\omega_5 = x + \dfrac{N_1 N_4}{N_3 N_5} y$    (G-5)

**Type H**

$\omega_1 = x + y$    (H-1)

$\omega_2 = x$    (H-2)

$\omega_3 = x + \dfrac{N_1}{N_3} y$    (H-3)

$\omega_4 = x + \dfrac{N_1}{N_3} y$    (H-4)

$\omega_5 = x + \dfrac{N_1 N_4}{N_3 N_5} y$    (H-5)

**Figure 6.14**   *(Continued)*

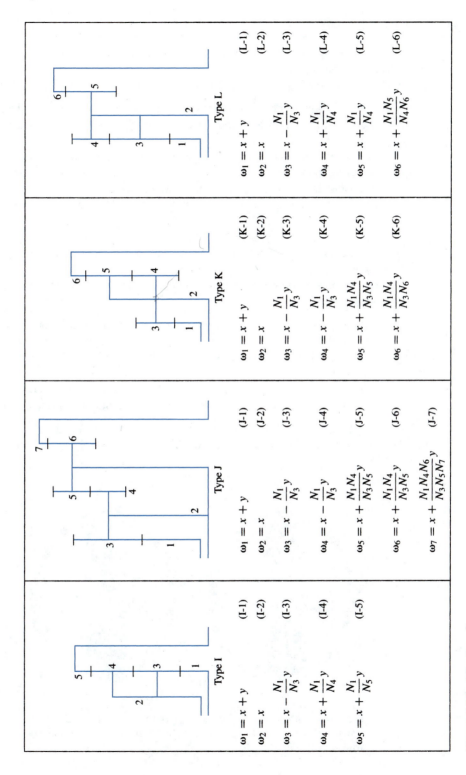

**Type I**

$\omega_1 = x + y$    (I-1)

$\omega_2 = x$    (I-2)

$\omega_3 = x - \dfrac{N_1}{N_3} y$    (I-3)

$\omega_4 = x + \dfrac{N_1}{N_4} y$    (I-4)

$\omega_5 = x + \dfrac{N_1}{N_5} y$    (I-5)

**Type J**

$\omega_1 = x + y$    (J-1)

$\omega_2 = x$    (J-2)

$\omega_3 = x - \dfrac{N_1}{N_3} y$    (J-3)

$\omega_4 = x - \dfrac{N_1}{N_3} y$    (J-4)

$\omega_5 = x + \dfrac{N_1 N_4}{N_3 N_5} y$    (J-5)

$\omega_6 = x + \dfrac{N_1 N_4}{N_3 N_5} y$    (J-6)

$\omega_7 = x + \dfrac{N_1 N_4 N_6}{N_3 N_5 N_7} y$    (J-7)

**Type K**

$\omega_1 = x + y$    (K-1)

$\omega_2 = x$    (K-2)

$\omega_3 = x - \dfrac{N_1}{N_3} y$    (K-3)

$\omega_4 = x - \dfrac{N_1}{N_3} y$    (K-4)

$\omega_5 = x + \dfrac{N_1 N_4}{N_3 N_5} y$    (K-5)

$\omega_6 = x + \dfrac{N_1 N_4}{N_3 N_6} y$    (K-6)

**Type L**

$\omega_1 = x + y$    (L-1)

$\omega_2 = x$    (L-2)

$\omega_3 = x - \dfrac{N_1}{N_3} y$    (L-3)

$\omega_4 = x + \dfrac{N_1}{N_4} y$    (L-4)

$\omega_5 = x + \dfrac{N_1}{N_4} y$    (L-5)

$\omega_6 = x + \dfrac{N_1 N_5}{N_4 N_6} y$    (L-6)

**Figure 6.14** *(Continued)*

## 6.4 TABULAR ANALYSIS OF PLANETARY GEAR TRAINS

In this section, the *tabular method* is presented to determine speed ratios and speeds of components of planetary gear trains.

A tabular method of a planetary gear train involves the following general steps:

1. Assume all components of the gear train rotate at $x$ rpm about the central axis. This motion constitutes rigid-body rotation of the entire gear train, and there is no meshing of the gear teeth.
2. Assume the crank is not permitted to move. Ignore all other motion constraints and choose a component of the gear train other than the crank, such as a sun or ring gear, and rotate it at $y$ rpm. Determine the rotational speeds of all components of the gear train. By holding the crank fixed, the planetary gear train becomes an ordinary gear train (Section 6.2).
3. Superimpose the two motions considered in steps 1 and 2 by summing the rotational speeds, that is, the rotational speeds of the components are expressed in terms of $x$ and $y$.
4. Equate expression(s) for rotational speed(s) found in step 3 to the specified input speed(s).

   (a) If the rotational speed of one link is specified, then express rotational speeds of all the components in terms of $x$ (or $y$). Use appropriate expressions to determine the speed ratio between two links, as required, and thereby eliminate $x$ (or $y$).
   (b) If the rotational speeds of two links are specified, then generate two equations and solve for $x$ and $y$. Use the result to determine rotational speeds, as required.

Figures 6.15(a) and 6.15(b) portray motions described in steps 1 and 2 as they apply to the planetary gear train illustrated in Figure 6.13.

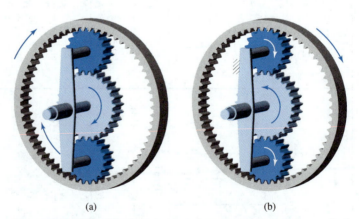

(a)                                    (b)

**Figure 6.15** Example of motions employed using the tabular method [Model 6.13]: (a) all links with same rotational speed, (b) motion of links with respect to fixed crank.

**TABLE 6.2** Tabular Analysis of the Planetary Gear Train Shown in Figure 6.13

| Component → Operation ↓ | Gear 1 $\omega_1$ | Crank $\omega_2$ | Gear 3 $\omega_3$ | Gear 4 $\omega_4$ |
|---|---|---|---|---|
| All Components Turn at x rpm | $x$ | $x$ | $x$ | $x$ |
| Crank Is Fixed, Gear 1 Turns at y rpm | $y$ | $0$ | $-\dfrac{N_1}{N_3}y$ | $-\dfrac{N_1}{N_4}y$ |
| Absolute Rotational Speeds | $x+y$ | $x$ | $x-\dfrac{N_1}{N_3}y$ | $x-\dfrac{N_1}{N_4}y$ |

Table 6.2 summarizes the first three steps of the above procedure for the planetary gear train shown in Figure 6.13. The last row of the table is equivalent to the following equations:

$$\omega_1 = x + y; \qquad \omega_2 = x$$

$$\omega_3 = x - \frac{N_1}{N_3}y; \qquad \omega_4 = x - \frac{N_1}{N_4}y$$

If the rotational speeds of two of the components are specified, then it is possible to solve for variables $x$ and $y$. For instance, if the sun gear of the planetary gear train shown in Figure 6.13 turns at $\omega_1 = 400$ rpm CW, and the ring gear is held fixed (i.e., $\omega_4 = 0$), the following two simultaneous linear equations are obtained:

$$\omega_1 = x + y = 400 \text{ rpm CW} = -400 \text{ rpm}; \qquad \omega_4 = x - \frac{N_1}{N_4}y = 0$$

where it has been assumed that positive rotational speeds are in the counterclockwise direction.

Substituting the given numbers of teeth ($N_1 = 27$, $N_4 = 63$), and solving yields

$$x = -120 \text{ rpm}; \qquad y = -280 \text{ rpm}$$

The output speed becomes

$$\omega_2 = x = -120 = 120 \text{ rpm CW}$$

The speed ratio of the gear train is

$$e_{2/1} = \frac{\omega_2}{\omega_1} = \frac{x}{x+y} = \frac{(-120)}{(-120)+(-280)} = 0.30 \qquad (6.4\text{-}1)$$

The rotational speeds of any component in the planetary gear train can now also be determined. For example

$$\omega_{\text{planet}} = \omega_3 = x - \frac{N_1}{N_3}y$$

$$= (-120) - \frac{27}{18} \times (-280) = 300 = 300 \text{ rpm CCW}$$

The above calculation gives the absolute value of the rotational speed, that is, with respect to the fixed link, gear 4. The relative rotational speed of the planet gear with respect to its own axis of rotation is

$$\omega_{\text{planet, relative}} = \omega_3 - \omega_2 = x - \frac{N_1}{N_3}y - x = 420 = 420 \text{ rpm CCW}$$

## EXAMPLE 6.4  Speed Ratio of a Planetary Gear Train

For the gear train shown in Figure 6.16, gears 6 and 7 are mounted on the input shaft and turn with the same speed ($\omega_6 = \omega_7$). These gears mesh with gears 2 and 5, which are connected to the crank (link 2) and sun gear (gear 1), respectively, and provide the two input motions to the planetary gear train. The output shaft is attached to the ring gear (gear 4). Gears 1 and 5 also turn at the same rate ($\omega_1 = \omega_5$). Determine the speed ratio of the gear train.

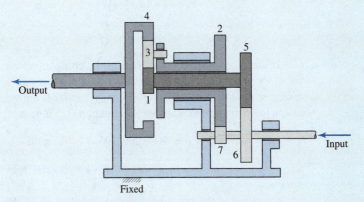

**Figure 6.16**  Planetary gear train.

SOLUTION

The two input rotational speeds to the planetary gear train are determined by first recognizing that

$$e_{5/6} = \frac{\omega_5}{\omega_6} = -\frac{N_6}{N_5}; \qquad e_{2/7} = \frac{\omega_2}{\omega_7} = -\frac{N_7}{N_2}$$

The rotational speeds of the sun gear and crank are therefore

$$\omega_1 = \omega_5 = e_{5/6} \times \omega_6 \tag{6.4-2}$$

$$\omega_2 = e_{2/7} \times \omega_7 \tag{6.4-3}$$

Since the arrangement of gears of the planetary gear train in Figures 6.12 and 6.16 are the same, Table 6.2 can also be applied in this example. Equating the input rotational speeds given in Equations (6.4-2) and (6.4-3) with those given in Table 6.2,

$$\omega_1 = -\frac{N_6}{N_5}\omega_6 = x + y \tag{6.4-4}$$

$$\omega_2 = -\frac{N_7}{N_2}\omega_7 = x \tag{6.4-5}$$

Solving Equations (6.4-4) and (6.4-5) for $y$, and recognizing that

$$\omega_6 = \omega_7 \tag{6.4-6}$$

gives

$$y = -\omega_7\left(\frac{N_6}{N_5} - \frac{N_7}{N_2}\right) \tag{6.4-7}$$

The value of rotational speed of the ring gear (gear 4) is also obtained from Table 6.2:

$$\omega_4 = x - \frac{N_1}{N_4}y \tag{6.4-8}$$

Substituting Equations (6.4-5)–(6.4-7),

$$\omega_4 = \omega_7\left[-\frac{N_7}{N_2} + \frac{N_1}{N_4}\left(\frac{N_6}{N_5} - \frac{N_7}{N_2}\right)\right] \tag{6.4-9}$$

and the speed ratio of the gear train employing Equation (6.4-9) is

$$e_{4/7} = \frac{\omega_4}{\omega_7} = -\frac{N_7}{N_2} + \frac{N_1}{N_4}\left(\frac{N_6}{N_5} - \frac{N_7}{N_2}\right) \tag{6.4-10}$$

## EXAMPLE 6.5 Speed Ratios of an Automatic Transmission

Determine the speed ratios of the automatic transmission illustrated in Figure 6.17. The speed ratios are generated as follows:

- *reverse:* Gear 5 is held fixed.
- *first speed:* Gear 1 is held fixed.
- *second speed:* Gears 1 and 6 are locked together.

In all cases, the input is connected to gear 9, and the output is connected to crank 2. The numbers of teeth are

$$N_1 = 27; \quad N_3 = 15; \quad N_4 = 9; \quad N_5 = 57$$
$$N_6 = 32 = N_7; \quad N_8 = 16; \quad N_9 = 8$$

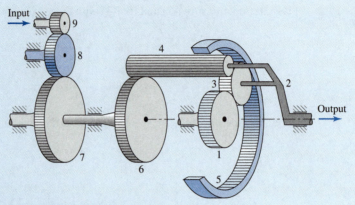

Input

9

8

4

3

2

Output

7

6

1

5

**Figure 6.17** Schematic of automatic transmission.

**SOLUTION**

Table 6.3 was created for this gear train. Using this table, the ratio of speeds between gear 6 and crank 2 is

$$e_{2/6} = \frac{\omega_2}{\omega_6} = \frac{x}{\left(x + \dfrac{N_5}{N_6}y\right)}$$

and therefore the speed ratio of the gear train is

$$e_{2/9} = \frac{\omega_2}{\omega_6} \times \frac{\omega_6}{\omega_9} = \frac{x}{\left(x + \dfrac{N_5}{N_6}y\right)} \times \frac{N_9}{N_7}$$  (6.4-11)

where it is recognized that $\omega_6 = \omega_7$.

**Reverse**

Since for this speed ratio gear 5 is held fixed, then from Table 6.3

$$\omega_5 = x + y = 0; \qquad y = -x$$  (6.4-12)

**TABLE 6.3** Tabular Analysis of the Gear Train Shown in Figure 6.17

| Component →<br>Operation ↓ | Gear 1<br>$\omega_1$ | Crank<br>$\omega_2$ | Gear 5<br>$\omega_5$ | Gear 6<br>$\omega_6$ |
|---|---|---|---|---|
| All Components Turn at $x$ rpm | $x$ | $x$ | $x$ | $x$ |
| Crank Is Fixed, Gear 5 Turns at $y$ rpm | $-\dfrac{N_5}{N_1}y$ | $0$ | $y$ | $\dfrac{N_5}{N_6}y$ |
| Absolute Rotational Speeds | $x - \dfrac{N_5}{N_1}y$ | $x$ | $x + y$ | $x + \dfrac{N_5}{N_6}y$ |

Combining Equations (6.4-11) and (6.4-12), the speed ratio is

$$e_{2/9} = \frac{\left(\dfrac{N_9}{N_7}x\right)}{\left(x + \dfrac{N_5}{N_6}y\right)} = \frac{\left(\dfrac{N_9}{N_7}x\right)}{x\left(1 - \dfrac{N_5}{N_6}\right)}$$

$$= \frac{\left(\dfrac{N_9}{N_7}\right)}{\left(1 - \dfrac{N_5}{N_6}\right)} = \frac{\left(\dfrac{8}{32}\right)}{\left(1 - \dfrac{57}{32}\right)} \tag{6.4-13}$$

$$= -0.320$$

**First Speed**

Since for this speed ratio gear 1 is held fixed, then from Table 6.3

$$\omega_1 = x - \frac{N_5}{N_1}y = 0; \qquad y = \frac{N_1}{N_5}x \tag{6.4-14}$$

Combining Equations (6.4-11) and (6.4-14), the speed ratio is

$$e_{2/9} = \frac{\left(\dfrac{N_9}{N_7}x\right)}{\left(x + \dfrac{N_5}{N_6}y\right)} = \frac{\left(\dfrac{N_9}{N_7}x\right)}{x\left(1 + \dfrac{N_5}{N_6}\dfrac{N_1}{N_5}\right)}$$

$$= \frac{\left(\dfrac{N_9}{N_7}\right)}{\left(1 + \dfrac{N_1}{N_6}\right)} = \frac{\left(\dfrac{8}{32}\right)}{\left(1 + \dfrac{27}{32}\right)} \tag{6.4-15}$$

$$= 0.136$$

**Second Speed**

Since for this speed ratio gears 1 and 6 are held together, then from Table 6.3

$$\omega_1 = \omega_6$$

$$x - \frac{N_5}{N_1}y = x + \frac{N_5}{N_6}y \tag{6.4-16}$$

which can only be satisfied if

$$y = 0 \tag{6.4-17}$$

Combining Equations (6.4-11) and (6.4-17), the speed ratio is

$$e_{2/9} = \frac{\left(\dfrac{N_9}{N_7}x\right)}{\left(x + \dfrac{N_5}{N_6}y\right)} = \frac{\left(\dfrac{N_9}{N_7}x\right)}{x} = \frac{8}{32} = 0.25 \qquad (6.4\text{-}18)$$

Figure 6.18 is an illustration of this gear train.

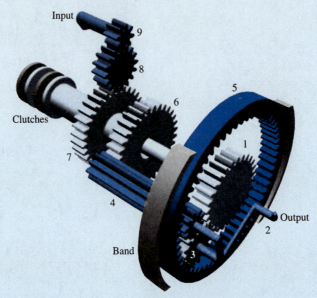

**Figure 6.18** Two-speed automatic transmission [Model 6.18].

## EXAMPLE 6.6  Analysis of a Wankel Engine

A Wankel engine, as shown in Figure 6.19, is a planetary gear train for which the internal planet gear, having $N_3$ teeth, is part of the rotor. The sun gear, with $N_1$ teeth, is fixed. Determine the speed ratio of the crank, link 2, with respect to the planet gear, link 3, if

$$\frac{N_1}{N_3} = \frac{2}{3} \qquad (6.4\text{-}19)$$

### SOLUTION

Table 6.4 was created for this gear train. Recognizing that sun gear 1 is fixed, its rotational speed, $\omega_1$, is

$$\omega_1 = x + y = 0; \qquad y = -x \qquad (6.4\text{-}20)$$

Substituting Equation (6.4-20) in the expression for $\omega_3$ obtained from Table 6.4 allows the rotational speed to be expressed in terms of one variable, $x$, as follows:

$$\omega_3 = x + \frac{N_1}{N_3}y = x\left(1 - \frac{N_1}{N_3}\right) \qquad (6.4\text{-}21)$$

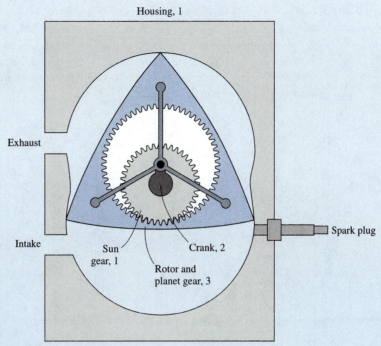

**Figure 6.19** Wankel engine.

Employing Equations (6.4-19) and (6.4-21), the speed ratio of the crank with respect to the planet gear is

$$e_{2/3} = \frac{\omega_2}{\omega_3} = \frac{x}{x\left(1 - \dfrac{N_1}{N_3}\right)} = \frac{1}{\left(1 - \dfrac{N_1}{N_3}\right)} = \frac{1}{\left(1 - \dfrac{2}{3}\right)} = 3 \qquad (6.4\text{-}22)$$

that is, for each rotation of the crank, the planet gear (i.e., rotor) makes one-third of a revolution in the same direction.

**TABLE 6.4** Tabular Analysis of the Wankel Engine Shown in Figure 6.19

| Component → <br> Operation ↓ | Gear 1 <br> $\omega_1$ | Crank <br> $\omega_2$ | Gear 3 <br> $\omega_3$ |
|---|---|---|---|
| All Components Turn at $x$ rpm | $x$ | $x$ | $x$ |
| Crank Is Fixed, Gear 1 Turns at $y$ rpm | $y$ | $0$ | $\dfrac{N_1}{N_3}y$ |
| Absolute Rotational Speeds | $x + y$ | $x$ | $x + \dfrac{N_1}{N_3}y$ |

## EXAMPLE 6.7  Speed Ratio of a Planetary Gear Train

For the gear train shown in Figure 6.20, planet gears 3, 4, and 5 are rigidly connected to a common concentric shaft and turn at the same rotational speed, $\omega_3 = \omega_4 = \omega_5$. Gear 1 is connected to the input, gear 7 is connected to the output, and gear 6 is fixed. The planet carrier, component 2, is connected to neither an input nor an output. Determine the speed ratio between the input and the output.

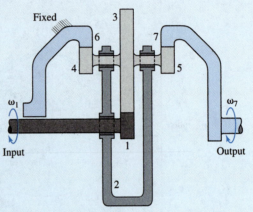

**Figure 6.20**  Planetary gear train.

SOLUTION

Table 6.5 was created for this gear train. Recognizing that ring gear 6 is fixed, its rotational speed, $\omega_6$, is

$$\omega_6 = x + y = 0; \qquad y = -x \tag{6.4-23}$$

Substituting Equation (6.4-23) in the expressions for $\omega_1$ and $\omega_7$ obtained from Table 6.5 allows the rotational speeds to be expressed in terms of one variable, $x$, as follows:

$$\omega_1 = x - \frac{N_6 N_3}{N_4 N_1} y = x \left( 1 + \frac{N_6 N_3}{N_4 N_1} \right) \tag{6.4-24}$$

$$\omega_7 = x + \frac{N_6 N_5}{N_4 N_7} y = x \left( 1 - \frac{N_6 N_5}{N_4 N_7} \right) \tag{6.4-25}$$

The speed ratio for this gear train is therefore

$$e_{7/1} = \frac{\omega_7}{\omega_1} = \frac{\left( 1 - \dfrac{N_6 N_5}{N_4 N_7} \right)}{\left( 1 + \dfrac{N_6 N_3}{N_4 N_1} \right)} \tag{6.4-26}$$

**TABLE 6.5** Tabular Analysis of the Planetary Gear Train Shown in Figure 6.20

| Component → Operation ↓ | Gear 1 $\omega_1$ | Gear 6 $\omega_6$ | Gear 7 $\omega_7$ |
|---|---|---|---|
| All Components Turn at $x$ rpm | $x$ | $x$ | $x$ |
| Crank Is Fixed, Gear 6 Turns at $y$ rpm | $-\dfrac{N_6 N_3}{N_4 N_1}y$ | $y$ | $\dfrac{N_6 N_5}{N_4 N_7}y$ |
| Absolute Rotational Speeds | $x-\dfrac{N_6 N_3}{N_4 N_1}y$ | $x+y$ | $x+\dfrac{N_6 N_5}{N_4 N_7}y$ |

When setting up the tabular method, it is not always necessary to include all components of the planetary gear train in the table. Unless the rotational speed of a planet gear is required, it need not be included in the table. However, components connected to the output(s) and the input(s) must always be included. This is necessary even when an input is fixed. For Example 6.7, since the planet carrier is not connected to either the inputs or the output, it has been left out of the table.

Figure 6.14 shows twelve basic types of planetary gear trains. Also listed in this figure are equations for the rotational speeds of the components obtained by performing a tabular analysis on each [7]. These equations can not only be applied to the gear trains shown, but are also useful in the analysis of other more complicated gear trains, which are combinations of the twelve basic types. The analysis of such gear trains is presented in Section 6.5.

## 6.5 KINEMATIC ANALYSIS OF MULTIPLE-STAGE PLANETARY GEAR TRAINS

It is possible to apply results presented in Figure 6.14 to more complicated systems, made up of combinations of the basic types. Each basic type of gear train that becomes part of the more complicated system will be considered as one *stage* of the gear train. A gear train that comprises a combination of more than one of the twelve basic types is referred to as a *multiple-stage planetary gear train*. Figure 6.21 shows a planetary gear train that is equivalent to a combination of type K and type L gear trains. These basic types of gear trains which make up this system are referred to as stage 1 and stage 2. Table 6.6 compares the numbering schemes used for the planetary gear train shown in Figure 6.21, with the two basic types it comprises (Figure 6.14).

Determining the speeds of the components of a planetary gear train with $k$ stages requires solving $2k$ equations containing $2k$ unknowns. In a single-stage planetary gear train ($k = 1$), there are two equations with two unknowns, which are normally based on the inputs to the gear train. Solving planetary gear trains for $k > 1$ requires not only that equations be generated based on the inputs, but also that constraint equations be obtained by considering connections of components between stages of the gear train. A total of $2k$ equations need to be generated.

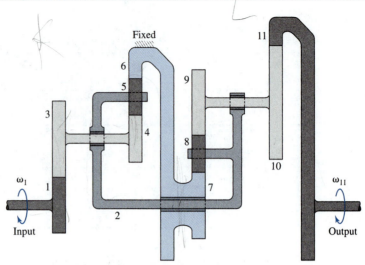

**Figure 6.21** Two-stage planetary gear train incorporating the basic planetary gear train types K and L.

**TABLE 6.6** Comparison of the Component Numbers in Figures 6.21 and 6.14

| Numbering of Components of Two-Stage Planetary Gear Train (Figure 6.21) | Numbering of Components of Individual Stages (Figure 6.14) | |
|---|---|---|
| | Stage 1 (Type K) | Stage 2 (Type L) |
| 1 | 1 | — |
| 2 | 2 | 2 |
| 3 | 3 | — |
| 4 | 4 | — |
| 5 | 5 | — |
| 6 | 6 | — |
| 7 | — | 1 |
| 8 | — | 3 |
| 9 | — | 4 |
| 10 | — | 5 |
| 11 | — | 6 |

For the two-stage planetary gear train in Figure 6.21, the numbers of teeth are

$$N_1 = 20; \quad N_3 = 25; \quad N_4 = 16; \quad N_5 = 17; \quad N_6 = 95$$
$$N_7 = 20; \quad N_8 = 17; \quad N_9 = 22; \quad N_{10} = 40; \quad N_{11} = 116$$

and the specified input speeds are

$$(\omega_1)_1 = 200 \text{ rpm} \tag{6.5-1}$$
$$(\omega_6)_1 = 0 \tag{6.5-2}$$

where subscript 1 outside of the parentheses indicates the stage number.

Additional constraints are obtained by equating the rotational speeds of those components connecting the two stages. The ring gear (component 6) of stage 1 is joined to the sun gear (component 1) of stage 2, giving

$$(\omega_6)_1 = (\omega_1)_2 \tag{6.5-3}$$

and the crank (component 2) of stage 1 is coupled to the crank (component 2) of stage 2, giving

$$(\omega_2)_1 = (\omega_2)_2 \tag{6.5-4}$$

Equations (6.5-1)–(6.5-4) may be expressed in terms of variables $x$ and $y$ using the appropriate equations listed in Figure 6.14, giving

$$x_1 + y_1 = 200 \tag{6.5-5}$$

$$x_1 + \left(\frac{N_1 N_4}{N_3 N_6}\right)_1 y_1 = 0 \tag{6.5-6}$$

$$x_1 + \left(\frac{N_1 N_4}{N_3 N_6}\right)_1 y_1 = x_2 + y_2 \tag{6.5-7}$$

$$x_1 = x_2 \tag{6.5-8}$$

where subscripts of $x$ and $y$ and those outside the parentheses indicate the associated stage number.

Employing Table 6.6, the numbers of the gears may be expressed in accordance with those given in Figure 6.21. Also, substituting the given number of gear teeth, Equations (6.5-6) and (6.5-7) become

$$x_1 + \left(\frac{N_1 N_4}{N_3 N_6}\right)_1 y_1 = x_1 + \left(\frac{N_1 N_4}{N_3 N_6}\right) y_1 = x_1 + \left(\frac{20 \times 16}{25 \times 95}\right) y_1 = x_1 + 0.1347 y_1 = 0 \tag{6.5-9}$$

$$x_1 + \left(\frac{N_1 N_4}{N_3 N_6}\right) y_1 = x_1 + \left(\frac{20 \times 16}{25 \times 95}\right) y_1 = x_1 + 0.1347 y_1 = x_2 + y_2 \tag{6.5-10}$$

Solving Equations (6.5-5), (6.5-8), (6.5-9), and (6.5-10) gives

$$x_1 = x_2 = -31.14; \qquad y_1 = 231.14; \qquad y_2 = 31.14$$

The rotational speed of every component in the planetary gear train may now be calculated by substituting values of $x_1$, $x_2$, $y_1$, and $y_2$ back in the appropriate equations listed in Figure 6.14. The speed ratio of this planetary gear train is

$$e_{11/1} = \frac{\omega_{11}}{\omega_1} = \frac{(\omega_6)_2}{(\omega_1)_1} = \frac{x_2 + \left(\dfrac{N_1 N_5}{N_4 N_6}\right)_2 y_2}{x_1 + y_1} = \frac{x_2 + \left(\dfrac{N_7 N_{10}}{N_9 N_{11}}\right) y_2}{x_1 + y_1} = 0.107$$

In general, for a planetary gear train with $k$ stages, a set of $2k$ equations can be generated and expressed in the form

$$[A]\{x\} = \{B\} \tag{6.5-11}$$

where

$$\{x\} = [x_1, y_1, x_2, y_2, \ldots, x_k, y_k]^T \tag{6.5-12}$$

Equation (6.5-11) is then solved for $\{x\}$.

The equations given in Figure 6.14 are ideally suited for obtaining solutions using a computer. All the types of planetary gear trains shown in Figure 6.14 can be stored in the computer's memory. Upon specification of a planetary gear train type, the computer will work with the appropriate set of equations. As each constraint is entered, matrix $[A]$ and vector $\{B\}$ of Equation (6.5-11) may be constructed.

## EXAMPLE 6.8  Speed Ratio of a Planetary Gear Train

Determine the speed ratio of the planetary gear train shown in Figure 6.20 by combining the basic types of planetary gear trains shown in Figure 6.14.

### SOLUTION

The gear train is equivalent to combining two basic planetary gear train stages, each of type B, shown in Figure 6.14. Components 1, 2, and 3 are shared by both stages. Therefore, they could be considered as linked components between the stages. Table 6.7 compares the numbering schemes presented in Figures 6.14 and 6.20.

The following equations are generated by considering two specified input rotational speeds, one of which is zero, and constraints imposed by the connected components of the planetary gear train:

$$(\omega_1)_1 = \omega_{\text{input}} = \omega_1 \tag{6.5-13}$$

$$\omega_6 = (\omega_5)_1 = 0 \tag{6.5-14}$$

$$(\omega_1)_1 = (\omega_1)_2 \tag{6.5-15}$$

$$(\omega_2)_1 = (\omega_2)_2 \tag{6.5-16}$$

$$(\omega_3)_1 = (\omega_3)_2 \tag{6.5-17}$$

**TABLE 6.7**  Comparison of the Component Numbers in Figures 6.20 and 6.14

| Numbering of Components of Two-Stage Planetary Gear Train (Figure 6.20) | Numbering of Components of Individual Stages (Figure 6.14) | |
|---|---|---|
| | Stage 1 (Type B) | Stage 2 (Type B) |
| 1 | 1 | 1 |
| 2 | 2 | 2 |
| 3 | 3 | 3 |
| 4 | 4 | — |
| 5 | — | 4 |
| 6 | 5 | — |
| 7 | — | 5 |

Expressing Equations (6.5-13)–(6.5-17) in terms of the unknowns given in Figure 6.14,

$$x_1 + y_1 = \omega_1 \tag{6.5-18}$$

$$x_1 - \left(\frac{N_1 N_4}{N_3 N_5}\right)_1 y_1 = 0 \quad \text{or} \quad x_1 - \frac{N_1 N_4}{N_3 N_6} y_1 = 0 \tag{6.5-19}$$

$$x_1 + y_1 = x_2 + y_2 \tag{6.5-20}$$

$$x_1 = x_2 \tag{6.5-21}$$

$$x_1 - \frac{N_1}{N_3} y_1 = x_2 - \frac{N_1}{N_3} y_2 \tag{6.5-22}$$

However, Equation (6.5-22) is a linear combination of Equations (6.5-20) and (6.5-21). Therefore, we only need to consider solving Equations (6.5-18)–(6.5-21) with unknowns $x_1$, $x_2$, $y_1$, and $y_2$, and then Equation (6.5-22) will be automatically satisfied.

Combining Equations (6.5-18)–(6.5-21) in matrix form,

$$\begin{bmatrix} 1 & 1 & 0 & 0 \\ 1 & \left(-\dfrac{N_1 N_4}{N_3 N_6}\right) & 0 & 0 \\ 1 & 1 & -1 & -1 \\ 1 & 0 & -1 & 0 \end{bmatrix} \begin{Bmatrix} x_1 \\ y_1 \\ x_2 \\ y_2 \end{Bmatrix} = \begin{Bmatrix} \omega_1 \\ 0 \\ 0 \\ 0 \end{Bmatrix} \tag{6.5-23}$$

Solving Equation (6.5-23) gives

$$x_1 = x_2 = \omega_1 \frac{N_1 N_4}{N_3 N_6 + N_1 N_4} \tag{6.5-24}$$

$$y_1 = y_2 = \omega_1 \frac{N_3 N_6}{N_3 N_6 + N_1 N_4} \tag{6.5-25}$$

The output rotational speed of the gear train is

$$\omega_7 = (\omega_5)_2 = x_2 - \left(\frac{N_1 N_4}{N_3 N_5}\right)_2 y_2 \quad \text{or} \quad \omega_7 = x_2 - \frac{N_1 N_5}{N_3 N_7} y_2 \tag{6.5-26}$$

Substituting Equations (6.5-24) and (6.5-25) in (6.5-26),

$$\omega_7 = \omega_1 \frac{N_1 N_4}{N_3 N_6 + N_1 N_4} - \frac{N_1 N_5}{N_3 N_7} \omega_1 \frac{N_3 N_6}{N_3 N_6 + N_1 N_4} \tag{6.5-27}$$

The speed ratio of the gear train is obtained from Equation (6.5-27) and yields after simplification

$$e_{7/1} = \frac{\omega_7}{\omega_1} = \frac{\left(1 - \dfrac{N_6 N_5}{N_4 N_7}\right)}{\left(1 + \dfrac{N_6 N_3}{N_4 N_1}\right)}$$  (6.5-28)

which is identical to the result obtained in Example 6.7.

## EXAMPLE 6.9  Kinematic Analysis of a Two-Stage Planetary Gear Train

Figure 6.22(a) shows the skeleton representation of a two-stage planetary gear train. The numbers of teeth are

$$N_1 = 24; \qquad N_2 = 60; \qquad N_3 = 12$$
$$N_4 = 48; \qquad N_5 = 12; \qquad N_6 = 24$$

**Figure 6.22** Two-stage planetary gear train: (a) skeleton representation, (b) mechanism [Model 6.22B].

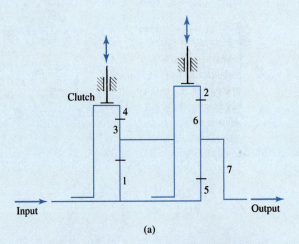

(a)

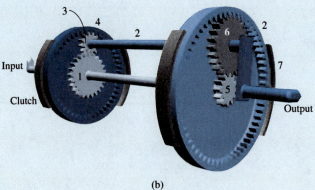

(b)

Determine the speed ratios of all components with respect to gear 1 when

(a) gear 2 is held fixed
(b) gear 4 is held fixed

### SOLUTION

The gear train is equivalent to combining two basic planetary gear train stages, each of type A, shown in Figure 6.14. The sun gears of both stages share the same rotational speed, and the crank of stage 1 is rigidly connected to the ring gear of stage 2.

Table 6.8 compares the numbering schemes presented in Figures 6.14 and 6.22(a).

**TABLE 6.8** Comparison of the Component Numbers in Figures 6.22(a) and 6.14

| Numbering of Components of Two-Stage Planetary Gear Train (Figure 6.22(a)) | Numbering of Components of Individual Stages (Figure 6.14) | |
| --- | --- | --- |
| | Stage 1 (Type A) | Stage 2 (Type A) |
| 1 | 1 | — |
| 2 | 2 | 4 |
| 3 | 3 | — |
| 4 | 4 | — |
| 5 | — | 1 |
| 6 | — | 3 |
| 7 | — | 2 |

Using the equations given in Figure 6.14, and the numbering scheme provided in Table 6.8, the rotational speeds of the components may be expressed as

$$\omega_1 = (\omega_1)_1 = x_1 + y_1 \tag{6.5-29}$$

$$\omega_2 = (\omega_2)_1 = x_1 \tag{6.5-30}$$

$$\omega_3 = (\omega_3)_1 = x_1 - \left(\frac{N_1}{N_3}\right)_1 y_1 = x_1 - \frac{N_1}{N_3} y_1 \tag{6.5-31}$$

$$\omega_4 = (\omega_4)_1 = x_1 - \left(\frac{N_1}{N_4}\right)_1 y_1 = x_1 - \frac{N_1}{N_4} y_1 \tag{6.5-32}$$

$$\omega_5 = (\omega_1)_2 = x_2 + y_2 \tag{6.5-33}$$

$$\omega_6 = (\omega_3)_2 = x_2 - \left(\frac{N_1}{N_3}\right)_2 y_2 = x_2 - \frac{N_5}{N_6} y_2 \tag{6.5-34}$$

$$\omega_7 = (\omega_2)_2 = x_2 \tag{6.5-35}$$

$$\omega_2 = (\omega_4)_2 = x_2 - \left(\frac{N_1}{N_4}\right)_2 y_2 = x_2 - \frac{N_5}{N_2} y_2 \tag{6.5-36}$$

Speed ratios of components of the gear train relative to gear 1 may be expressed as

$$e_{j/1} = \frac{\omega_j}{\omega_1}, \qquad j = 2, \ldots, 7 \tag{6.5-37}$$

Considering the case of $j = 7$, and substituting Equations (6.5-29) and (6.5-35) in Equation (6.5-37),

$$e_{7/1} = \frac{\omega_7}{\omega_1} = \frac{x_2}{x_1 + y_1} \tag{6.5-38}$$

In addition, the following equations are generated by considering the connections of the components between the stages:

- connection of the sun gears

$$\omega_1 = \omega_5 \quad \text{or} \quad (\omega_1)_1 = (\omega_1)_2 \quad \text{or} \quad x_1 + y_1 = x_2 + y_2 \tag{6.5-39}$$

- connection of the crank of stage 1 to the ring gear of stage 2

$$\dot{\omega_2} = (\omega_2)_1 = (\omega_4)_2 \quad \text{or} \quad x_1 = x_2 - \frac{N_5}{N_2}y_2 \tag{6.5-40}$$

Equations (6.5-29)–(6.5-40) are valid for parts (a) and (b) of this problem. We now proceed to calculate the speed ratios when either gear 2 or gear 4 is held fixed.

(a) If gear 2 is fixed, then from Equation (6.5-30)

$$\omega_2 = x_1 = 0 \tag{6.5-41}$$

Solving Equations (6.5-39)–(6.5-41) for $y_1$ and $y_2$ in terms of $x_2$ gives

$$y_1 = \frac{(N_2 + N_5)}{N_5}x_2; \qquad y_2 = \frac{N_2}{N_5}x_2 \tag{6.5-42}$$

Substituting Equations (6.5-41) and (6.5-42) in Equation (6.5-38) gives

$$e_{7/1} = \frac{x_2}{x_1 + y_1} = \frac{x_2}{\left[0 + \dfrac{(N_2 + N_5)}{N_5}x_2\right]} \tag{6.5-43}$$

$$= \frac{N_5}{N_2 + N_5} = \frac{12}{60 + 12} = \frac{1}{6}$$

Similar results of analyses for other components of the gear train are given in Table 6.9.

(b) If gear 4 is fixed, then from Equation (6.5-32)

$$\omega_4 = x_1 - \frac{N_1}{N_4}y_1 = 0 \tag{6.5-44}$$

TABLE 6.9 Speed Ratios of the Planetary Gear Train with Respect to Gear 1 in Figure 6.22(a)

| Condition →  Speed Ratio ↓ | Part (a)  (Gear 2 Fixed) | Part (b)  (Gear 4 Fixed) |
|---|---|---|
| $e_{2/1}$ | 0 | $\dfrac{N_1}{N_1 + N_4} = \dfrac{1}{3}$ |
| $e_{3/1}$ | $-\dfrac{N_1}{N_3} = -2$ | $-\dfrac{N_1(N_3 - N_4)}{N_3(N_1 + N_4)} = -1$ |
| $e_{4/1}$ | $-\dfrac{N_1}{N_4} = -\dfrac{1}{2}$ | 0 |
| $e_{5/1}$ | 1 | 1 |
| $e_{6/1}$ | $\dfrac{N_5(N_6 - N_2)}{N_6(N_2 + N_5)} = -\dfrac{1}{4}$ | $\dfrac{N_5 N_6(N_1 + N_4) + N_2(N_1 N_6 - N_4 N_5)}{N_6(N_1 + N_4)(N_2 + N_5)} = \dfrac{1}{6}$ |
| $e_{7/1}$ | $\dfrac{N_5}{N_2 + N_5} = \dfrac{1}{6}$ | $\dfrac{N_4 N_5 + N_1(N_2 + N_5)}{(N_1 + N_4)(N_2 + N_5)} = \dfrac{4}{9}$ |

Solving Equations (6.5-39), (6.5-40), and (6.5-44) for $x_2$, $y_1$, and $y_2$ in terms of $x_1$ gives

$$x_2 = \left[1 + \frac{N_4 N_5}{N_1(N_2 + N_5)}\right] x_1; \qquad y_1 = \frac{N_4}{N_1} x_1$$

$$y_2 = \frac{N_2 N_4}{N_1(N_2 + N_5)} x_1$$

(6.5-45)

Substituting Equations (6.5-45) in Equation (6.5-38) gives

$$e_{7/1} = \frac{x_2}{x_1 + y_1} = \frac{\left[1 + \dfrac{N_4 N_5}{N_1(N_2 + N_5)}\right] x_1}{\left(x_1 + \dfrac{N_4}{N_1} x_1\right)}$$

(6.5-46)

$$= \frac{N_4 N_5 + N_1(N_2 + N_5)}{(N_1 + N_4)(N_2 + N_5)} = \frac{48 \times 12 + 24(60 + 12)}{(24 + 48)(60 + 12)} = \frac{4}{9}$$

**Model 6.22B**
Two-Stage
Planetary Gear
Train

Similar results of analyses for other components of the gear train are given in Table 6.9. Figure 6.22(b) shows another illustration of the same gear train.

# 6.6 DIFFERENTIALS

**Model 6.23**
Differential,
Bevel Gears

A *differential* is a mechanism that permits transmission of input power to two separate outputs at unequal rotational speeds. A planetary gear train may be employed to create such a mechanism and can incorporate either spur gears or bevel gears. Consider the gear train illustrated in Figure 6.23. A corresponding sectional top view is given in Figure 6.24.

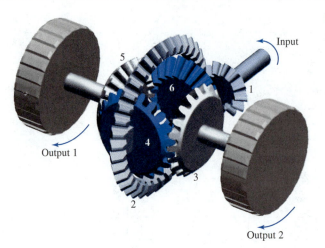

**Figure 6.23** Differential gear train [Model 6.23].

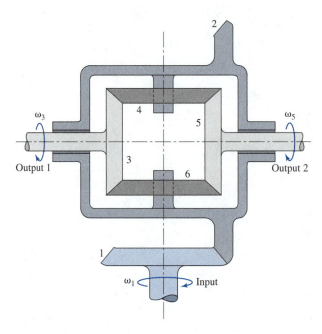

**Figure 6.24** Top view of differential gear train.

Included in these figures are four bevel gears, numbered 3 through 6, which combine to form a differential.

Gear trains of the configuration shown in Figure 6.23 are commonly employed in drive trains of vehicles. Gear 1, referred to as the *pinion*, is the input to the gear train. Ring gear 2 is attached to the planet carrier. Gears 1 and 2 may be plain bevel gears as shown. Alternatively, these gears could be either a spiral bevel set or a hypoid gear set, as illustrated in Figures 5.9(b) and 5.10, respectively. Gears 4 and 6 are planet gears because their axes of rotation are not fixed. These planet gears are allowed to rotate with respect to gear 2, and they revolve about the axis of rotation of gear 2. Gears 3 and 5 are attached to the same shafts as

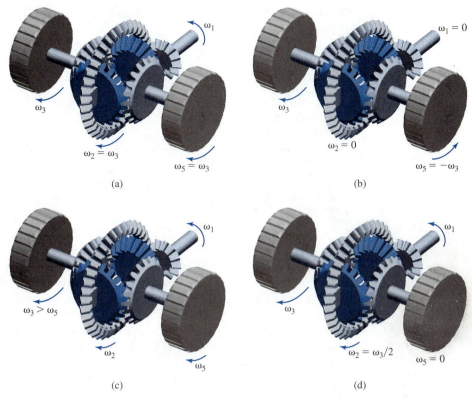

**Figure 6.25** Motions of a differential gear train [Model 6.23]: (a) equal outputs (no slip), (b) wheels rotate in opposite directions (zero input), (c) cornering or partial slip, (d) full slip.

the two drive wheels and share the same axis of rotation as gear 2. Gears 3 and 5 have the same size, and so are gears 4 and 6.

The differential, as part of the gear train, allows the drive wheels to rotate at either the same or unequal speeds. Four typical motions that may be generated are shown in Figure 6.25. In all of these cases, it is considered that there is one input and two outputs. When a vehicle travels straight ahead, as illustrated in Figure 6.25(a), both drive wheels, as well as gears 3 and 5, have the same rotational speed. In this case, gears 2, 3, 4, 5, and 6 all rotate about a common axis as though they formed a rigid body, and there is no meshing between these gears. For the three other cases shown in Figure 6.25, there is meshing between gears 3, 4, 5, and 6. Figure 6.25(b) shows the case when gear 1 is stationary, and gears 3 and 5 rotate in opposite directions. Figure 6.25(c) corresponds to the case where the vehicle is turning to the right. Here, the drive wheel connected to gear 5 moves on a larger radius of curvature than the drive wheel connected to gear 3. To avoid slippage of either wheel with the ground, gear 5 must turn at a faster rate than gear 3. The differential permits this difference of the two output speeds. Figure 6.25(d) illustrates the case where one of the outputs is stopped, while the other output continues to rotate.

The tabular method presented in Section 6.4 is now employed to determine the magnitudes of speeds under different driving conditions. Following the first three general steps

**TABLE 6.10** Tabular Analysis of the Bevel Differential Gear Train Shown in Figure 6.23

| Component → <br> Operation ↓ | Gear 1 <br> $\omega_1$ | Gear 2 <br> $\omega_2$ | Gear 3 <br> $\omega_3$ | Gear 5 <br> $\omega_5$ |
|---|---|---|---|---|
| Gears 4 and 6 Are Stationary with Respect to Gear 2, Gear 1 Turns at $x$ rpm | $x$ | $\dfrac{N_1}{N_2}x$ | $\dfrac{N_1}{N_2}x$ | $\dfrac{N_1}{N_2}x$ |
| Gear 1 Is Fixed, Gear 5 Turns at $y$ rpm | 0 | 0 | $-y$ | $y$ |
| Absolute Rotational Speeds | $x$ | $\dfrac{N_1}{N_2}x$ | $\dfrac{N_1}{N_2}x - y$ | $\dfrac{N_1}{N_2}x + y$ |

for a tabular analysis, Table 6.10 was constructed. In this table, gear 1 is listed along with the other gears, even though their axes of rotation are perpendicular to one another. When comparing the speeds of gear 1 with those of others in the gear train, only the magnitudes should be considered.

When the vehicle is travelling along a straight line, the two output rotational speeds are equal. This requires that $y = 0$, and since $x = \omega_1$, then

$$\omega_3 = \omega_5 = \frac{N_1}{N_2}x = \frac{N_1}{N_2}\omega_1 = \omega_2 \qquad (6.6\text{-}1)$$

When the vehicle is in a turn, the two drive wheels no longer have the same rotational speed. Gear 2 rotates at the average value of the two output speeds, and one wheel, say, connected to gear 5, has a greater speed, while the other, connected to gear 3, has a lower speed. Here, variable $y$ takes on a finite value, and $2y$ is the difference of the rotational speeds of the two outputs. If gear 3 stops turning, then

$$\omega_3 = 0 = \frac{N_1}{N_2}x - y \qquad \text{or} \qquad y = \frac{N_1}{N_2}x \qquad (6.6\text{-}2)$$

The rotational speed of gear 5 is then

$$\omega_5 = \frac{N_1}{N_2}x + y = 2\frac{N_1}{N_2}x = 2\frac{N_1}{N_2}\omega_1 = 2\omega_2 \qquad (6.6\text{-}3)$$

As indicated, gear 5 rotates at twice the average speed of gear 2.

The condition of having one of the outputs stationary can occur when one drive wheel, say, connected to gear 3, is on dry pavement, while the other is freely spinning on a slippery surface such as ice. In this instance, minimal torque resistance is offered by the spinning wheel, and negligible torque can be transmitted to the other drive wheel. The result is that the wheel on ice spins with twice the average speed of gear 2, while the other wheel remains stationary.

The above problem can be relieved somewhat by employing a *limited-slip differential*. A schematic of such a system is shown in Figure 6.26. The arrangement of gears is the same as a conventional differential (Figure 6.24). However, in addition there are friction

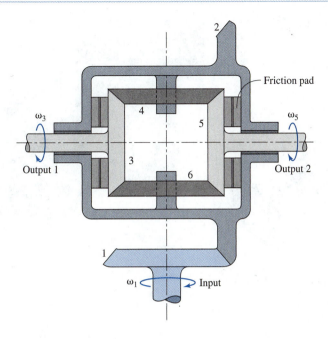

**Figure 6.26** Limited-slip differential.

pads placed between the planet carrier (component 2) and gears 3 and 5. In this instance, when one of the drive wheels is on ice, a limited amount of torque may still be transmitted to the other wheel. The amount of torque that can be transmitted to one wheel when the other has no traction is dependent on the frictional torque generated through the friction pads. A disadvantage associated with a limited-slip differential is that energy is lost, and wear occurs in the friction pads when relative motion occurs between the output gears and the carrier, that is, each time the vehicle changes direction during normal operation.

The arrangement of gears illustrated in Figure 6.23 is employed for rear wheel drive vehicles, where the input connected to gear 1 is perpendicular to the two outputs, gears 3 and 5. Alternatively, Figure 6.27 shows the arrangement commonly used in front wheel drive vehicles, where the engine and transmission (not shown) drive sun gear 1 of the planetary gear train while ring gear 4 is held fixed. This Type A planetary gear train (Figure 6.14) provides one final speed reduction prior to delivering motion to the differential. Component 2 is both the crank of the Type A planetary gear train and the planet carrier of the differential. It rotates at the average of the two output speeds of the differential. Gears 6 and 8 are planet gears of the differential, and gears 5 and 7 are connected to the two output wheels.

A differential may also be created by using a Type I planetary gear train (Figure 6.14) for which

**Model 6.27**
Front Wheel
Drive Gear Train

$$N_5 = 2N_1 \qquad (6.6\text{-}4)$$

Employing Equation (6.6-4) in the equations provided in Figure 6.14,

$$\omega_1 = x + y \qquad (6.6\text{-}5)$$

$$\omega_2 = x \qquad (6.6\text{-}6)$$

$$\omega_5 = x + \frac{N_1}{N_5}y = x + \frac{N_1}{2N_1}y = x + \tfrac{1}{2}y \qquad (6.6\text{-}7)$$

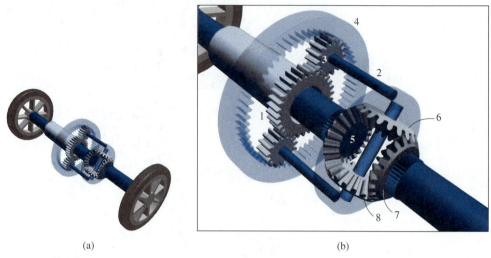

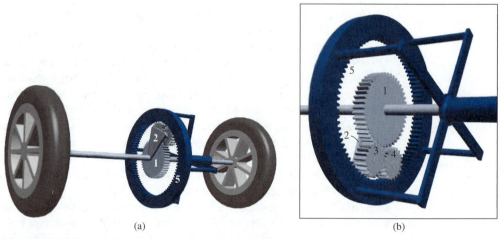

(a)                                    (b)

**Figure 6.27** Front wheel drive gear train [Model 6.27].

(a)                                    (b)

**Figure 6.28** Differential incorporating spur gears [Model 6.28].

Therefore, if the input to this gear train is supplied to gear 5, then as required, its value of rotational speed is the average of the two outputs, $\omega_1$ and $\omega_2$.

Figure 6.28 shows such a differential using spur gears. Differentials that incorporate spur gears instead of bevel gears have the advantage of a smaller width of the gear train. For this reason, they are often employed in four wheel drive vehicles, which require a differential for the front wheels, a differential for the rear pair of wheels, and the option to have differential action between the front and rear wheel pairs.

**Model 6.28**
Differential,
Spur Gears

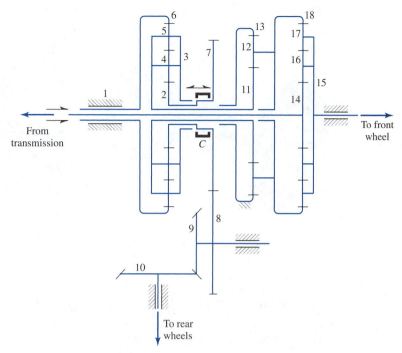

**Figure 6.29** Four wheel drive gear train.

Figure 6.29 shows the skeleton form of an all-time four wheel drive gear train. Input from the transmission is supplied to gear 6. One differential, consisting of gears 2, 4, 5, and 6 and crank arm 3, allows a differential action between the front pair and rear pair of wheels. A second differential, consisting of gears 14, 16, 17, and 18 and crank arm 15, allows differential motion between the two front wheels. A third differential (not shown) is employed for the two rear wheels. Gears 11, 12, and 13 are part of a planetary gear train that provides a speed reduction to the front pair of wheels. Gears 7, 8, 9, 10, and the gears at the rear differential provide the same speed reduction to the rear wheels.

By moving collar $C$ from the position shown and locking together component 3 and gear 7, there can be no differential action between the front and rear pairs of wheels. In this instance, if neither of the front wheels were able to provide traction, power could still be transmitted to the rear pair of wheels.

**Model 6.30**
Torsen
Differential

Another type of differential is the *Torsen*™ *differential*. It is shown in Figure 6.30. With this differential it is possible to provide improved driving torque to each of the driving wheels without the use of friction pads. Two or more planet gears are in mesh with the central helical gears, also called the *side gears*. The planet gears are interconnected by means of straight spur gears. The number of planet gears employed in a specific design is a function of the required torque capacity. A comprehensive description of the operation of the Torsen differential was prepared by Chocholek [8].

(a)                                                              (b)

**Figure 6.30**  Torsen differential [Model 6.30].

## EXAMPLE 6.10  Speed Ratio of a Gear Train Incorporating Bevel Gears

In the planetary gear train shown in Figure 6.31, the input speed is $\omega_2 = 100$ rpm CCW. Determine the magnitude and direction of the output speed $\omega_{10}$. The numbers of teeth are

$$N_2 = 40; \quad N_4 = 30; \quad N_5 = 25; \quad N_6 = 120$$
$$N_7 = 50; \quad N_8 = 20; \quad N_9 = 70; \quad N_{10} = 20$$

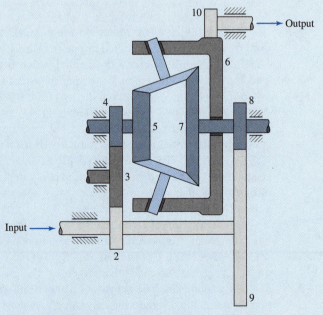

**Figure 6.31**  Planetary gear train incorporating bevel gears.

SOLUTION

The input shaft drives the two gears of the planetary gear train. Their rotational speeds are

$$\omega_5 = \omega_4 = \frac{N_2}{N_4}\omega_2 = \frac{40}{30} \times 100 = 133 = 133 \text{ rpm CCW}$$

$$\omega_7 = \omega_8 = -\frac{N_9}{N_8}\omega_2 = -\frac{70}{20} \times 100 = -350 = 350 \text{ rpm CW}$$

where it has been assumed that positive rotational speeds are in the counterclockwise direction.

Employing the first three steps for the tabular method, Table 6.11 was constructed (note that if the carrier is fixed, gears 5 and 7 turn in opposite directions).

**TABLE 6.11** Tabular Analysis of the Bevel Differential Gear Train Shown in Figure 6.31

| Component → <br> Operation ↓ | Gear 5 <br> $\omega_5$ | Gear 7 <br> $\omega_7$ | Gear 6 <br> $\omega_6$ |
|---|---|---|---|
| Gears 5, 6, and 7 Turn at $x$ rpm | $x$ | $x$ | $x$ |
| Gear 6 Is Fixed, Gear 5 Turns at $y$ rpm | $y$ | $-\dfrac{N_5}{N_7}y$ | $0$ |
| Absolute Rotational Speeds | $x + y$ | $x - \dfrac{N_5}{N_7}y$ | $x$ |

Substituting values from Table 6.11,

$$\omega_5 = 133 = x + y$$

$$\omega_7 = -350 = x - \frac{N_5}{N_7}y = -350 = x - \frac{25}{50}y$$

Solving the above two equations for $x$ and $y$,

$$x = -189; \qquad y = 322$$

and therefore

$$\omega_6 = x = -189 = 189 \text{ rpm CW}$$

The output rotational speed is then

$$\omega_{10} = -\frac{N_6}{N_{10}}\omega_6 = -\frac{120}{20} \times (-189) = 1{,}134 = 1{,}134 \text{ rpm CCW}$$

# 6.7 HARMONIC DRIVES

A *harmonic drive* in exploded form is shown in Figure 6.32. This drive incorporates a unique means of power transmission in which flexibility of one of its components is utilized. Large reductions or increases of rotational speed are possible with such drives. Due

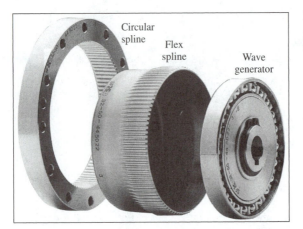

**Figure 6.32** Harmonic drive (courtesy of HD Systems, Inc., Hauppauge, NY).

Circular spline

Flex spline

Wave generator

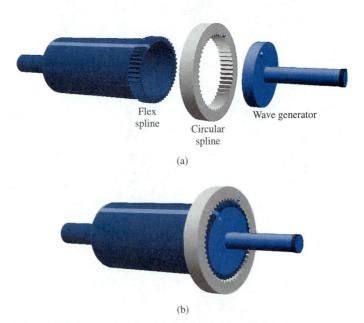

Flex spline

Circular spline

Wave generator

(a)

(b)

**Figure 6.33** Harmonic drive [Model 6.33]: (a) exploded view, (b) assembled unit.

**Model 6.33**
Harmonic Drive

to their relatively small, in-line configuration between the driver and driven components, they have been useful in a variety of applications, including drives of the axes in robots.

Figure 6.33 shows another illustration of a harmonic drive. The *wave generator* has a surface in the shape of an ellipse. The elliptical shape is inserted into a flexible component, called the *flex spline*, which is forced to be deflected into the same shape. Through these deflections, external teeth on the flex spline mesh with a rigid internal circular gear known as the *circular spline*. Teeth on the flex spline and circular spline simultaneously mesh at two locations, 180° apart from one another. This arrangement leads to a balanced torque reaction between the input and output about the central axis of the drive. Also, with this form of meshing, a precisely machined unit can be made to operate with essentially zero backlash.

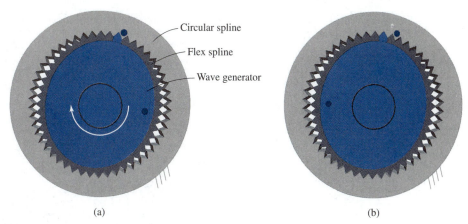

**Figure 6.34** Harmonic drive [Model 6.33].

Although teeth on both splines have the same pitch, the circular spline has two more teeth than the flex spline, that is,

$$N_c = N_f + 2 \qquad (6.7\text{-}1)$$

where $N_c$ and $N_f$ are the numbers of teeth on the circular spline and flex spline, respectively.

The animation provided through **[Model 6.33]** has two distinct motions. For the first motion, the wave generator is the input component, the flex spline is the output component, and the circular spline is held fixed. As the wave generator rotates, the flex spline forms a travelling deflection wave. Figure 6.34 shows two configurations of the harmonic drive. In Figure 6.34(b), the wave generator has rotated 180° clockwise from the position shown in Figure 6.34(a). Notice that the marking on the flex spline has shifted one tooth with respect to the circular spline as a result of rotation of the wave generator. The rotational direction of shift is opposite to that of the travelling deflection wave, that is, for each rotation of the wave generator in the clockwise direction, the flex spline rotates an equivalent of two teeth in the counterclockwise direction. Therefore, the first speed ratio generated is

$$e_{f/w} = \frac{\omega_f}{\omega_w} = -\frac{2}{N_f} \qquad (6.7\text{-}2)$$

where $\omega_f$ and $\omega_w$ are rotational speeds of the flex spline and wave generator, respectively, and the negative sign indicates that the flex spline and wave generator turn in opposite directions.

For the harmonic drive illustrated in Figure 6.33

$$N_f = 48; \qquad N_c = 50$$

and thus for this unit

$$e_{f/w} = -\frac{2}{N_f} = -\frac{1}{24}$$

Any one of the three components of a harmonic drive can be selected for the input, the output, or the fixed component. Once the selection has been made, the corresponding speed ratio may be determined using a tabular method, similar to that employed for planetary

**TABLE 6.12** Tabular Analysis of the Harmonic Drive Shown in Figure 6.33

| Component → <br><br> Operation ↓ | Wave Generator <br> $\omega_w$ | Flex Spline <br> $\omega_f$ | Circular Spline <br> $\omega_c$ |
|---|---|---|---|
| All Components Turn at $x$ rpm | $x$ | $x$ | $x$ |
| Circular Spline Is Fixed, Wave Generator Turns at $y$ rpm | $y$ | $-\dfrac{2}{N_f}y$ | $0$ |
| Absolute Rotational Speeds | $x + y$ | $x - \dfrac{2}{N_f}y$ | $x$ |

gear trains. Table 6.12 provides such a table. Included in the table is a row for which all three components have the same rotational speed of $x$ rpm. In the subsequent row, the circular spline has zero rotational speed, the wave generator is given $y$ rpm, and the corresponding value of rotational speed of the flex spline is provided. Absolute values of rotational speeds that are a superposition of the two motions are also listed.

For the second portion of the animation provided through [**Model 6.33**], the flex spline is prevented from rotating, and hence, from Table 6.12,

$$\omega_f = x - \frac{2}{N_f}y = 0$$

Therefore

$$x = \frac{2}{N_f}y \qquad \text{or} \qquad y = \frac{N_f}{2}x \tag{6.7-3}$$

In addition, if the wave generator is the input component, and the circular spline is the output component, then the second speed ratio generated is

$$e_{c/w} = \frac{\omega_c}{\omega_w} = \frac{x}{x+y} \tag{6.7-4}$$

Combining Equations (6.7-4), (6.7-3), and (6.7-1),

$$e_{c/w} = \frac{x}{x+y} = \frac{x}{\left(x + \dfrac{N_f}{2}x\right)} = \frac{2}{2+N_f} = \frac{2}{N_c} \tag{6.7-5}$$

For the harmonic drive shown in Figure 6.33

$$e_{c/w} = \frac{2}{N_c} = \frac{2}{50} = \frac{1}{25}$$

Here, the circular spline and wave generator rotate in the same direction.

Harmonic drives may be combined in multiple stages, similar to planetary gear trains, to provide even greater speed ratios.

## EXAMPLE 6.11 Analysis of a Two-Stage Harmonic Drive

Figure 6.35 shows the cross section of a two-stage harmonic drive. The numbers of teeth are

$$N_{f_1} = 104; \qquad N_{c_1} = 106; \qquad N_{f_2} = 108; \qquad N_{c_2} = 110$$

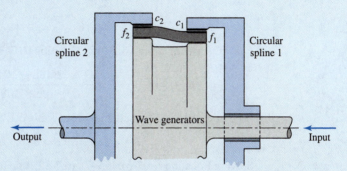

**Figure 6.35** Two-stage harmonic drive.

Determine the speed ratio of

(a) circular spline 1 with respect to the wave generator if circular spline 2 is held fixed
(b) circular spline 2 with respect to the wave generator if circular spline 1 is held fixed

### SOLUTION

The drive is equivalent to combining two harmonic drives. The wave generators of both stages share the same rotational speed, and the flex splines of both stages are rigidly connected. We employ Table 6.12 to generate the following equations:

- connection of wave generators:

$$x_1 + y_1 = x_2 + y_2 \qquad (6.7\text{-}6)$$

- connection of flex splines:

$$x_1 - \frac{2}{N_{f_1}} y_1 = x_2 - \frac{2}{N_{f_2}} y_2 \qquad (6.7\text{-}7)$$

where subscripts 1 and 2 indicate the stage number. Equations (6.7-6) and (6.7-7) are valid for parts (a) and (b) of this problem. We now proceed to calculate the speed ratios when either circular spline 1 or circular spline 2 is held fixed.

(a) If circular spline 2 is fixed, then from Table 6.12

$$\omega_{c_2} = x_2 = 0 \qquad (6.7\text{-}8)$$

Solving Equations (6.7-6)–(6.7-8) and (6.7-1) for $x_1$ and $y_1$ in terms of $y_2$,

$$x_1 = \frac{2(N_{f_2} - N_{f_1})}{N_{c_1} N_{f_2}} y_2; \qquad y_1 = \frac{N_{c_1} N_{f_2}}{N_{c_1} N_{f_2}} y_2 \qquad (6.7\text{-}9)$$

The speed ratio may be expressed as

$$e_{c_1/w} = \frac{\omega_{c_1}}{\omega_w} = \frac{x_1}{x_1 + y_1} \qquad (6.7\text{-}10)$$

Substituting Equations (6.7-9) in Equation (6.7-10), and simplifying,

$$e_{c_1/w} = \frac{2(N_{f_2} - N_{f_1})}{N_{c_1} N_{f_2}} = \frac{2(108 - 104)}{106 \times 108} = \frac{1}{1,431} \qquad (6.7\text{-}11)$$

(b)  If circular spline 1 is fixed, then from Table 6.12

$$\omega_{c_1} = x_1 = 0 \qquad (6.7\text{-}12)$$

Solving Equations (6.7-6), (6.7-7), (6.7-12), and (6.7-1) for $x_2$ and $y_1$ in terms of $y_2$,

$$x_2 = \frac{2(N_{f_1} - N_{f_2})}{N_{c_1} N_{f_2}} y_2; \qquad y_1 = \frac{N_{f_1} N_{c_2}}{N_{c_1} N_{f_2}} y_2 \qquad (6.7\text{-}13)$$

The speed ratio may be expressed as

$$e_{c_2/w} = \frac{\omega_{c_2}}{\omega_w} = \frac{x_2}{x_1 + y_1} \qquad (6.7\text{-}14)$$

Substituting Equations (6.7-12) and (6.7-13) in Equation (6.7-14), and simplifying,

$$e_{c_2/w} = \frac{2(N_{f_1} - N_{f_2})}{N_{f_1} N_{c_2}} = \frac{2(104 - 108)}{104 \times 110} = -\frac{1}{1,430} \qquad (6.7\text{-}15)$$

## 6.8  TORQUE RELATIONS IN GEARBOXES

In a gearbox, torque is transmitted from the input to the output shaft, and if there is a difference in rotational speed between input and output, there will be a related change in torque. This means that the input and output torques are unequal, and to maintain static equilibrium, a reaction torque must be applied to the gearbox housing. Figure 6.36 shows the torques acting on a typical gearbox. The *input torque* and *output torque* are designated as $T_i$ and $T_o$, respectively, and the input and output rotational speeds are $\omega_i$ and $\omega_o$. The input and output powers (see Appendix D, Section D.10.3) to and from the gearbox are

$$\boxed{P_i = T_i \omega_i} \qquad (6.8\text{-}1)$$

$$\boxed{P_o = -T_o \omega_o} \qquad (6.8\text{-}2)$$

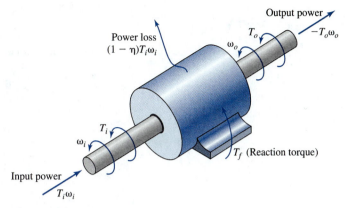

Figure 6.36 Torques on a gearbox.

where it has been recognized that the output torque and output rotational speed must have opposite signs.

Due to frictional losses, the power output is less than the power input to the gear train. The *mechanical efficiency* of the gearbox, $\eta$, is defined as the ratio of the output power to the input power:

$$\eta = \frac{P_o}{P_i} = -\frac{T_o \omega_o}{T_i \omega_i}$$

or

$$\boxed{T_o \omega_o = -\eta T_i \omega_i} \tag{6.8-3}$$

In addition, the sum of the moments applied to the gearbox must be zero, that is,

$$\boxed{T_i + T_o + T_f = 0} \tag{6.8-4}$$

where $T_f$ is the *reaction torque* acting on the gearbox, which is required to keep the gearbox housing stationary. The following example illustrates the use of the above equations.

## EXAMPLE 6.12 Torque Analysis of a Gear Train

Determine the torque on the housing of the gearbox in Figure 6.13, with the ring gear fixed, while under the following conditions:

$$\text{input power} = P_i = 30{,}000 \text{ N-m/sec} = 30{,}000 \text{ watts} = 30 \text{ kW}$$

$$\omega_i = \omega_1 = 2{,}000 \text{ rpm}; \qquad \eta = 0.98$$

The speed ratio of the gearbox was previously determined (see Equation (6.4-1)) to be

$$e_{o/i} = \frac{\omega_o}{\omega_i} = e_{2/1} = 0.30$$

From Equation (6.8-1)

$$T_i = \frac{P_i}{\omega_i} = \frac{30,000 \text{ N-m/sec}}{\left(2,000 \times \dfrac{2\pi}{60} \text{ rad/sec}\right)} = 143 \text{ N-m}$$

From Equation (6.8-3)

$$T_o\omega_o = T_o e_{o/i}\omega_i = -\eta T_i \omega_i$$

and therefore

$$T_o = -\frac{\eta T_i}{e_{o/i}} = -\frac{0.98 \times 143}{0.30} = -467 \text{ N-m}$$

Combining the above results with Equation (6.8-4),

$$T_f = T_4 = -T_o - T_i = 467 - 143 = 324 \text{ N-m}$$

Since $T_f$ is positive, the torque on link 4 is in the same direction as the input rotational speed.

# PROBLEMS

**P6.1**  For the gear train shown in Figure P6.1, determine
  (a) the speed ratio $e_{6/2}$
  (b) the output rotational speed, $\omega_6$, if the input rotational speed, $\omega_2$, is 75 rpm CW

  $N_2 = 15$;     $N_3 = 25$;     $N_4 = 20$

  $N_5 = 15$;     $N_6 = 15$

**P6.2**  For a reverted gear train (Figure 5.33), gears 2 and 3 have a module of 5 mm, and gears 4 and 5 have a module of 3 mm. All gears are straight spur. Determine suitable tooth numbers for the gears if the speed ratio is to be
  (a) 0.25
  (b) approximately 0.42

**P6.3**  Specify the numbers of teeth for the gears shown in Figure 6.8, having speed ratios of approximately 0.8,

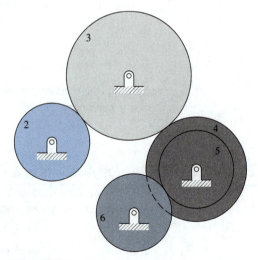

**Figure P6.1**

0.5, and 0.3. The smallest gear must have at least 20 teeth, and the module of all teeth is 4 mm.

**P6.4** Figure P6.4 illustrates a three-speed transmission. Gears 4, 5, and 6 are free to spin about their shaft in the configuration shown, while gears 3, 7, 8, 9, 10, and 11 are keyed to their respective shafts. Gear 10 is an idler gear whose center is not in line with those of gears 6 and 11. $C_1$ and $C_2$ are synchronizers that fix one free gear to the shaft each time, so that power is transmitted from the input shaft to the output shaft. The input and output shafts are collinear. Using gears of diametral pitch 12 in$^{-1}$, and using a centre-to-center distance between the input shaft and countershaft of 4 inches, specify the number of teeth of all gears to obtain the following speed ratios: $+1$, $+0.5$, $+0.2$, $-0.2$ (reverse), subject to the constraint that the minimum number of teeth in any gear is 16.

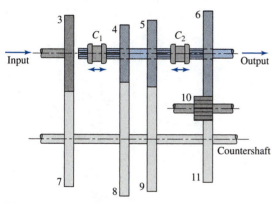

**Figure P6.4**

**P6.5** Using the tabular method, derive the expressions for the component speeds for a type I planetary gear train as given in Figure 6.14.

**P6.6** Using the tabular method, derive the expressions for the component speeds in a type J planetary gear train as given in Figure 6.14.

**P6.7** Using the tabular method, derive the expressions for the component speeds in a type K planetary gear train as given in Figure 6.14.

**P6.8** For the gear train of Figure P6.8, calculate the speed ratio and determine the speed of rotation of output gear 10, given that gear 3 is driven at 200 rpm in the direction shown.

$$N_2 = 20; \qquad N_3 = 10; \qquad N_4 = 60$$

$$N_5 = 72; \qquad N_6 = 15; \qquad N_7 = 22$$
$$N_8 = 16; \qquad N_9 = 12; \qquad N_{10} = 10$$

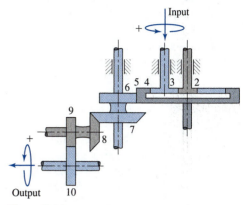

**Figure P6.8**

**P6.9** Determine the expression for the speed ratio of each of the following planetary gear trains (refer to Figure 6.14):

(a) type D, $\omega_1 = 0$, $\omega_2 = $ input, $\omega_7 = $ output
(b) type J, $\omega_7 = 0$, $\omega_1 = $ input, $\omega_2 = $ output
(c) type C, $\omega_7 = 0$, $\omega_1 = $ input, $\omega_2 = $ output
(d) type L, $\omega_1 = 0$, $\omega_2 = $ input, $\omega_6 = $ output

**P6.10** In the gear train of Figure P6.10, shaft $A$ rotates at 200 rpm and shaft $B$ rotates at 300 rpm in the directions indicated. Determine the speed of shaft $C$ and its direction of rotation.

$$N_2 = 35; \qquad N_3 = 25; \qquad N_4 = 14$$
$$N_5 = 46; \qquad N_6 = 20; \qquad N_7 = 16$$

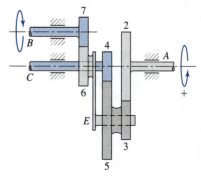

**Figure P6.10**

**P6.11** For a planetary gear train type A, the sun and ring gears have $N_1 = 20$ and $N_4 = 70$ teeth, respectively.

(a) If the speed of the ring gear is 500 rpm CW, at what speed must the sun gear be driven if the crank is to rotate at

(i)  75 rpm CW

(ii) 75 rpm CCW

(b) Determine the maximum number of equally spaced planet gears that can be employed in this application.

**P6.12** Figure P6.12 shows an epicyclic gear train called *Ferguson's paradox*. Gears 2, 3, and 4 are loosely attached to their respective shafts while gear 5 is fixed.

(a) Find the number of rotations the crank has to turn so that gear 4 rotates five times in the direction shown. How many times does gear 3 rotate, and in which direction?

(b) Determine the number of turns that gear 2 makes about its own shaft.

$$N_2 = 15; \qquad N_3 = 80$$
$$N_4 = 81; \qquad N_5 = 82$$

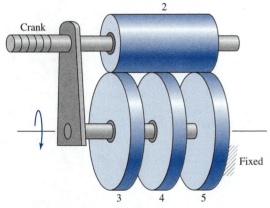

**Figure P6.12**

**P6.13** In Figure P6.13, $C$ and $D$ represent band brakes that can be used to stop the rotation of either arm $E$ or gear 4, one at a time. Determine the speed and direction of shaft $B$ when shaft $A$ is rotating at 1,000 rpm CW, while

(a) brake $C$ holds arm $E$ fixed

(b) brake $D$ holds gear 4 fixed

$$N_2 = 90; \qquad N_3 = 32$$
$$N_5 = 94; \qquad N_6 = 28$$

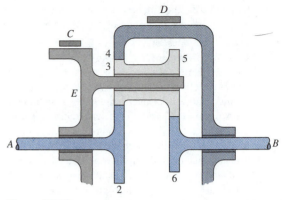

**Figure P6.13**

**P6.14** For the gear train shown in Figure P6.14, gear 2 has a module of 2.0 mm with 75 teeth; gear 5 has a module of 4.0 mm with 50 teeth; and gear 4 has 40 teeth. Determine the number of teeth on gear 3 and speed ratio of the gear train.

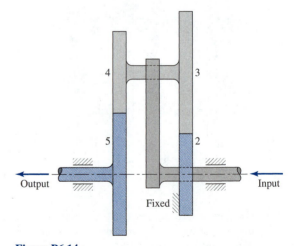

**Figure P6.14**

**P6.15** If an automobile is making a right-hand turn at 30 km/hr, determine the rotational speed of the differential carrier. The radius of curvature of the curve is 40 m to the center of the automobile, and the automobile tread (the distance between the two drive wheels) is 1.60 m. The outside diameter of the wheels is 850 mm. The differential gear train of the automobile is shown in Figure 6.24.

$$N_1 = 30; \qquad N_2 = 75$$
$$N_3 = N_5 = 30; \qquad N_4 = N_6 = 20$$

**P6.16** Determine the three speed ratios, $e_{2/1}, e_{5/1}$, and $e_{8/1}$, of the planetary gear train shown in Figure P6.16. Employ

(a) the tabular method, and obtain results with one table

(b) the combination of the basic types in Figure 6.14

$$N_1 = 15; \quad N_3 = 45; \quad N_4 = 105$$
$$N_5 = 13; \quad N_6 = 10; \quad N_7 = 27$$
$$N_8 = 87$$

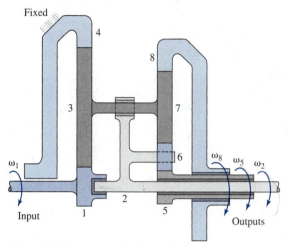

**Figure P6.16**

**P6.17** Figure P6.17 illustrates a two-speed transmission. The two output speeds are obtained by alternatively fixing gears 3 and 7 by using band brakes $C$ and $D$, respectively. Determine the expressions for the two possible values of the speed ratio, $e_{8/1}$.

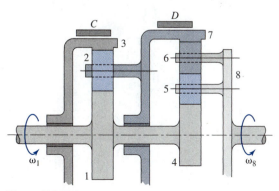

**Figure P6.17**

**P6.18** In the gear train of Figure P6.18, gears 1 and 2 are cut from the same casting and revolve freely about the arms of the planet carrier $C$. Gears 5 and 6 are also cut from the same casting and are freely attached to shaft $A$. Gears 3 and 4 are keyed to shaft $A$. For an output speed of 500 rpm in the direction shown, determine the sense and speed of the input shaft rotation.

$$N_1 = 42; \quad N_2 = 44; \quad N_3 = 40$$
$$N_4 = 14; \quad N_5 = 38; \quad N_6 = 150$$
$$N_7 = 50$$

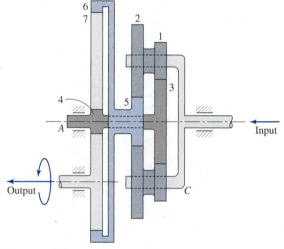

**Figure P6.18**

**P6.19** The three-stage planetary gear train illustrated in Figure P6.19 has the following two input rotational speeds: $\omega_1 = 100$ rpm CW, $\omega_{19} = 6,000$ rpm CW.

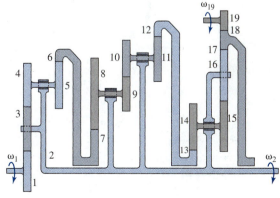

**Figure P6.19**

Determine the output rotational speed, $\omega_2$, using the tabular method.

$$N_1 = 20; \qquad N_3 = 23; \qquad N_4 = 21$$
$$N_5 = 25; \qquad N_6 = 115; \qquad N_7 = 43$$
$$N_8 = 38; \qquad N_9 = 17; \qquad N_{10} = 19$$
$$N_{11} = 23; \qquad N_{12} = 117; \qquad N_{13} = 38$$
$$N_{14} = 25; \qquad N_{15} = 23; \qquad N_{16} = 28$$
$$N_{17} = 110; \qquad N_{18} = 120; \qquad N_{19} = 10$$

**P6.20** A two-stage planetary gear train incorporates type A for the first stage and type H for the second. The ring gear of the first stage is connected to the crank of the second stage, while the crank of the first stage is connected to the ring gear 1 of the second stage. If ring gear 5 of the second stage is fixed, write down, but do not solve, the four governing equations for the motion in terms of $x_1$, $y_1$, $x_2$, and $y_2$ according to the equations given in Figure 6.14, and determine the condition for the mechanism to function with a nonzero mobility.

**P6.21** In the operation of the gear train shown in Figure P6.21, ring gear 1 is fixed. Determine a relation involving $N_6$ and $N_7$ to obtain a speed ratio of $-0.48$.

$$N_1 = 100; \qquad N_2 = 30; \qquad N_3 = 10$$
$$N_4 = 85; \qquad N_5 = 88; \qquad N_8 = 22$$

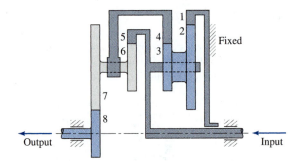

**Figure P6.21**

**P6.22** Determine the output rotational speed, $\omega_2$, for the gear train shown in Figure P6.22 by
(a) using the tabular method
(b) considering the gear train to be a combination of the basic types illustrated in Figure 6.14

$$N_1 = 20; \qquad N_3 = 25; \qquad N_4 = 70$$

$$N_5 = 15; \qquad N_6 = 15; \qquad N_7 = 35$$
$$N_8 = 80; \qquad \omega_1 = 200 \text{ rpm CW}$$

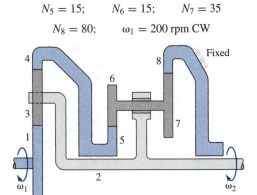

**Figure P6.22**

**P6.23** The driveshaft of an automobile is turning at 1,500 rpm with output 1 on the garage floor and the right (i.e., output 2) jacked up. The bevel gear differential of Figure 6.24 is connected to the wheels. Determine the speed of output 2 and the speed of ring gear 2.

$$N_1 = 15; \qquad N_2 = 60$$
$$N_3 = N_5 = 20; \qquad N_4 = N_6 = 18$$

**P6.24** Refer to the Figure P6.24.
(a) Determine the rotation of the carrier when gear 1 makes 30 rotations clockwise and gear 2 makes 12 rotations counterclockwise.
Also, determine the angular displacement of gear 3 about its own axis.
(b) If gear 1 rotates at $f$ rpm and gear 2 at $g$ rpm, determine the rotational speed of the carrier in terms of $f$ and $g$.

$$N_1 = 40; \qquad N_2 = 30; \qquad N_3 = 24$$

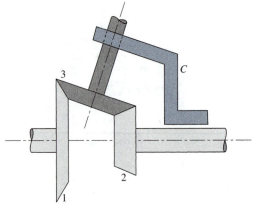

**Figure P6.24**

**P6.25** For a harmonic drive similar to that shown in Figure 6.32, determine the rotational speed of the flex spline.

$$N_f = 100; \qquad N_c = 102$$

$$\omega_w = 70 \text{ rpm CW}; \qquad \omega_c = 0$$

**P6.26** Refer to Figure P6.26. Ring gear 1 is driven at 500 rpm in the direction shown, while gear 4 is held stationary.

(a) Determine the speed ratio of the gear train and calculate the sense and rotational speed of carrier $C$.

(b) If 30 kW of power is obtained at the output, and the drive has a mechanical efficiency of 98 percent, calculate the torque required to hold gear 4 fixed.

$$N_1 = 500; \qquad N_2 = 60$$

$$N_3 = 60; \qquad N_4 = 503$$

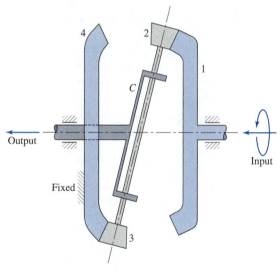

**Figure P6.26**

**P6.27** For an input speed $\omega_1 = 600$ rad/sec of the gear train shown in Figure P6.27, calculate the rotational speed of the planet carrier and the output rotational speed, $\omega_7$.

$$N_1 = 15; \qquad N_3 = 42; \qquad N_4 = 33$$

$$N_5 = 150; \qquad N_6 = 32; \qquad N_7 = 100$$

**P6.28** Determine the torque required to hold down the base of the gearbox casing shown on Figure 6.36 while under the following operating conditions:

$$\omega_i = 300 \text{ rpm}; \qquad T_i = 200 \text{ N-m}$$

$$\eta = 0.95; \qquad e_{o/i} = -0.60$$

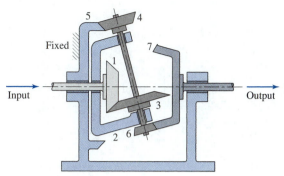

**Figure P6.27**

**P6.29** In the planetary gear train shown on Figure P6.29, internal gear 4 turns at 300 rpm in the direction shown. The diametral pitch of gear 1 is 4 in$^{-1}$.

(a) Determine the diametral pitch of gear 3 and its pitch diameter.

(b) Determine the speed and direction of rotation of the output shaft.

(c) If the input power is 20 kW, and the mechanical efficiency is 95 percent, determine the external torque on gear 1.

$$N_1 = 48; \qquad N_2 = 24$$

$$N_3 = 37; \qquad N_4 = 127$$

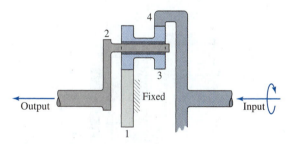

**Figure P6.29**

**P6.30** Refer to Figure P6.30. Determine the speed of rotation of shaft $A$ needed to produce the 300 rpm output rotation of shaft $B$ in the direction shown. If the mechanical efficiency of the gear train is 90 percent, determine the torque $T_f$ that must be applied to the gearbox to convert a 20 kW input to output at shaft $B$.

$$N_2 = 20; \qquad N_3 = 40; \qquad N_4 = 12$$

$$N_6 = 39; \qquad N_7 = 16; \qquad N_8 = 60$$

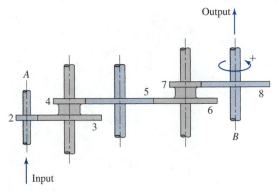

Figure P6.30

**P6.31** In the two-stage planetary gear train of Figure P6.31, gear 6 is fixed and crank arms $C$ and $D$ are attached to the output shaft. Gears 3 and 4 form a compound wheel that rotates freely about the output shaft. Determine

(a) the speed and direction of rotation of the output shaft

(b) the relative rotational speed of planet gear 2 with respect to crank arm $C$

(c) the maximum number of equally spaced planet gears 5 may be employed

$$N_1 = 20; \qquad N_3 = 72; \qquad N_4 = 24$$
$$N_6 = 64; \qquad \omega_1 = 300 \text{ rpm CW}$$

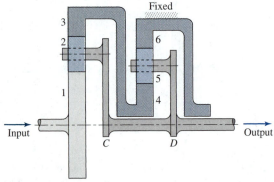

Figure P6.31

**P6.32** Figure P6.32 illustrates a planetary gear train. Points $A$, $C$, $D$, and $E$ are used to keep track of the angular positions of links 1, 3, 4, and 5, respectively. Starting from the position shown, gear 5 is rotated 270° counterclockwise (point $E$ moves to point $E'$), and gear 1

is rotated 90° clockwise (point $A$ moves to point $A'$). After completion of the motions of gears 1 and 5

(a) specify the angular position of the crank arm 2

(b) specify the angular position of planet gear 3 about its own axis

(c) specify the angular position of planet gear 4 about its own axis

(d) sketch the configuration of the gear train and indicate the positions of points $C$ and $D$ after the motions (indicate as $C'$ and $D'$, respectively)

$$N_1 = 54; \qquad N_3 = 20$$
$$N_4 = 24; \qquad N_5 = 108$$

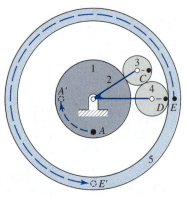

Figure P6.32

**P6.33** Information pertaining to the two-speed transmission shown in Figure P6.33 is given in Table P6.33.

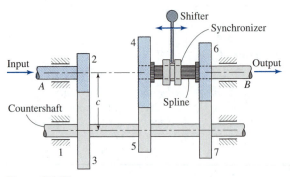

Figure P6.33

All gears are to be manufactured using a hob cutter having a 20° pressure angle and 0.500-inch hob pitch. All gears are to have full-depth teeth, each

**TABLE P6.33**

| Gear No. | No. of Teeth | Type of Gear |
|----------|--------------|--------------|
| 2 | 20 | Straight spur |
| 3 | 40 | — |
| 4 | — | Helical spur |
| 5 | — | — |
| 6 | — | — |
| 7 | — | — |

with more than 15 teeth and fewer than 45 teeth. Shafts $A$ and $B$ are collinear. The desired speed ratios are +38/156 (synchronizer connects gear 4 to shaft $B$), and +54/132 (synchronizer connects gear 6 to shaft $B$).

Determine

(a) the base circle radius of gear 2
(b) the center-to-center distance, $c$, between gears 2 and 3
(c) the contact ratio between gears 2 and 3
(d) the number of teeth on gears 4, 5, 6, and 7 that will generate exactly the desired speed ratios
(e) the circular pitch of gear 4
(f) the helix angle of gear 4
(g) the helix angle of gear 7
(h) the pitch circle diameter of gear 5
(i) the addendum circle diameter of gear 5

**P6.34** For the planetary gear train shown in Figure P6.34, all gears have straight spur teeth, full depth, 20° pressure angle, and module of 3 mm. Determine

(a) the speed and direction of rotation of the output shaft
(b) the relative rotational speed of the planet carrier with respect to gear 7
(c) the maximum number of equally spaced planet gears 7 that may be employed

$$N_1 = 35; \qquad N_2 = 21; \qquad N_3 = 27$$
$$N_5 = 56; \qquad N_6 = 100; \qquad N_7 = 50$$
$$N_8 = 180; \qquad \omega_1 = 500 \text{ rpm CCW}$$

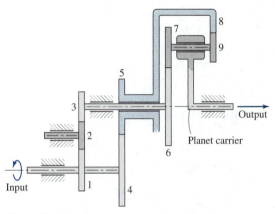

**Figure P6.34**

# 7 Cams

## 7.1 INTRODUCTION

A *cam* is a link of a mechanism. It transmits motion, by direct contact, to another link, called the *follower*. Some common types of cam mechanisms are illustrated in Figure 7.1, showing input and output motions, which may be translational or rotational.

Figure 7.2 shows a moving-headstock milling machine, used in machining a wide variety of small metal components employed in high precision mechanical devices. This machine is an excellent example of implementing different types of cam mechanisms. Here, the followers transmit motions to the cutting tools. Cam mechanisms often provide a simpler and less expensive method of achieving required motions than other types of mechanisms. Motions can more readily be modified by changing the shape of the cam. Figure 7.3 shows some of the *disc cam mechanisms* used in this machine. The photographs depict different types of followers. Figures 7.4 and 7.5 show a *cylindrical cam mechanism* and a *face cam mechanism*, respectively.

In this chapter, graphical and analytical methods of determining the shapes of disc cams are presented. In addition, the computer program entitled Cam Design is introduced, for use in the design of disc cam mechanisms.

## 7.2 DISC CAM MECHANISM NOMENCLATURE

Disc cams may be classified according to their type of follower. Figure 7.6 shows six cams and followers. Followers are classified according to their shape, location relative to the cam, and whether the follower motion is translational or rotational. Three of the followers shown in Figure 7.6 execute linear translation, whereas three execute rotational motion.

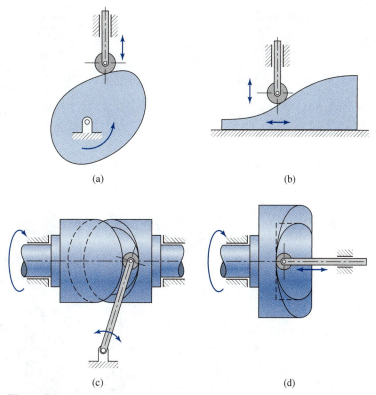

Figure 7.1 Common types of cams: (a) disc or plate, (b) wedge, (c) cylindrical, (d) end or face.

**Figure 7.2** Moving-headstock milling machine [Video 7.2].

Common definitions associated with disc cam mechanisms are

• *cam profile:* the surface of the cam contacted by the follower
• *trace point:* the point of contact of the knife-edge follower, or the center of the roller follower
• *pitch curve:* the path of the trace point with respect to the cam

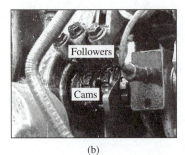
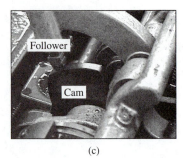

(a)                                (b)                                (c)

**Figure 7.3** Disc cam mechanisms: (a) single knife-edge follower [Video 7.3A], (b) multiple knife-edge followers [Video 7.3B], (c) roller follower [Video 7.3C].

**Figure 7.4** Cylindrical cam mechanism [Video 7.4].

**Figure 7.5** Face cam mechanism [Video 7.5].

- *base circle:* the smallest circle tangent to the cam profile, with its center on the axis of the camshaft
- *prime circle:* the smallest circle tangent to the pitch curve, with its center on the axis of the camshaft
- *pressure angle:* the angle between the normal to the pitch curve and the direction of motion of the trace point

This nomenclature for disc cam mechanisms having translating and pivoting followers is illustrated in Figures 7.7 and 7.8. Springs (not shown) between the follower and base link are employed to keep the follower in contact with the cam profile.

## 7.3 PRESSURE ANGLE

Pressure angle changes as the mechanism moves. Figures 7.7 and 7.8 illustrate pressure angles in mechanisms for given configurations.

The value of the pressure angle is a critical parameter in design of disc cam mechanisms and should be kept as small as possible. The practical maximum is 30°. Even if the follower and disc cam were a frictionless kinematic pair, there is still a component of

Translating Pivoting

Knife
edge

Flat
face

Roller

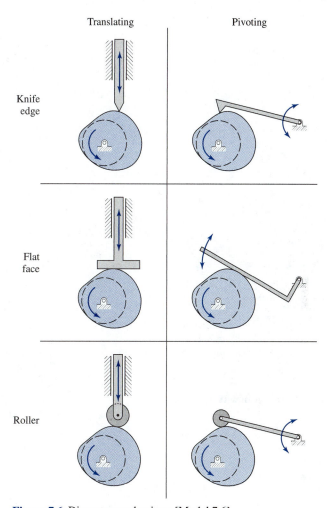

**Figure 7.6** Disc cam mechanisms [Model 7.6].

unwanted side thrust. This can be illustrated by a free body diagram as shown in Figure 7.9. Figure 7.9(a) shows a disc cam and roller follower. Figure 7.9(b) shows the roller follower with the guide of the follower removed and replaced by the equivalent forces. Larger values of pressure angle cause an increase in the unwanted side thrust on the follower stem. For excessive pressure angles, the mechanism will tend to bind.

Some means by which the pressure angle may be reduced are

- increasing the diameter of the base circle
- increasing the diameter of the roller follower
- changing the offset of the follower (see Figure 7.7)
- changing the motion of the follower

An illustration of the application of the above methods is provided later in this chapter by means of an example.

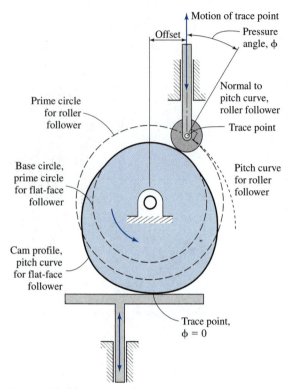

**Figure 7.7** Disc cam mechanism with translating followers.

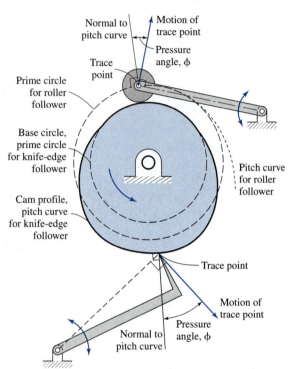

**Figure 7.8** Disc cam mechanism with pivoting followers.

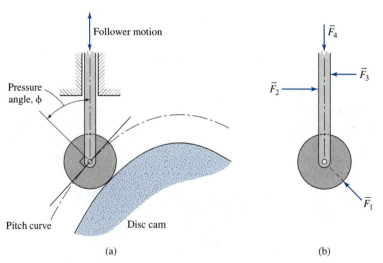

**Figure 7.9** Roller follower: (a) disc cam and follower, (b) free body diagram of follower.

## 7.4  THE DISPLACEMENT DIAGRAM

In designing a disc cam, one starts by specifying the required motion of the follower throughout a complete cycle. It may be expressed in the form

$$s = f(\theta), \qquad 0 \le \theta \le 2\pi \qquad\qquad (7.4\text{-}1)$$

A *displacement diagram* is a plot of follower displacement versus cam rotation. The motions of the follower may be grouped into the following three categories:

- *rise:* The follower is moving away from the center of the disc cam.
- *dwell:* The follower is at rest.
- *fall* or *return:* The follower is moving toward the center of the disc cam.

Figure 7.10 shows an example of a displacement diagram. It depicts portions of rise, dwell, and return. Rotation of the cam is usually specified in degrees. Displacement of the follower is plotted in terms of the amount of its linear movement or rotation, for translating and pivoting cam mechanisms, respectively.

## 7.5  TYPES OF FOLLOWER MOTIONS

A displacement diagram is generally made up of portions of standardized functions. Some common functions adapted for rise and fall portions of the displacement diagram are given below; all of which are of the form

$$s^* = g(\theta^*), \qquad 0 \le \theta^* \le \beta \qquad\qquad (7.5\text{-}1)$$

where

$s^* = $ displacement of the follower in the current portion
$\theta^* = $ rotation of the cam in the current portion
$\beta = $ total rotation in the current portion

Figure 7.11 gives an illustration of relating a standardized function to a displacement diagram. To compare various types of motion, we will consider dwell of the follower just prior to, and just after, its motion. Also, all motions will start at $\theta^* = 0$ and end at $\theta^* = \beta$.

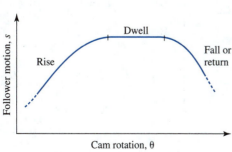

**Figure 7.10**  Typical displacement function.

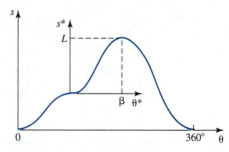

**Figure 7.11**  Incorporation of a standardized function in a displacement diagram.

The follower lift during the current portion is $L$, and the constant rotational speed of the cam is $\dot{\theta}$.

In the design of cam mechanisms, particularly for high-speed operation, time derivative quantities of the motions of the follower are important. The second time derivative of the displacement is the *acceleration*, which is proportional to the force applied to produce the motion. It is generally desirable to determine extremum values of the peak accelerations, and try to reduce their absolute values. It is often worthwhile to reduce the time rate of change of acceleration, referred to as the *jerk*. Higher values of jerk often correspond to increased noise levels under operating conditions.

## 7.5.1 Uniform Motion

*Uniform motion* is equivalent to *constant velocity*. Figure 7.12 shows a portion of a displacement diagram that incorporates uniform motion. The equations for follower motion during lift and return are given in the same figure.

Expressions for the velocity, acceleration, and jerk during uniform motion are

$$\dot{s}^* = \frac{L}{\beta}\dot{\theta}^* = \text{constant}; \qquad \ddot{s}^* = 0; \qquad \dddot{s}^* = 0, \qquad 0 < \theta^* < \beta \qquad (7.5\text{-}2)$$

The acceleration and jerk are expressed as zero. However, it must be kept in mind that when $\theta^* = 0$ and $\theta^* = \beta$, acceleration and the jerk are infinite. This is because at the start and end of the motion there is a step change of velocity. In theory, this would correspond to an infinite force and jerk. Because of this undesirable characteristic, uniform motion of the follower without proper blending with adjoining motions should be avoided in high-speed applications to prevent excessive accelerations.

To improve the dynamic characteristics, constant slope in the displacement profile is sometimes combined with other shapes at its ends to smooth out the sharp corners. As a target, infinite accelerations should be eliminated. Section 7.5.3 presents a cam motion that

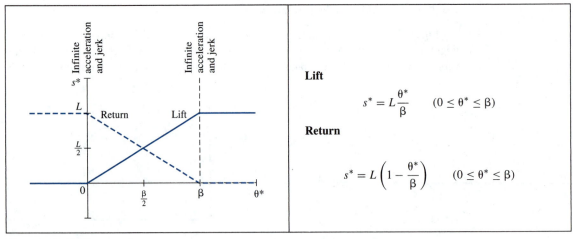

**Figure 7.12** Uniform motion.

incorporates a portion of uniform motion combined with smoothing segments at the beginning and end of the motion.

## 7.5.2 Parabolic Motion

*Parabolic motion* is also referred to as *constant acceleration*. Figure 7.13 shows a portion of a displacement diagram that incorporates parabolic motion for the lift and return. At the beginning and ending of the motion there is a step change of acceleration and an infinite value of jerk. Equations for follower displacement are given in the same figure.

The displacement curve shown is made up of two parabolas. For a lift motion, the first half of the motion (rotation 0 to β/2) has a constant positive acceleration, whereas the second

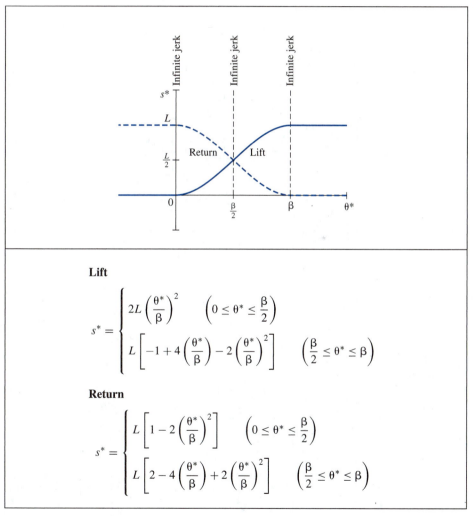

**Lift**

$$s^* = \begin{cases} 2L\left(\dfrac{\theta^*}{\beta}\right)^2 & \left(0 \le \theta^* \le \dfrac{\beta}{2}\right) \\[4mm] L\left[-1 + 4\left(\dfrac{\theta^*}{\beta}\right) - 2\left(\dfrac{\theta^*}{\beta}\right)^2\right] & \left(\dfrac{\beta}{2} \le \theta^* \le \beta\right) \end{cases}$$

**Return**

$$s^* = \begin{cases} L\left[1 - 2\left(\dfrac{\theta^*}{\beta}\right)^2\right] & \left(0 \le \theta^* \le \dfrac{\beta}{2}\right) \\[4mm] L\left[2 - 4\left(\dfrac{\theta^*}{\beta}\right) + 2\left(\dfrac{\theta^*}{\beta}\right)^2\right] & \left(\dfrac{\beta}{2} \le \theta^* \le \beta\right) \end{cases}$$

**Figure 7.13** Parabolic motion.

half of the motion (rotation $\beta/2$ to $\beta$) has a constant negative acceleration (i.e., a deceleration). At $\theta = \beta/2$, both the displacement and slope of the displacement curve are matched.

### 7.5.3 Modified Parabolic Motion

*Modified parabolic motion* is a combination of uniform motion and parabolic motion. Figure 7.14 shows a portion of a displacement diagram that incorporates modified

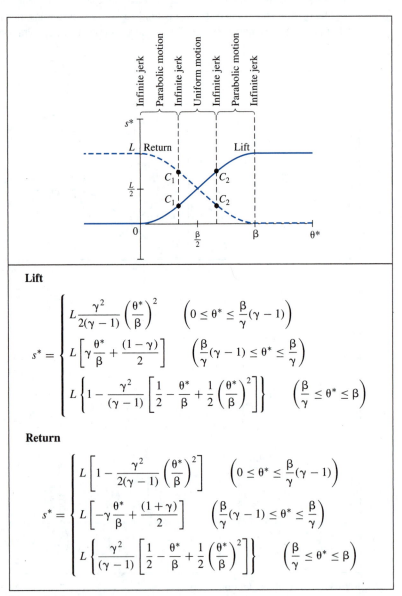

**Lift**

$$s^* = \begin{cases} L\dfrac{\gamma^2}{2(\gamma-1)}\left(\dfrac{\theta^*}{\beta}\right)^2 & \left(0 \le \theta^* \le \dfrac{\beta}{\gamma}(\gamma-1)\right) \\[3mm] L\left[\gamma\dfrac{\theta^*}{\beta} + \dfrac{(1-\gamma)}{2}\right] & \left(\dfrac{\beta}{\gamma}(\gamma-1) \le \theta^* \le \dfrac{\beta}{\gamma}\right) \\[3mm] L\left\{1 - \dfrac{\gamma^2}{(\gamma-1)}\left[\dfrac{1}{2} - \dfrac{\theta^*}{\beta} + \dfrac{1}{2}\left(\dfrac{\theta^*}{\beta}\right)^2\right]\right\} & \left(\dfrac{\beta}{\gamma} \le \theta^* \le \beta\right) \end{cases}$$

**Return**

$$s^* = \begin{cases} L\left[1 - \dfrac{\gamma^2}{2(\gamma-1)}\left(\dfrac{\theta^*}{\beta}\right)^2\right] & \left(0 \le \theta^* \le \dfrac{\beta}{\gamma}(\gamma-1)\right) \\[3mm] L\left[-\gamma\dfrac{\theta^*}{\beta} + \dfrac{(1+\gamma)}{2}\right] & \left(\dfrac{\beta}{\gamma}(\gamma-1) \le \theta^* \le \dfrac{\beta}{\gamma}\right) \\[3mm] L\left\{\dfrac{\gamma^2}{(\gamma-1)}\left[\dfrac{1}{2} - \dfrac{\theta^*}{\beta} + \dfrac{1}{2}\left(\dfrac{\theta^*}{\beta}\right)^2\right]\right\} & \left(\dfrac{\beta}{\gamma} \le \theta^* \le \beta\right) \end{cases}$$

**Figure 7.14** Modified parabolic motion.

parabolic motion for the lift and return. At the beginning and end of the rotation of the cam, the follower motion is parabolic. In the central region, the follower motion is uniform. At the transition points $C_1$ and $C_2$, between the uniform and parabolic motions, both the displacement and the slope of the displacement curve are matched. Thus, the acceleration remains finite at the transition points. Expressions for follower displacement are included in Figure 7.14.

The expressions for displacement given in Figure 7.14 include the parameter $\gamma$. It may take on values

$$1 < \gamma < 2$$

Cases where $\gamma = 1$ and $\gamma = 2$ correspond to uniform motion and parabolic motion, respectively.

### 7.5.4 Harmonic Motion

*Harmonic motion* incorporates a portion of a sine wave. A portion of a displacement diagram with harmonic motion is given in Figure 7.15. The equations for lift and return motions in this instance are listed in the same figure.

Acceleration has a finite value at the beginning and end of harmonic motion. Because there is a step change of acceleration, there is infinite jerk at these locations. Other than the start and end points, jerk remains finite.

### 7.5.5 Cycloidal Motion

Figure 7.16 shows a portion of a displacement diagram that incorporates *cycloidal motion*. The displacement functions for lift and return are also provided. For this motion, the expression for acceleration can be easily shown to be zero at the start and finish. As a result, jerk remains finite throughout the entire motion, including at the start and finish.

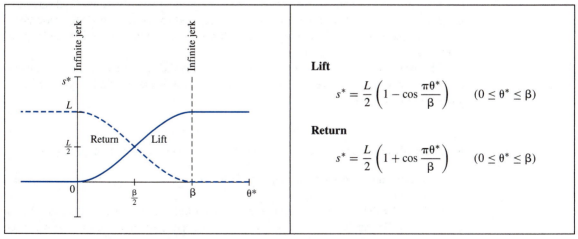

**Lift**

$$s^* = \frac{L}{2}\left(1 - \cos\frac{\pi\theta^*}{\beta}\right) \qquad (0 \le \theta^* \le \beta)$$

**Return**

$$s^* = \frac{L}{2}\left(1 + \cos\frac{\pi\theta^*}{\beta}\right) \qquad (0 \le \theta^* \le \beta)$$

**Figure 7.15** Harmonic motion.

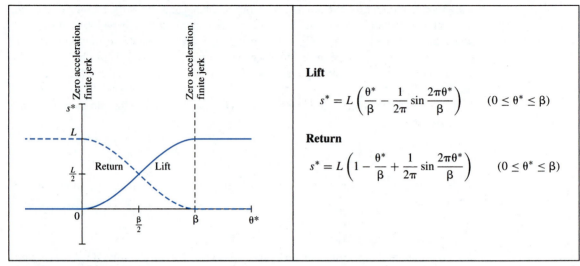

**Figure 7.16** Cycloidal motion.

The figure panel on the right contains:

**Lift**

$$s^* = L\left(\frac{\theta^*}{\beta} - \frac{1}{2\pi}\sin\frac{2\pi\theta^*}{\beta}\right) \qquad (0 \le \theta^* \le \beta)$$

**Return**

$$s^* = L\left(1 - \frac{\theta^*}{\beta} + \frac{1}{2\pi}\sin\frac{2\pi\theta^*}{\beta}\right) \qquad (0 \le \theta^* \le \beta)$$

## 7.6  COMPARISON OF FOLLOWER MOTIONS

A comparison of the four of the types of motion presented in Section 7.5 is given in Figure 7.17. The plot for each is nondimensional, that is, results are plotted as a unit lift taking place in one unit of time. For the displacements, only the plot of uniform motion can be immediately identified. However, there are significant differences for the velocities, accelerations, and jerks. Jerks for the parabolic motion are infinite at the beginning, middle, and end of the motion cycle, and zero everywhere else.

## 7.7  DETERMINATION OF DISC CAM PROFILE

The basic problem of cam design is to find the cam profile that will produce a desired follower motion. The cam profile depends on the required follower motion, the base circle diameter of the cam, the follower type and its dimensions, and the position of the follower relative to the cam.

Cam profiles may be generated graphically or by evaluating analytical expressions for the particular type of follower under consideration. Both of these techniques will be considered in this section.

### 7.7.1  Graphical Determination of Disc Cam Profile

To graphically construct a disc cam profile, the following general steps are required:

1. Specify the displacement diagram, base circle diameter, and follower (type and dimensions).
2. Consider the cam as fixed and move the other links with respect to the cam; that is, invert the mechanism. In order to obtain the same relative motions between the

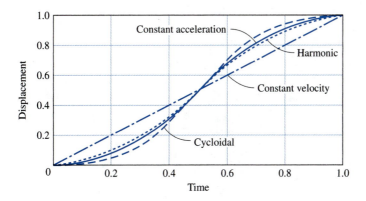

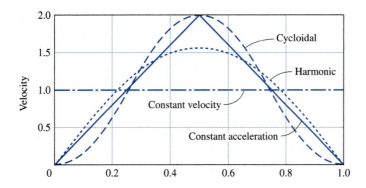

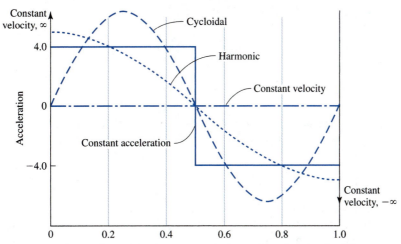

**Figure 7.17** Comparison of displacement, velocity, and acceleration for follower motions.

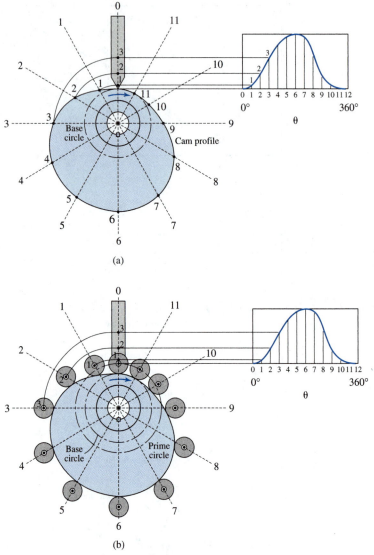

(a)

(b)

**Figure 7.18** Graphical construction of a disc cam: (a) translating knife-edge follower, (b) translating roller follower, zero offset.

links, the base link and follower are rotated about the cam in the *opposite direction* to cam rotation. The follower is then drawn in numerous positions with respect to the cam throughout a cycle.

3. Draw the cam profile inside the envelope of the follower positions.

Figure 7.18 illustrates this procedure for six types of followers.

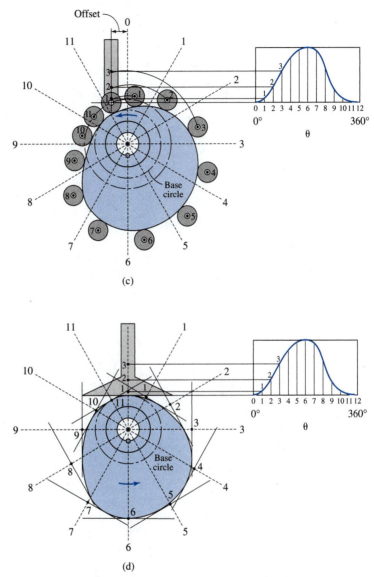

**Figure 7.18** (*Continued*) Graphical construction of a disc cam: (c) translating roller follower, offset, (d) translating flat-face follower.

## 7.7.2 Analytical Determination of Disc Cam Profile

The points on a disc cam may be determined analytically, using the governing mathematical equations that are presented in reference [5] and that are summarized in Figure 7.19. In these equations, quantities $x_c$ and $y_c$ are the coordinates of a point of contact between the

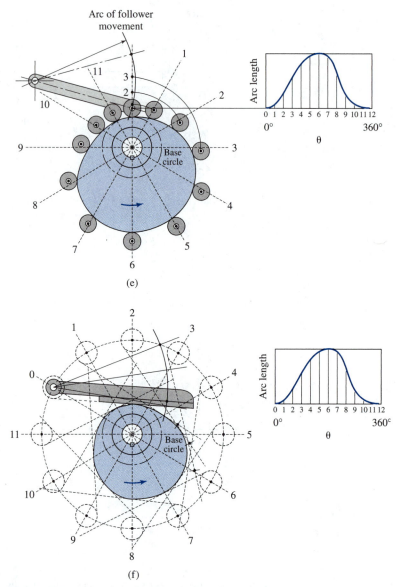

**Figure 7.18** (*Continued*) Graphical construction of a disc cam: (e) pivoting roller follower, (f) pivoting flat-face follower.

cam and the follower for a given rotation of the cam. Similar to the graphical procedure used in drawing the cam in one position, it is necessary to rotate these calculated points around the center of rotation of the cam, in a direction *opposite* to that of the operational direction of rotation of the cam. The use of these equations is illustrated by the following example.

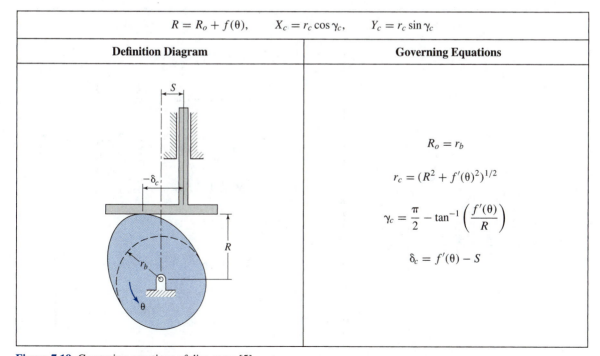

$$R = R_o + f(\theta), \qquad X_c = r_c \cos\gamma_c, \qquad Y_c = r_c \sin\gamma_c$$

| Definition Diagram | Governing Equations |
|---|---|
| | $$R_o = \left(r_b^2 - S^2\right)^{1/2}$$ $$r_c = (S^2 + R^2)^{1/2}$$ $$\gamma_c = \frac{\pi}{2} - \tan^{-1}(S/R)$$ $$\phi = \frac{\pi}{2} - \gamma_c - \tan^{-1}\left(\frac{Rf'(\theta)}{r_c{}^2}\right)$$ |

$$R = R_o + f(\theta), \qquad X_c = r_c \cos\gamma_c, \qquad Y_c = r_c \sin\gamma_c$$

| Definition Diagram | Governing Equations |
|---|---|
| | $$R_o = r_b$$ $$r_c = (R^2 + f'(\theta)^2)^{1/2}$$ $$\gamma_c = \frac{\pi}{2} - \tan^{-1}\left(\frac{f'(\theta)}{R}\right)$$ $$\delta_c = f'(\theta) - S$$ |

**Figure 7.19** Governing equations of disc cams [5].

$$R = R_o + f(\theta), \qquad X_c = r_c \cos \gamma_c, \qquad Y_c = r_c \sin \gamma_c$$

| Definition Diagram | Governing Equations |
|---|---|
| 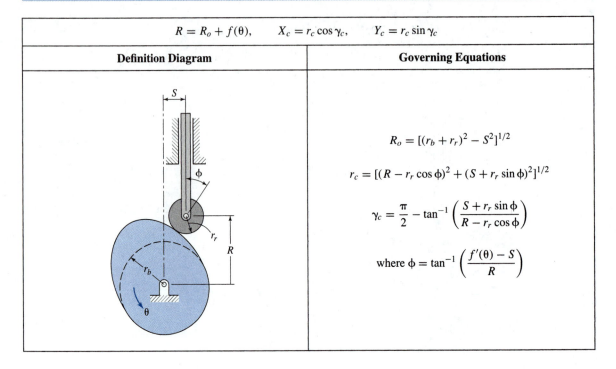 | $$R_o = [(r_b + r_r)^2 - S^2]^{1/2}$$ $$r_c = [(R - r_r \cos \phi)^2 + (S + r_r \sin \phi)^2]^{1/2}$$ $$\gamma_c = \frac{\pi}{2} - \tan^{-1}\left(\frac{S + r_r \sin \phi}{R - r_r \cos \phi}\right)$$ where $\phi = \tan^{-1}\left(\dfrac{f'(\theta) - S}{R}\right)$ |

$$\alpha = \alpha_o + f(\theta), \qquad r_c = \left(X_c^2 + Y_c^2\right)^{1/2}, \qquad \gamma_c = \tan^{-1}(Y_c/X_c)$$

| Definition Diagram | Governing Equations |
|---|---|
| 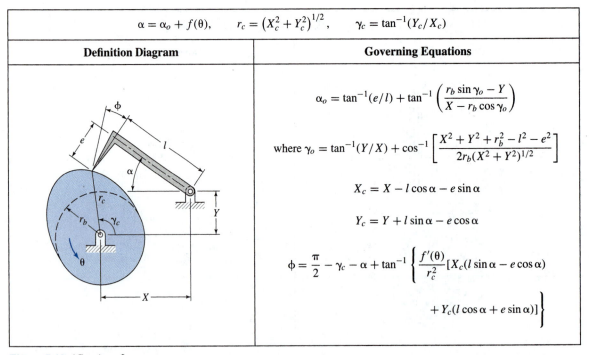 | $$\alpha_o = \tan^{-1}(e/l) + \tan^{-1}\left(\frac{r_b \sin \gamma_o - Y}{X - r_b \cos \gamma_o}\right)$$ where $\gamma_o = \tan^{-1}(Y/X) + \cos^{-1}\left[\dfrac{X^2 + Y^2 + r_b^2 - l^2 - e^2}{2r_b(X^2 + Y^2)^{1/2}}\right]$ $$X_c = X - l \cos \alpha - e \sin \alpha$$ $$Y_c = Y + l \sin \alpha - e \cos \alpha$$ $$\phi = \frac{\pi}{2} - \gamma_c - \alpha + \tan^{-1}\left\{\frac{f'(\theta)}{r_c^2}[X_c(l \sin \alpha - e \cos \alpha)\right.$$ $$\left. + Y_c(l \cos \alpha + e \sin \alpha)]\right\}$$ |

**Figure 7.19** (*Continued*)

$$\alpha = \alpha_o + f(\theta), \qquad r_c = \left(X_c^2 + Y_c^2\right)^{1/2}, \qquad \gamma_c = \tan^{-1}(Y_c/X_c)$$

| Definition Diagram | Governing Equations |
|---|---|
| 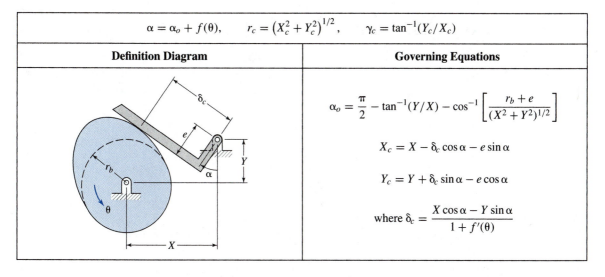 | $$\alpha_o = \frac{\pi}{2} - \tan^{-1}(Y/X) - \cos^{-1}\left[\frac{r_b + e}{(X^2 + Y^2)^{1/2}}\right]$$ $$X_c = X - \delta_c \cos\alpha - e \sin\alpha$$ $$Y_c = Y + \delta_c \sin\alpha - e \cos\alpha$$ where $$\delta_c = \frac{X\cos\alpha - Y\sin\alpha}{1 + f'(\theta)}$$ |

$$\alpha = \alpha_o + f(\theta), \qquad r_c = \left(X_c^2 + Y_c^2\right)^{1/2}, \qquad \gamma_c = \tan^{-1}(Y_c/X_c)$$

| Definition Diagram | Governing Equations |
|---|---|
| 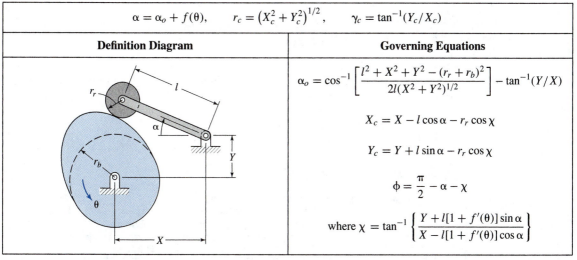 | $$\alpha_o = \cos^{-1}\left[\frac{l^2 + X^2 + Y^2 - (r_r + r_b)^2}{2l(X^2 + Y^2)^{1/2}}\right] - \tan^{-1}(Y/X)$$ $$X_c = X - l\cos\alpha - r_r \cos\chi$$ $$Y_c = Y + l\sin\alpha - r_r \cos\chi$$ $$\phi = \frac{\pi}{2} - \alpha - \chi$$ where $$\chi = \tan^{-1}\left\{\frac{Y + l[1 + f'(\theta)]\sin\alpha}{X - l[1 + f'(\theta)]\cos\alpha}\right\}$$ |

**Figure 7.19** (*Continued*)

## EXAMPLE 7.1 Determination of a Disc Cam Profile

Consider the following the input parameters of a disc cam to be manufactured:

- translating roller follower
- base circle diameter $d_b = 3.0$ cm
- roller circle diameter $d_r = 1.0$ cm
- offset $S = 0.6$ cm

$$r_b = \frac{d_b}{2} = \frac{3.0}{2} = 1.5 \text{ cm}; \qquad r_r = \frac{d_r}{2} = \frac{1.0}{2} = 0.5 \text{ cm}$$

A layout of these dimensions is provided in Figure 7.20.

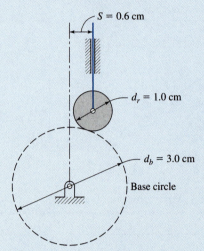

**Figure 7.20** Prescribed dimensions of a disc cam.

The displacement parameters are given in Table 7.1, and the corresponding displacement diagram is illustrated in Figure 7.21. Determine

(a) the cam profile
(b) the pressure angle as a function of cam rotation

**TABLE 7.1** Follower Motion

| Interval ↓   Parameter → | Initial Angle for Interval (degrees) | Final Angle for Interval (degrees) | Type of Motion | Rise (+) or Return (−) (cm) |
|:---:|:---:|:---:|:---|:---:|
| 1 | 0 | 100 | Harmonic | 2.5 |
| 2 | 100 | 160 | Dwell | 0 |
| 3 | 160 | 300 | Cycloidal | −1.5 |
| 4 | 300 | 360 | Parabolic | −1.0 |

## SOLUTION

(a) We apply the equations as listed in Figure 7.19. A typical calculated value of the pressure angle is made for $\theta = 30°$. At this rotation, the follower is undergoing lift with harmonic motion, with a lift of 2.5 cm in the interval

$$0 \le \theta \le 100°$$

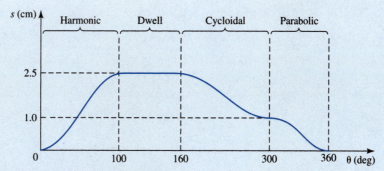

**Figure 7.21** Displacement diagram.

Therefore

$$L = 2.5 \text{ cm}; \qquad \beta = 100° - 0° = 100°$$

Using the expressions from Figure 7.15, and recognizing in this instance that yields

$$\theta = \theta^*$$

$$f(\theta) = s^* = \frac{L}{2}\left(1 - \cos\frac{\pi\theta}{\beta}\right); \qquad f(30°) = 0.515 \text{ cm}$$

$$f'(\theta) = \frac{df(\theta)}{d\theta} = \frac{\pi L}{2\beta}\sin\frac{\pi\theta}{\beta}; \qquad f'(30°) = 1.82 \text{ cm}$$

$$R_o = [(r_b + r_r)^2 - S^2]^{1/2} = 1.908 \text{ cm}$$

$$R(\theta) = R_o + f(\theta); \qquad R(30°) = 2.423 \text{ cm}$$

$$\phi(\theta) = \tan^{-1}\left(\frac{f'(\theta) - S}{R(\theta)}\right); \qquad \phi(30°) = 26.7°$$

$$\gamma_c(\theta) = \frac{\pi}{2} - \tan^{-1}\left[\frac{S + r_r\sin(\phi(\theta))}{R(\theta) - r_r\sin(\phi(\theta))}\right]; \qquad \gamma_c(30°) = 67.3°$$

$$r_c(\theta) = \{[R(\theta) - r_r\cos(\phi(\theta))]^2 + [S + r_r\sin(\phi(\theta))]^2\}^{1/2}; \qquad r_c(30°) = 2.142 \text{ cm}$$

$$x_c(\theta) = r_c\cos(\phi(\theta)); \qquad x_c(30°) = 0.825 \text{ cm}$$

$$y_c(\theta) = r_c\sin(\phi(\theta)); \qquad y_c(30°) = 1.98 \text{ cm}$$

The calculated coordinates are rotated about the center of rotation of the cam in an opposite direction to that of the rotation of the cam to locate a point on the cam profile. This is illustrated in Figure 7.22(a). Figure 7.22(b) shows the disc cam mechanism.

As a second point on the profile, consider the cam rotation of $\theta = 200°$. At this position, the follower is undergoing return with cycloidal motion. This interval of

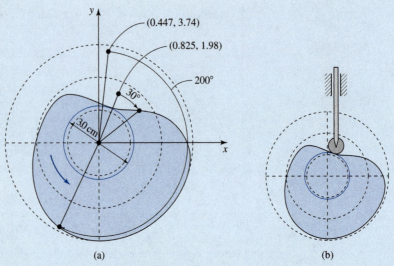

**Figure 7.22** (a) Development of disc cam profile. (b) Mechanism.

motion starts at 160° and ends at 300°. The return during the interval is 1.5 cm. Therefore

$$L = 1.5 \text{ cm}; \qquad \beta = 300° - 160° = 140°$$

$$\theta^* = \theta - 160° = 200° - 160° = 40°$$

$$f(200°) = f(300°) + s^*(40°) = 2.30 \text{ cm}$$

Employing the equations from Figures 7.16 and 7.19, in this instance

$$x_c(200°) = 0.447 \text{ cm}; \qquad y_c(200°) = 3.74 \text{ cm}; \qquad \phi(200°) = -17.8°$$

The point is added to Figure 7.22.

(b) The pressure angle was included in calculating part (a) of this problem. Figure 7.23 shows a plot of pressure angle as a function of cam rotation.

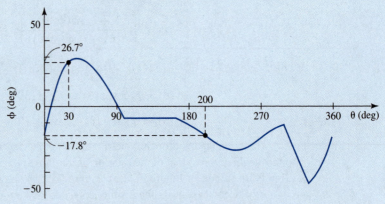

**Figure 7.23** Pressure angle as a function of cam rotation.

The cam profile and pressure angle presented in the above example may also be obtained using the program Cam Design (see Section 7.10).

## 7.8  UNDERCUTTING OF A DISC CAM PROFILE

On occasion, for a given geometry of a mechanism and specified follower displacement, it may be impossible to generate the cam profile. Consider employing the same displacement diagram as was employed in Example 7.1. If we now incorporate a flat-face follower and employ the procedure presented in Section 7.7, then the result is shown in Figure 7.24. Part of the cam profile doubles back on itself. This portion of the cam profile is impossible to manufacture, and is said to be *undercut*. This has resulted from attempting to achieve too great a lift of the follower, within inadequate cam rotation, and the base circle of the cam is too small relative to the rise required.

Undercutting may be eliminated by employing one or more of the following methods:

- increasing the diameter of the base circle
- incorporating a different type of follower
- modifying the motion of the follower

This is illustrated by an example later in this chapter.

## 7.9  POSITIVE-MOTION CAM MECHANISMS

**Model 7.25**
Positive-Motion
Cam Mechanism

*Positive-motion cam mechanisms* are those where the cam exerts affirmative control of motion of the follower during a complete cycle. The cylindrical cams shown in Figures 7.1 and 7.4 are also positive-motion cams. Such cams do not require gravity or spring force to ensure contact between the cam and follower.

Figure 7.25 shows another type of positive-motion cam mechanism. It is very similar to a disc cam mechanism with a roller follower. However, in this instance, a groove of

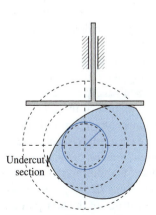

**Figure 7.24** Undercut disc cam.

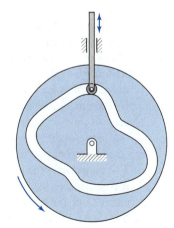

**Figure 7.25** Positive-motion cam mechanism [Model 7.25].

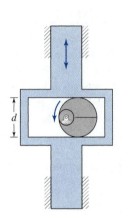

**Figure 7.26** Constant-breadth cam mechanism [Model 7.26].

**Figure 7.27** Movie projector mechanism.

constant width is machined in the cam. As the cam rotates, the roller follower is constrained to move in this groove.

Figure 7.26 shows a *constant-breadth cam mechanism*, which is another type of positive-motion cam mechanism. The cam surface is always in contact with the two parallel surfaces of the translating follower. The distance between parallel surfaces, $d$, equals the base circle diameter plus the maximum follower rise. In order to be a constant-breadth cam, the lengths of all diagonal lines through the center of the cam to the cam profile must be equal. Constant-breadth cams may be designed similar to disc cams with a translating flat face, provided that the rise during 180° of cam rotation is the mirror image of the return during the remainder of the cycle.

Figure 7.27 shows an example of a constant-breadth cam employed in a movie projector mechanism. Figure 7.28 gives another illustration of the same system. The function of this mechanism is to intermittently advance film, one frame at a time. While a frame is stationary, light is permitted to shine through the film and momentarily project a stationary image. While the film is moving, the light must be blocked. This mechanism incorporates two positive-motion cam mechanisms. This includes the constant-breadth cam mechanism used to provide up and down motion of the transport arm. The constant-breadth cam is located on shaft $B$, as shown in Figure 7.28(a). Shaft $A$ is geared to shaft $B$. A cylindrical cam is attached to shaft $A$. One end of the transport arm is in the slot of the cylindrical cam. This cam causes the intermittent rotation of the transport arm about the vertical axis $C$. The rotation causes the other end of the transport arm to enter perforations of the film. While entered into the perforations (Figure 7.28(b)), the arm moves vertically, advancing the film one frame. Every third upward stroke of the transport arm, the end of the arm enters perforations in the sides of the film. A *shutter* (shown in animation, but not in Figure 7.28) is attached to shaft $B$.

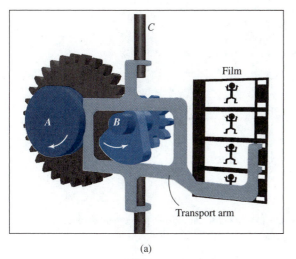 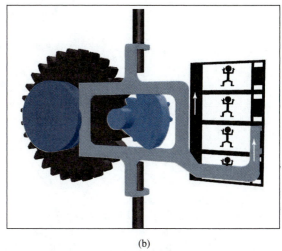

(a)                                          (b)

**Figure 7.28** Movie projector mechanism [Model 7.28].

## 7.10 PROGRAM CAM DESIGN

The program Cam Design is included on the CD-ROM accompanying this textbook. With this program it is possible to design disc cams having either translating or pivoting followers. The face of the follower may be a knife edge, roller, or flat face. The disc cam profile may be plotted, and an animated motion of the cam mechanism may be displayed. Any of the motions presented in Section 7.4 may be incorporated in the displacement diagram. Values of pressure angle may be displayed. Also, by plotting the profile, we may determine whether or not undercutting is encountered.

As an illustration, consider the following input parameters of a disc cam mechanism to be analyzed:

- translating roller follower
- base circle diameter: 4.0 cm
- roller circle diameter: 1.0 cm
- offset: 0.6 cm
- cam rotational speed: 20 rad/sec CCW

Table 7.2 lists the required motions of the follower. Figure 7.29(a) illustrates the above input parameters, and Figure 7.29(b) shows the displacement diagram.

**TABLE 7.2** Follower Motion

| Parameter →<br>Interval ↓ | Initial Angle<br>for Interval<br>(degrees) | Final Angle<br>for Interval<br>(degrees) | Type of<br>Motion | Rise (+) or<br>Return (−)<br>(cm) |
|---|---|---|---|---|
| 1 | 0 | 100 | Parabolic | 0.8 |
| 2 | 100 | 260 | Dwell | 0 |
| 3 | 260 | 360 | Harmonic | −0.8 |

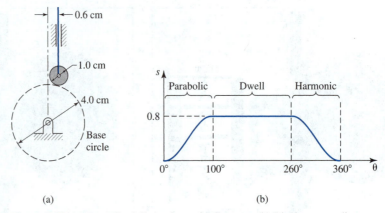

(a)                                                    (b)

**Figure 7.29** (a) Prescribed dimensions of a disc cam. (b) Displacement diagram.

**Figure 7.30** Cam Design example—introductory logo.

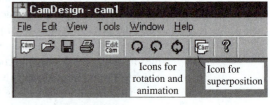

**Figure 7.31** Cam Design—menu icons.

After initiating the program, we obtain the opening screen logo as shown in Figure 7.30. By selecting the **File** menu (Figure 7.31), we have the option to either open an existing file or create a new one. If we select a **New** file, a wizard dialogue box appears. It includes three property pages: **Design Page, Motion Page,** and **Display Page.** Each is described separately below.

- **Design Page** (Figure 7.32)

    Figure 7.32(a) shows an illustration of the **Design Page** containing default parameters of a cam mechanism. For this illustration, we enter the following modifications: replace the knife-edge follower with a roller follower, and replace the values of the base circle diameter, follower offset, and roller diameter. After entering these data, the screen appears as shown in Figure 7.32(b). Then selecting **Next >**, we obtain the **Motion Page.**

- **Motion Page** (Figure 7.33)

    The default **Motion Page** is shown in Figure 7.33(a). Default parameters are replaced by those to be employed for this illustration. This includes the cam rotational speed, along with the number of intervals of motion during a complete cycle (changed from one to three—see Table 7.2). Furthermore, by selecting each interval, we input the

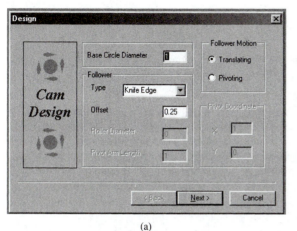
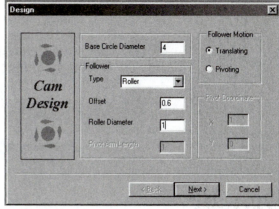

(a)                                              (b)

**Figure 7.32**  Cam Design example—Design Page.

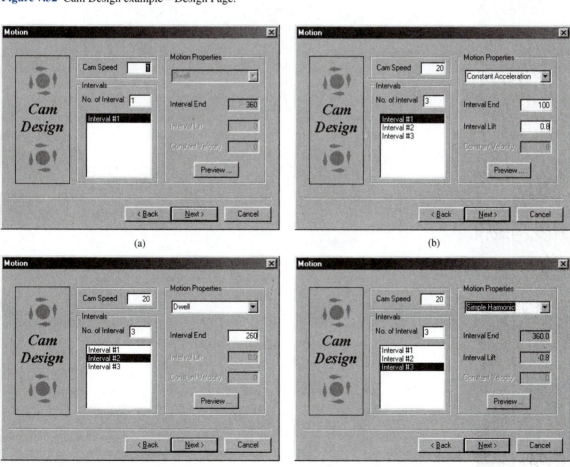

(a)                                              (b)

(c)                                              (d)

**Figure 7.33**  Cam Design example—Motion Page.

cam rotation at the end of the interval, the type of motion, and if needed, the rise or fall of the follower during the interval. For this illustration, we input the parameters listed in Table 7.2. Figures 7.33(b), 7.33(c) and 7.33(d) illustrate the **Motion Page** after the parameters for each interval have been entered. Then, selecting **Next >**, we obtain the **Display Page.**

• **Display Page** (Figure 7.34)
This page outlines a variety of display modes that may be obtained in either tabular or graphical form. Figure 7.34(a) shows the default **Display Page,** which specifies a

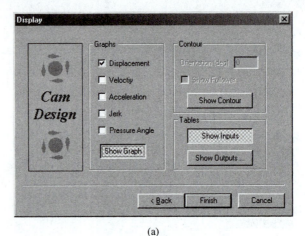

(a)

Absolute Maximum Values

| Function | Absolute Maximum | Maximum at (deg) |
|---|---|---|
| Follower Displacement | 8.000e−001 | 100.0 |
| Follower Velocity | 1.833e+001 | 50.0 |
| Follower Acceleration | 5.184e+002 | 260.0 |
| Follower Jerk | 1.866e+004 | 310.0 |
| Pressure Angle | 2.544e+001 | 318.0 |

Output Graphs (normalized with the maximum values)

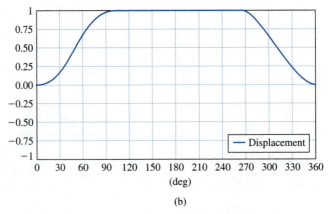

(b)

**Figure 7.34** Cam Design example—display or displacement diagram: (a) Display Page, (b) displacement diagram.

displacement diagram. By clicking on **Finish,** the wizard dialogue box is closed, and the screen as shown in Figure 7.34(b) appears. It includes a displacement diagram in nondimensional form. Maximum absolute values of displacement, velocity, acceleration, jerk, and pressure angle throughout a cycle are also provided in a table contained within the figure.

By clicking the **Edit Cam** icon (see Figure 7.31), the **Properties Sheet** appears, as shown in Figure 7.35(a). We now have the choice to return to the **Design Page,** the **Motion Page,** or the **Display Page.** It is possible to modify the parameters or repeat the displays. We have the option to change the type and dimensions of the cam mechanism through the **Design Page**. Alternatively, by selecting the **Motion Page**, we may alter the follower motions. A further option is to select the **Display Page**, and either review or select outputs. We may go to the **Display Page,** deselect the **Show Graph** button, and select the **Show Contour** button. We also have the option of including the follower with the cam in the illustration by clicking on the appropriate box (Figure 7.35(b)). Then, clicking **Apply** gives the result as shown in Figure 7.36. Once again clicking the **Edit Cam** icon, then the **Display**

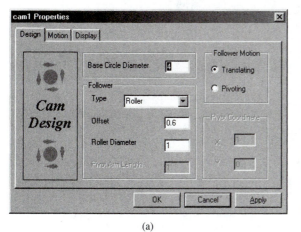

(a)

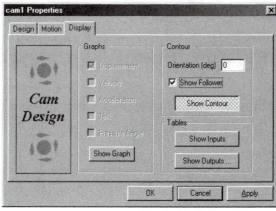

(b)

**Figure 7.35** Cam Design example—Properties Sheet.

**Figure 7.36** Cam Design example—cam and follower.

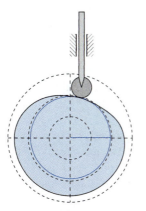

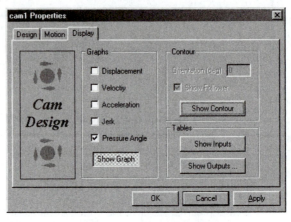

(a)

| Function | Absolute Maximum | Maximum at (deg) |
|---|---|---|
| Follower Displacement | 8.000e−001 | 100.0 |
| Follower Velocity | 1.833e+001 | 50.0 |
| Follower Acceleration | 5.184e+002 | 260.0 |
| Follower Jerk | 1.866e+004 | 310.0 |
| Pressure Angle | 2.544e+001 | 318.0 |

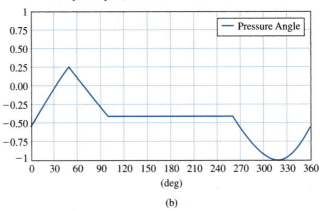

(b)

**Figure 7.37** Cam Design example—display of pressure angle.

**Page** button, deselecting the **Show Contour** button, clicking on the **Show Graph** button, selecting the **Pressure Angle** graph as shown in Figure 7.37(a), and clicking **Apply** gives the result shown in Figure 7.37(b).

An animated motion of the cam mechanism may also be obtained. Animations may be generated with or without the follower. For animations, it is necessary to deselect all other inputs and outputs, except for **Show Contour** (see Figure 7.38(a)). Then, using the icons shown in Figure 7.31, we have the choice to step the motion in increments of 5° of

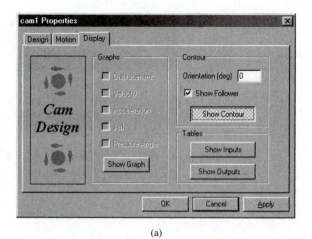

(a)

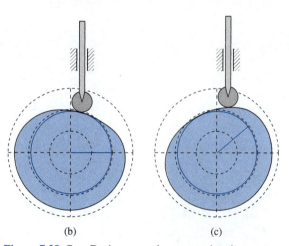

(b)                                        (c)

**Figure 7.38** Cam Design example—cam animation.

cam rotation in either the clockwise or counterclockwise direction. Alternatively, the motion may be animated. Figures 7.38(b) and 7.38(c) illustrate two positions of the cam mechanism.

Under the **File** menu, we may save the above parameters using a desired file name. The above cam mechanism is saved as **cam1.** Then we proceed to return to the **Motion Page** and change the displacement diagram. The rotation of the cam dividing the second and third intervals is altered from 260° to 300° (Figure 7.39(a)). This modified cam mechanism and displacement diagram are shown in Figures 7.39(b) and 7.39(c), respectively. It is then saved as **cam2.**

Two or more files may be opened simultaneously. Figure 7.40(a) shows the case when both **cam1** and **cam2** are opened at the same time. When more than one file is open, it is possible to superimpose graphs of displacement and the time derivatives of displacement, as well as the cam contours. The icon shown in Figure 7.31 is used to activate the

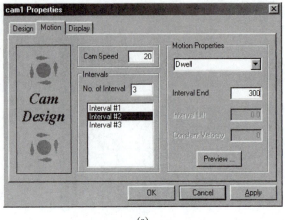

(a)

| Function | Absolute Maximum | Maximum at (deg) |
|---|---|---|
| Follower Displacement | 8.000e−001 | 100.0 |
| Follower Velocity | 2.400e+001 | 330.0 |
| Follower Acceleration | 1.440e+003 | 300.0 |
| Follower Jerk | 8.640e+004 | 330.0 |
| Pressure Angle | 3.288e+001 | 334.0 |

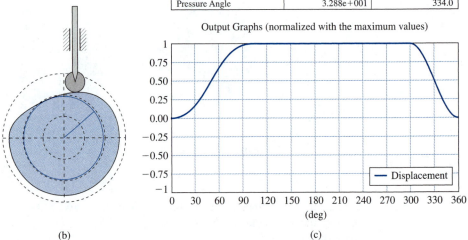

(b)                                          (c)

**Figure 7.39** Cam Design example—modification of displacement diagram.

superimpose option. Then, the screen as shown in Figure 7.40(b) appears. We then indicate the desired files to be superimposed by first highlighting them and then clicking the **Add >>** button. If **cam1** and **cam2** are added, we obtain the screen as shown in Figure 7.40(c). If we select to superimpose graphs by clicking **OK,** then we obtain the result shown in Figure 7.40(d). Alternatively, if we select to superimpose the contours, we obtain the result shown in Figure 7.40(e).

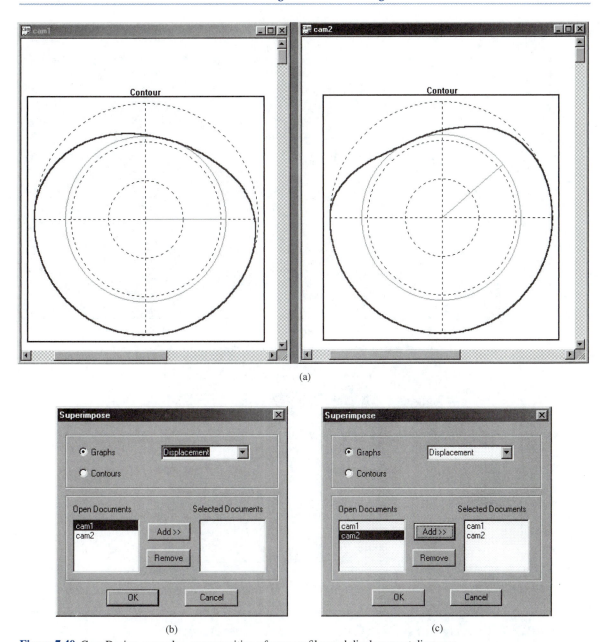

Figure 7.40 Cam Design example—superposition of cam profiles and displacement diagrams.

This program can be used effectively as a design tool to search for a cam mechanism that has a prescribed motion while subject to a variety of constraints, such as space and maximum pressure angle. It may also be used to check for the existence of undercutting, and to design a cam mechanism where undercutting does not exist.

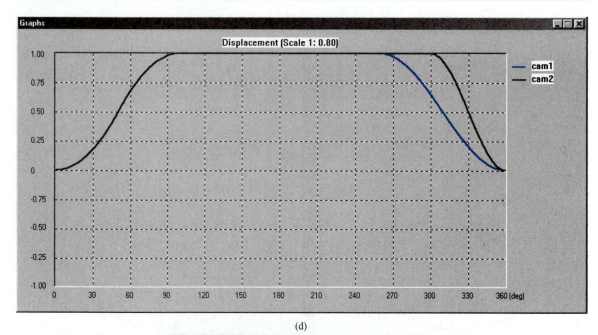

(d)

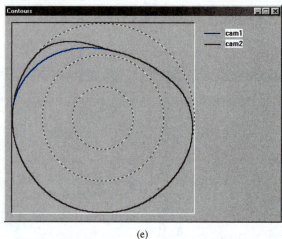

(e)

**Figure 7.40**  (*Continued*)

## EXAMPLE 7.2  Design of a Disc Cam to Reduce Pressure Angle

Consider the displacement parameters given in Table 7.3. The corresponding displacement diagram is illustrated in Figure 7.41.

In addition, we have the following parameters:

- translating roller follower
- base circle diameter: 4.0 cm
- roller circle diameter: 1.0 cm
- offset: 0.0 cm

**TABLE 7.3** Follower Motion

| Parameter →<br>Interval ↓ | Initial Angle for Interval (degrees) | Final Angle for Interval (degrees) | Type of Motion | Rise (+) or Return (−) (cm) |
|---|---|---|---|---|
| 1 | 0 | 50 | Dwell | 0 |
| 2 | 50 | 120 | Parabolic | 1.5 |
| 3 | 120 | 160 | Dwell | 0 |
| 4 | 160 | 270 | Harmonic | −0.7 |
| 5 | 270 | 360 | Cycloidal | −0.8 |

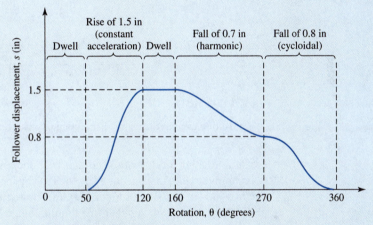

**Figure 7.41** Displacement diagram.

Based on the above parameters and motions, Figure 7.42(a) shows the corresponding cam mechanism, and Figure 7.42(b) gives a plot of the pressure angle as a function of cam rotation. The maximum pressure angle during a cycle is 37.1°. Figure 7.42(a) illustrates the configuration of the cam mechanism where the pressure angle is maximum.

Design a cam mechanism for which the pressure angle is always less than 30° through changing

(a) the diameter of the base circle
(b) the diameter of the roller follower
(c) the offset of the follower
(d) the displacement diagram

Employ the program Cam Design.

**SOLUTION**

We consider the disc cam profile shown in Figure 7.42(a) as the *original profile*.

(a) Increasing the diameter of the base circle will generally decrease the maximum value of pressure angle. However, this will be accomplished at the expense of

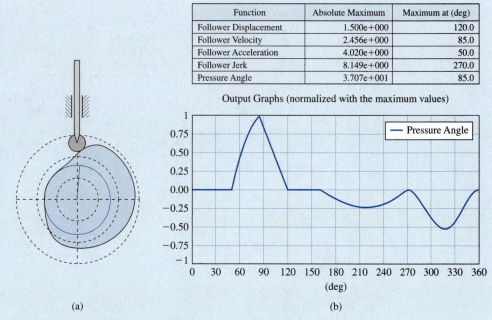

| Function | Absolute Maximum | Maximum at (deg) |
|---|---|---|
| Follower Displacement | 1.500e+000 | 120.0 |
| Follower Velocity | 2.456e+000 | 85.0 |
| Follower Acceleration | 4.020e+000 | 50.0 |
| Follower Jerk | 8.149e+000 | 270.0 |
| Pressure Angle | 3.707e+001 | 85.0 |

(a)                                          (b)

**Figure 7.42**  Original disc cam mechanism—distribution of pressure angle.

creating a larger mechanism. Through trial and error we are able to reduce the maximum pressure angle from 37.1° to 30.0° by increasing the base circle diameter from 4.0 cm to 6.01 cm. Figure 7.43(a) shows the configuration of the mechanism where the pressure angle is maximum. Figure 7.43(b) shows a superposition of the original and modified cam profiles.

(b) We return the diameter of the base circle back to 4.0 cm and now increase the diameter of the roller follower. Through trial and error, by increasing the roller diameter from 1.0 cm to 3.01 cm, the maximum pressure angle has been reduced from 37.1° to 30.0°. Figure 7.44(a) illustrates the disc cam mechanism in the configuration of maximum pressure angle. Figure 7.44(b) shows a superposition of the original and modified cam profiles. The greatest difference between the profiles occurs during the steep rise of the follower.

(c) We return the diameter of the roller follower back to 1.0 cm. Several values of offset are now tried in a search for the minimum value of the maximum pressure angle throughout a cycle. The optimum offset is 0.606 cm. Figure 7.45(a) shows the corresponding cam mechanism. The maximum pressure angle has been reduced from 37.1° to 30.2°. Figure 7.45(b) shows a superposition of the original and modified profiles. In this instance, it was not possible to reduce the maximum pressure angle to 30.0° solely by changing the follower offset. However, a designer has the option to vary more than one parameter in order to achieve a desired result. By increasing the base circle diameter from 4.0 cm to 4.5 cm, with zero offset, the maximum pressure angle reduces to 35.1°. Then, through trial and error,

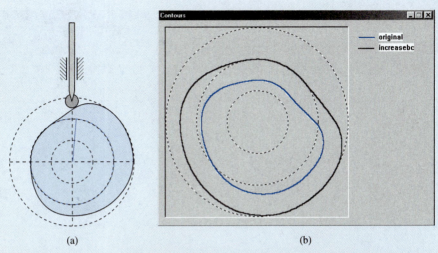

**Figure 7.43** Disc cam mechanism with increased base circle diameter: (a) configuration of maximum pressure angle, (b) superposition of original and modified cam profiles.

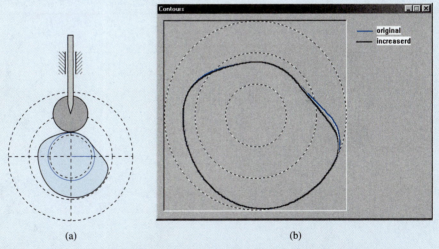

**Figure 7.44** Disc cam mechanism with increased roller diameter: (a) configuration of maximum pressure angle, (b) superposition of original and modified cam profiles.

the pressure angle may be reduced to 30.0° by incorporating an offset of 0.45 cm. This result is illustrated in Figure 7.46.

(d) We return the offset back to zero and the base circle diameter to 4.0 cm. For the results presented in Figure 7.42(b), the configuration corresponding to the largest value of pressure angle occurs near the steepest slope of the displacement

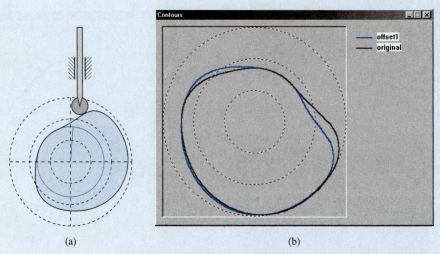

(a)                                        (b)

**Figure 7.45** Disc cam mechanism with offset: (a) configuration of maximum pressure angle, (b) superposition of original and modified cam profiles.

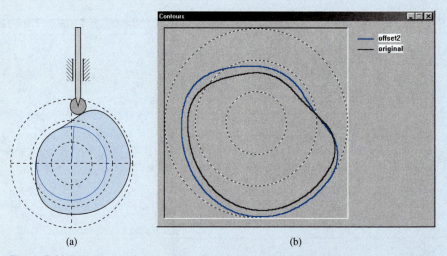

(a)                                        (b)

**Figure 7.46** Disc cam mechanism with offset and increased base circle diameter: (a) configuration of maximum pressure angle, (b) superposition of original and modified cam profiles.

diagram. Thus, in order to reduce this large pressure angle, we may reduce the slope of the displacement diagram. This may be accomplished by reducing the initial dwell, which spans 0° to 50°. Through trial and error, if the initial dwell is reduced, to span 0° to 29.0°, then the maximum pressure angle is reduced to 30.0°. Figure 7.47 shows the modified displacement diagram. Figure 7.48(a) shows the cam mechanism in the configuration of maximum pressure angle. Figure 7.48(b) shows a superposition of the new and original profiles.

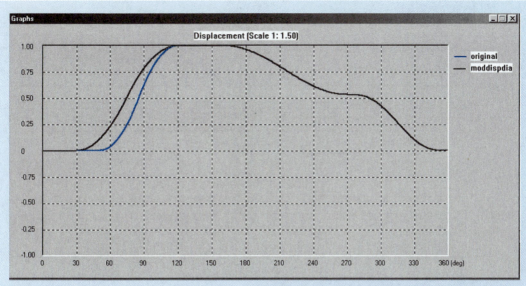

**Figure 7.47** Modified displacement diagram.

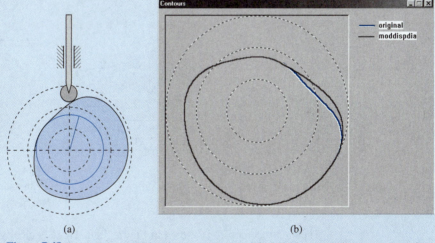

(a)                                                                      (b)

**Figure 7.48** Disc cam mechanism incorporating modified displacement diagram: (a) configuration of maximum pressure angle, (b) superposition of original and modified cam profiles.

## EXAMPLE 7.3  Design of a Disc Cam to Eliminate Undercutting

Consider the displacement parameters given in Table 7.4. The corresponding displacement diagram is illustrated in Figure 7.49.

In addition, we have the following parameters:

- translating flat-face follower
- base circle diameter: 15 cm

**TABLE 7.4**  Follower Motion

| Interval ↓ \ Parameter → | Initial Angle for Interval (degrees) | Final Angle for Interval (degrees) | Type of Motion | Rise (+) or Return (−) (cm) |
|---|---|---|---|---|
| 1 | 0 | 50 | Dwell | 0 |
| 2 | 50 | 100 | Harmonic | 2.5 |
| 3 | 100 | 360 | Parabolic | −2.5 |

| Function | Absolute Maximum | Maximum at (deg) |
|---|---|---|
| Follower Displacement | 2.500e+000 | 100.0 |
| Follower Velocity | 4.500e+000 | 75.0 |
| Follower Acceleration | 1.620e+001 | 50.0 |
| Follower Jerk | 5.832e+001 | 75.0 |
| Length | 4.500e+000 | 75.0 |

Output Graphs (normalized with the maximum values)

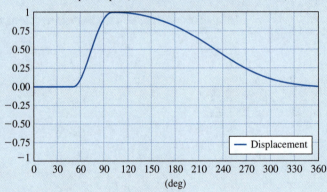

**Figure 7.49**  Displacement diagram.

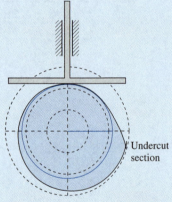

**Figure 7.50**  Disc cam mechanism with undercutting.

The cam mechanism is shown in Figure 7.50. There is clearly a section of the profile that has been undercut.

Design a cam for which undercutting is eliminated by changing

(a) the diameter of the base circle
(b) the type of follower

Employ the program Cam Design.

SOLUTION

(a) Figure 7.51 shows the result when the base circle diameter has been increased to 25 cm. The undercut section has been completely eliminated.

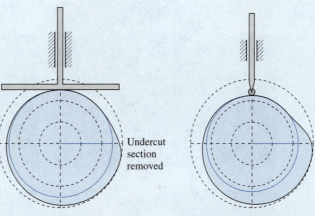

**Figure 7.51** Disc cam with increased base circle diameter.

**Figure 7.52** Disc cam mechanism implementing roller follower.

(b) We return the value of the base circle back to 15 cm and now replace the flat-face follower with a roller follower. The roller diameter is 1.0 cm. Figure 7.52 shows the cam mechanism with no undercutting.

## 7.11 MANUFACTURING OF CAMS

Cams are manufactured using a wide range of methods. Depending on the quantity to be produced, speed of operation, load rating, and required accuracy of the motion, cams are made from a variety of materials and may be fabricated by hand, or molded, or machined. This section presents two methods of manufacturing cams; one for low-volume production and the other for mass production.

For low-volume production of disc cams, the shape of the cam profile may be transcribed by hand onto a cam blank. This technique is used for the fabrication of cams used in moving-headstock milling machines (Figure 7.2). The steps are illustrated in Figure 7.53. First, radial lines are drawn from a center on the blank to segregate portions of motions and dwells of the follower (Figure 7.53(a)). Circular arcs are then scribed, using the common center. These arcs correspond to the periods of dwell. The contours for the rises and falls are then added (Figure 7.53(b)). This is followed by completing a rough cut of the profile using a band saw (Figure 7.53(c)). Then a fine cut is taken using a shaping machine (Figure 7.53(d)). Finally, the cam profile is smoothed as necessary by hand.

Video 7.53A1,
Video 7.53A2,
Video 7.53B,
Video 7.53C,
Video 7.53D
Manufacture of a
Disc Cam

Cams for internal combustion engines are usually machined from a solid shaft, so all cams on the shaft are rigidly connected and will be driven at the same speed. Figure 1.1 shows a one-cylinder engine with two cams on the camshaft. Most internal combustion engines are designed with two cams for each cylinder.

In a modern facility for the mass production of camshafts, the entire process is highly automated. The process starts with a solid cylindrical shaft, which is accurately cut to length by machining the two end surfaces. Typically four holes are then drilled in one end of the shaft, all offset from its centerline. Three of the holes are tapped and will be used

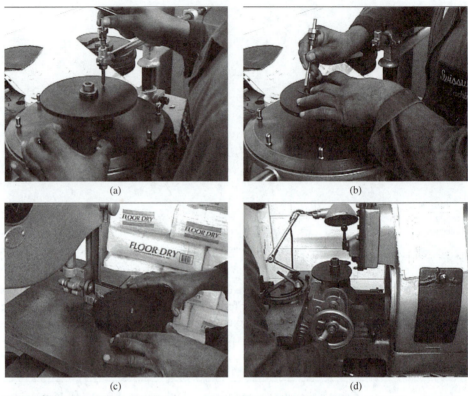

**Figure 7.53** Manufacture of a disc cam: (a) scribing radial lines [Video 7.53A1] and circular arcs [Video 7.53A2], (b) scribing rises and falls [Video 7.53B], (c) rough cut using band saw [Video 7.53C], (d) fine cut using shaper [Video 7.53D].

later for attaching a pulley or timing gear to drive the camshaft. A tapered pin is press-fit in the hole that is not tapped. This pin will be used for controlling the rotations of the shaft in the subsequent manufacturing operations.

Multiple annular slots are cut on the circular surface, spaced along the length of the shaft. These slots will become the gaps between the cams. The cam profiles are then machined. In machining each cam profile, the shaft is given one revolution at a constant rate, taking approximately two seconds. During rotation, a milling cutter machines a cam profile. The milling cutter, rotating on its own axis, is actuated to move radially with respect to the centerline of the camshaft. This radial motion is computer controlled and mimics the motion of the follower in the engine. This process is carried out for all cams on the camshaft. Between the machining of cam profiles, the milling cutter is repositioned by moving it parallel to the axis of the camshaft.

The camshafts are automatically transferred to other machines for the finish grinding of the cam profiles and camshaft bearings. There is a separate grinding wheel for each cam profile and bearing. The grinding wheels are also computer controlled, similar to the radial movements used for the milling operation.

At a subsequent station, the cam profiles only are heat treated. The camshafts are later fed through a machine where they are wire brushed to remove burs and scaling caused by the heat treatment process. The camshafts are then transferred to a polishing machine. Here, a polishing compound is applied, and the camshaft is rotated several revolutions in both directions. The polishing is carried out using long bands of polishing cloth that are applied to the working surfaces. In the next machine, the camshafts are washed to remove the polishing compound and subjected to final inspection.

# PROBLEMS

**P7.1**  A disc cam's translating roller follower with zero off-set is to rise 3.0 cm with simple harmonic motion in 160° of cam rotation and return with simple harmonic motion in the remaining 200°. If the roller radius is 0.50 cm, and the prime circle radius is 3 cm, construct the displacement diagram, pitch curve, and cam profile for counterclockwise cam rotation.

**P7.2**  The follower shown in Figure P7.2 is to be pivoted 35° clockwise with harmonic motion during a 120° counterclockwise turn of the cam, then allowed to dwell during a 30° counterclockwise turn, then return to the starting point shown, with uniformly accelerated and decelerated motion. Sketch the cam profile surface during the rise motion of the follower, by considering six equally spaced rotations of the cam during the rise.

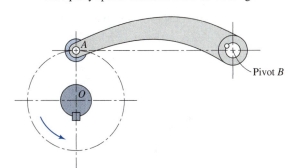

**Figure P7.2**

**P7.3**  A disc cam with a translating flat-face follower is to have the same motion as in Problem 7.1. The prime circle is to have a radius of 3.0 cm, and the cam is to rotate clockwise. Construct the displacement diagram and the cam profile, offsetting the follower stem by 1.0 cm in the direction that reduces the bending stress in the follower during the rise.

**P7.4**  Construct the displacement diagram and the cam profile for the disc cam with a pivoting flat-face follower that rises through 10° with cycloidal motion in 150° of counterclockwise cam rotation, then

dwells for 30°, returns with cycloidal motion in 120°, and dwells for 60°. Graphically determine the necessary length of the follower face, allowing 5.0 mm clearance at each end. The prime circle is to have a radius of 40 mm. The follower pivot arm is 120 mm to the right. The cam rotation is counterclockwise.

**P7.5**  The motion of the mechanism shown in Figure P7.5 is to be generated by means of a disc cam, rotating counterclockwise about point $O_6$. Starting from the configuration shown, link 4 is to move from position $A$ to position $B$ during 120° of cam rotation. Each number on the path of link 4 represents 30° of cam rotation. After link 4 reaches $B$, it is to remain

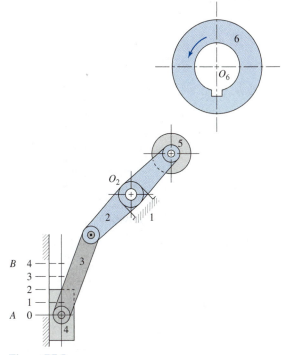

**Figure P7.5**

stationary for 60° of cam rotation, and then return as quickly as possible to position $A$, where it stays during the remainder of the cycle.

(a) Construct the cam profile.
(b) Determine the pressure angles of the cam follower corresponding to cam rotations of 30°, 60°, and 90°.

**P7.6**  A disc cam mechanism is to be used for a vibration platform. The follower motion is to have rise of 2.0 cm with constant acceleration in 0.5 seconds, dwell for 0.3 seconds, and fall with constant acceleration in 0.7 seconds. There is a translating roller follower with zero offset.

(a) Determine the rotational speed of the disc.
(b) Plot the displacement diagram.
(c) Graphically construct the cam profile.
(d) Determine analytically the maximum follower velocity and acceleration.

$$d_b = 3.0 \text{ cm}; \qquad d_r = 1.0 \text{ cm}$$

**P7.7**  A disc cam mechanism is to provide the motion given in Table P7.7. There is a translating roller follower with zero offset.

(a) Plot the displacement diagram.
(b) Graphically construct the cam profile.

$$d_b = 4.0 \text{ cm}; \qquad d_r = 1.0 \text{ cm}$$

**P7.8**  A disc cam mechanism is to provide the motion given in Table P7.8. There is a translating flat-face follower.

(a) Plot the displacement diagram.
(b) Graphically construct the cam profile.

$$d_b = 4.0 \text{ cm}$$

**P7.9**  A disc cam mechanism is to provide the motion given in Table P7.9. There is a translating roller follower with offset.

(a) Plot the displacement diagram.
(b) Graphically construct the cam profile.

$$d_b = 4.0 \text{ cm}; \qquad d_r = 1.0 \text{ cm}; \qquad S = 0.5 \text{ cm}$$

**P7.10**  A disc cam mechanism is to provide the motion given in Table P7.10. There is a translating roller follower. The nominal dimensions are provided. Employ the program Cam Design and reduce the peak value of pressure angle to 30° by

(a) increasing the base circle diameter
(b) increasing the roller diameter
(c) adjusting the offset

$$d_b = 2.0 \text{ cm}; \qquad d_r = 1.0 \text{ cm}; \qquad S = 0$$

**P7.11**  A disc cam mechanism is to provide the motion given in Table P7.11. There is a translating flat-face follower. The nominal base circle diameter is provided. For these parameters there is undercutting of

**TABLE P7.7**  Follower Motion

| Parameter →<br>Interval ↓ | Initial Angle for Interval (degrees) | Final Angle for Interval (degrees) | Type of Motion | Rise (+) or Return (−) (cm) |
|---|---|---|---|---|
| 1 | 0 | 100 | Harmonic | 1.2 |
| 2 | 100 | 240 | Dwell | 0 |
| 3 | 240 | 360 | Parabolic | −1.2 |

**TABLE P7.8**  Follower Motion

| Parameter →<br>Interval ↓ | Initial Angle for Interval (degrees) | Final Angle for Interval (degrees) | Type of Motion | Rise (+) or Return (−) (cm) |
|---|---|---|---|---|
| 1 | 0 | 120 | Cycloidal | 1.2 |
| 2 | 120 | 200 | Dwell | 0 |
| 3 | 200 | 360 | Uniform | −1.2 |

**TABLE P7.9**  Follower Motion

| Parameter →<br>Interval ↓ | Initial Angle<br>for Interval<br>(degrees) | Final Angle<br>for Interval<br>(degrees) | Type of<br>Motion | Rise (+) or<br>Return (−)<br>(cm) |
|---|---|---|---|---|
| 1 | 0 | 80 | Harmonic | 1.5 |
| 2 | 80 | 200 | Dwell | 0 |
| 3 | 200 | 360 | Parabolic | −1.5 |

**TABLE P7.10**  Follower Motion

| Parameter →<br>Interval ↓ | Initial Angle<br>for Interval<br>(degrees) | Final Angle<br>for Interval<br>(degrees) | Type of<br>Motion | Rise (+) or<br>Return (−)<br>(cm) |
|---|---|---|---|---|
| 1 | 0 | 100 | Harmonic | 1.4 |
| 2 | 100 | 220 | Dwell | 0 |
| 3 | 220 | 360 | Uniform | −1.4 |

**TABLE P7.11**  Follower Motion

| Parameter →<br>Interval ↓ | Initial Angle<br>for Interval<br>(degrees) | Final Angle<br>for Interval<br>(degrees) | Type of<br>Motion | Rise (+) or<br>Return (−)<br>(cm) |
|---|---|---|---|---|
| 1 | 0 | 130 | Harmonic | 3.6 |
| 2 | 130 | 280 | Dwell | 0 |
| 3 | 280 | 360 | Harmonic | −3.6 |

**TABLE P7.12**  Follower Motion

| Parameter →<br>Interval ↓ | Initial Angle<br>for Interval<br>(degrees) | Final Angle<br>for Interval<br>(degrees) | Type of<br>Motion | Rise (+) or<br>Return (−)<br>(cm) |
|---|---|---|---|---|
| 1 | 0 | 90 | Harmonic | 1.3 |
| 2 | 90 | 200 | Dwell | 0 |
| 3 | 200 | 360 | Uniform | −1.3 |

the cam profile. Employ the program Cam Design and eliminate the undercutting by increasing the base circle diameter.

$$d_b = 4.0 \text{ cm}$$

**P7.12** A disc cam mechanism is to provide the motion given in Table P7.12. There is a translating knife-edge follower. The nominal dimensions are provided. Employ the program Cam Design and reduce the peak value of pressure angle to 30° by

(a) increasing the base circle diameter.
(b) adjusting the offset.

$$d_b = 3.0 \text{ cm}; \qquad S = 0$$

# 8 Graphical Force Analysis of Planar Mechanisms

## 8.1 INTRODUCTION

This chapter presents a method of completing force analyses of mechanisms under static or dynamic conditions. It is possible to determine the driving force or torque required either to hold the mechanism in a stationary configuration or to generate a specified motion. In addition, internal loads in links, and forces across kinematic pairs, may be determined.

To make a dynamic analysis, the mechanism will be considered to be in a state of dynamic equilibrium. Thus, similar procedures may be employed for static and dynamic analyses.

If a mechanism is in static or dynamic equilibrium, every link within the mechanism must also be in equilibrium. Sections 8.2 and 8.3 describe two basic cases for which links of a mechanism are in equilibrium. These basic cases are subsequently used in graphical force analyses of mechanical systems.

## 8.2 TWO-FORCE MEMBER

A link acted on by only two forces is known as a *two-force member*. In order for a link to qualify as a two-force member, no moments can be applied. For instance, Figure 8.1(a) illustrates a four-bar mechanism in which moments $M_{O_2}$ and $M_{O_4}$ are applied from the base link to links 2 and 4, respectively. Therefore, these two links do not qualify as two-force members. However, neglecting gravity, inertia, and friction in the turning pairs at $A$ and $B$, link 3 can only be subjected to forces applied through the turning pairs. A free body diagram of the link is shown in Figure 8.1(b). It is subjected to forces $\overline{F}_1$ and $\overline{F}_2$. In this figure, the directions and magnitudes of these forces have been arbitrarily depicted. However, in order for link 3 to remain in equilibrium, the sum of the applied forces and sum of the

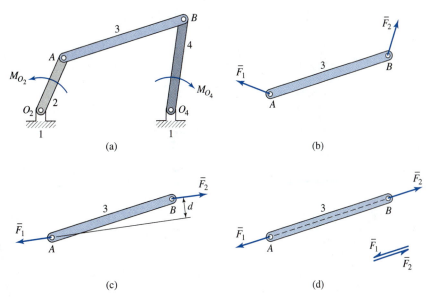

**Figure 8.1** Two-force member analysis.

applied moments must be zero. The sum of the applied forces is

$$\bar{F}_1 + \bar{F}_2 = \bar{0} \tag{8.2-1}$$

or

$$\bar{F}_1 = -\bar{F}_2 \tag{8.2-2}$$

that is, the two applied forces are equal in magnitude and opposite in direction. Figure 8.1(c) illustrates such a condition. However, as shown, the two forces in Figure 8.1(c) provide a couple of magnitude $F_1 d$, where $d$ is the perpendicular distance between the lines of action. For the net moment to be zero, the distance $d$ must be zero; thus, the two forces must be collinear. Figure 8.1(d) shows the case where the two forces are applied such that the net force and the moment acting on the link are zero.

In summary, for a member to be in equilibrium when acted on by two forces, the two forces must be equal in magnitude, collinear, and opposite in direction.

## 8.3 THREE-FORCE MEMBER

A link acted on by only three forces is known as a *three-force member*. No moments can be applied. Consider the mechanism shown in Figure 8.2(a). Links 2 and 4 do not qualify as three-force members, because moments are applied to them. However, neglecting gravity, inertia, and friction in the kinematic pairs, link 3 is subjected to only three forces. A free body diagram of the link is shown in Figure 8.2(b). One of the forces, $\bar{F}_3$, is assumed known and is applied through point $C$. The direction and magnitude of $\bar{F}_1$ and $\bar{F}_2$ have

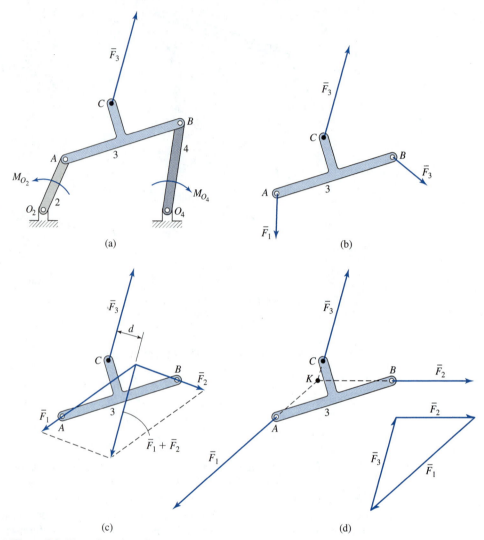

**Figure 8.2** Three-force member analysis.

been arbitrarily chosen. For this body to remain in equilibrium, the sum of applied forces and the sum of applied moments must be zero. Summing the applied forces to zero gives

$$\bar{F}_1 + \bar{F}_2 + \bar{F}_3 = \bar{0} \tag{8.3-1}$$

or

$$\bar{F}_3 = -(\bar{F}_1 + \bar{F}_2) \tag{8.3-2}$$

Figure 8.2(c) illustrates a case that satisfies Equation (8.3-2). However, as shown, the three forces provide a force couple of magnitude $F_3 d$, and thus link 3 is not in equilibrium.

In order for the force couple to vanish, distance $d$ must equal zero, which is accomplished by having the three forces concurrent at some point. Figure 8.2(d) shows the three lines of action concurrent at point $K$. As shown in this example, the point of concurrency need not lie within the physical bounds of the link.

In summary, for a member acted on by three forces to be in equilibrium, the forces must add vectorially to zero and must have their lines of action concurrent at some point.

## 8.4 FORCE TRANSMISSION IN FRICTIONLESS KINEMATIC PAIRS

Forces and moments in a mechanism are transmitted through kinematic pairs of adjoining links. In this chapter, we restrict ourselves to frictionless kinematic pairs. In such instances, forces are transmitted perpendicular to the surfaces of members in contact. If friction and gravity are neglected for a sliding pair, the force transmitted between the members is perpendicular to the direction of the slide. Figure 8.3(a) illustrates a slider crank mechanism with a force, $\bar{P}$, applied to the slider and a moment, $M$, applied to the crank. A free body diagram of the slider is shown in Figure 8.3(b). The line of action of the force *from* link 3 *on* link 4, designated as $\bar{F}_{34}$, is known since link 3 is a two-force member. Force $\bar{F}_{14}$ is drawn perpendicular to the surfaces in contact, and thus is also perpendicular to the direction of the slide.

A frictionless rolling pair is shown in Figure 8.4(a). Here, the transmitted force is perpendicular to the surfaces of the links at the point of contact. A free body diagram of the roller is shown in Figure 8.4(b). Neglecting rolling friction, the roller is a two-force member.

For a turning pair, the direction of the force transmitted is dependent on the relative position of the other kinematic pairs in the kinematic chain, and cannot be determined by examining the turning pair alone. Considering the mechanism shown in Figure 8.4, there is a turning pair between links 2 and 3. The direction of the force transmitted between these two links depends on the direction of the force between links 2 and 4. As shown, the two forces acting on link 2 must be collinear for the link to remain in static equilibrium.

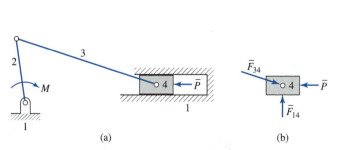

Figure 8.3  Free body diagram of a frictionless sliding pair:
(a) mechanism, (b) free body diagram.

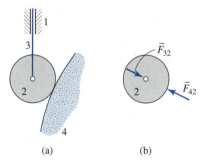

Figure 8.4  Free body diagram of a frictionless rolling pair.

## 8.5 FORCE POLYGONS

For each three-force member within a mechanism, a *force polygon* may be constructed as an aid in determining either directions or magnitudes of forces. As an example, consider the three-force member shown in Figure 8.5(a). Let points $A$, $B$, and $C$ represent kinematic pairs with adjacent links of the mechanism, or locations of externally applied loads to the mechanism. At $B$, both the direction and magnitude of the applied force are known. Also, the line of action of the force at $C$ is given. It is now required to determine the remaining information about the forces at $A$ and $C$. This consists of the force magnitude and direction at $A$ and $C$.

We begin by extending the known lines of action of forces through $B$ and $C$. These lines intersect at point $K$ as shown in Figure 8.5(b). Since this is a three-force member, the line of action of the force at $A$ must pass through the same intersection point. Now that we have all three lines of action, we can construct the force polygon by first drawing in the known force through point $B$ using an appropriate scale. At the head and tail of this vector,

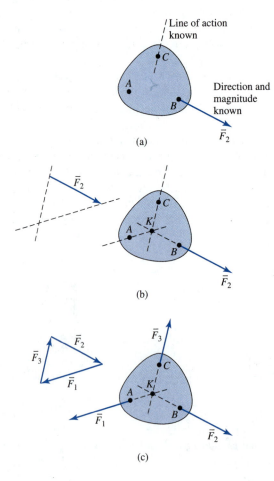

**Figure 8.5** Construction of a force polygon associated with a three-force member: (a) given information, (b) construction of force polygon, (c) final result.

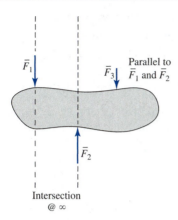

**Figure 8.6**

we proceed to draw the other two lines of action. The intersection point of these lines of action dictates the shape of the force polygon. The polygon is completed, as shown in Figure 8.5(c), by labelling the vectors, all head to tail to ensure that

$$\bar{F}_1 + \bar{F}_2 + \bar{F}_3 = \bar{0} \tag{8.5-1}$$

Note that for a force polygon, unlike velocity and acceleration polygons, there is no pole point.

A special case occurs when two of the forces acting on a three-force member are parallel, as shown in Figure 8.6, where their intersection occurs at infinity. It follows that the third force acting on the member must be parallel to the other two. In this instance, it is not possible to draw a force polygon to determine the unknown quantities. Instead, it is necessary to apply Equations (2.12-2) and (2.12-3) directly.

## 8.6 STATIC FORCE ANALYSIS USING FORCE POLYGON METHOD

Utilizing force polygons as presented in Section 8.5, it is possible to determine the static forces transmitted throughout an entire mechanism. The force polygon method is suitable for only one configuration.

The method consists of the following general steps:

1. Draw a diagram of the mechanism to scale, in the configuration for which an analysis is required.
2. Break the mechanism into a suitable number of subsystems (either individual links or groups of links) and, if required for clarification, draw a free body diagram for each.
3. Complete the force polygons for those subsystems which can be analyzed immediately (such as those subjected to known external forces).
4. Use results from step 3 to construct the remaining force polygons by examining the forces transmitted between subsystems.

# EXAMPLE 8.1  Static Force Analysis of a Slider Crank Mechanism

Consider a slider crank mechanism shown in Figure 8.7 having zero offset and with

$$\theta_2 = 120°; \qquad r_2 = 1.2 \text{ in}; \qquad r_3 = 4.8 \text{ in}$$

A 250 lb force is applied to the slider as shown. Determine the moment required from the base link onto link 2 to keep the mechanism in static equilibrium.

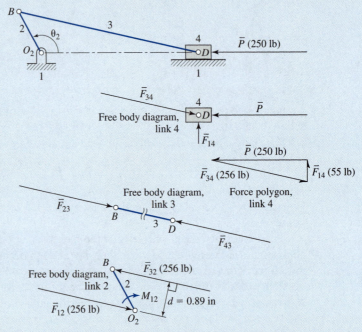

**Figure 8.7**  Static force analysis of a slider crank mechanism.

### SOLUTION

Free body diagrams of links 3 and 4 are shown in Figure 8.7. We note that link 3 is a two-force member, and link 4 is a three-force member. Therefore, for link 4

$$\overline{P} + \overline{F}_{14} + \overline{F}_{34} = \overline{0} \qquad (8.6\text{-}1)$$

Assuming frictionless kinematic pairs, $\overline{F}_{14}$ is perpendicular to the direction of the slide. Also, the force from link 3 is along the direction defined by points $B$ and $D$. For link 4, we construct the force polygon, shown in Figure 8.7.

We now consider the analysis of link 3, and note that at point $D$, based on Newton's third law of mechanics (Appendix D),

$$\overline{F}_{43} = -\overline{F}_{34} \qquad (8.6\text{-}2)$$

Since link 3 is a two-force member,

$$\overline{F}_{23} = -\overline{F}_{43} \tag{8.6-3}$$

Now consider the force transmitted from link 3 onto link 2. We note that

$$\overline{F}_{32} = -\overline{F}_{23} \tag{8.6-4}$$

To keep link 2 in static equilibrium, the sum of the forces and sum of the moments acting on this link must be zero. The sum of the forces being zero is accomplished by setting

$$\overline{F}_{12} = -\overline{F}_{32} \tag{8.6-5}$$

However, $\overline{F}_{12}$ and $\overline{F}_{32}$ are not collinear. Therefore, link 2 is not a two-force member. Thus, in addition to the forces, an external moment must be applied to the link to counteract the couple produced by the forces. The magnitude of the moment is

$$M_{12} = |\overline{F}_{32}|d \tag{8.6-6}$$

where $d$ is the perpendicular distance between the two lines of action. This moment must be applied in the clockwise direction to counteract the counterclockwise moment created by the force couple. Therefore

$$M_{12} = |\overline{F}_{32}|d = 256 \times 0.89 = 227 \, \text{in-lb CW}$$

## EXAMPLE 8.2 Static Force Analysis of a Lever Mechanism

For the mechanism shown in Figure 8.8(a), determine the magnitude and sense of the torque to be applied to link 2 by the base link to keep the mechanism in static equilibrium while being subjected to a force of magnitude 50 N.

$$r_{O_2B} = 3.0 \, \text{cm}$$

### SOLUTION

The lines of action of the forces are shown in Figure 8.8(b). For link 5, the line of action of force $\overline{F}$ and the line of action of the force from link 6 intersect at $K_1$. Then the line of action of the third force on link 5, through the turning pair at $C$, is drawn through this intersection point. For link 3, the line of action of the force through $C$ intersects the line of action of the force acting on link 3 from link 4 at $K_2$. The line of action of the third force acting on link 3, through the turning pair at $B$, is drawn through this intersection point.

Force polygons of links 5 and 3 are also shown in Figure 8.8(b). Since

$$\overline{F}_{32} = -\overline{F}_{23}$$

then we construct a free body diagram of link 2 as shown in Figure 8.8(b). The moment that must be supplied to link 2 is

$$M_{12} = |\overline{F}_{32}|d = 26 \times 2.8 = 73 \, \text{N-cm CW}$$

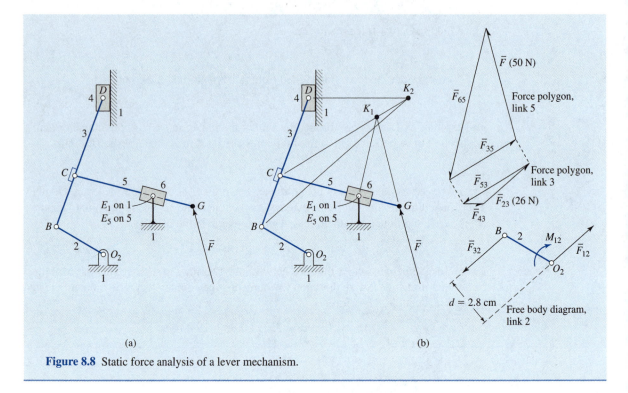

(a)    (b)

**Figure 8.8** Static force analysis of a lever mechanism.

# EXAMPLE 8.3  Static Force Analysis of a Front-End Loader

Consider the front-end loader mechanism shown in Figure 8.9. The mobility of this mechanism is equal to two. The two inputs are provided by the linear actuators, links 5 and 7. For static analysis, it is permissible to designate the linear actuators as two-force members.

For the position shown, a load of 5,000 lb is applied to the bucket, link 6. Determine

    (a)  the magnitude of the force in the actuators, links 5 and 7
    (b)  the magnitude of the bending moment at $B$

$$r_{AB} = 32 \, \text{in}$$

### SOLUTION

(a)  Link 2 is a two-force member. Links 4 and 6 are three-force members.

We begin the analysis by considering link 6. Both the magnitude and direction of the load are known. Since link 7 is a two-force member, the line of action of the force at point $F$ is known. The lines of action of these two forces intersect at point $K_1$. The line of action of the force through $D$ must also pass through $K_1$. Using the three lines of action, we complete the first force polygon associated with link 6.

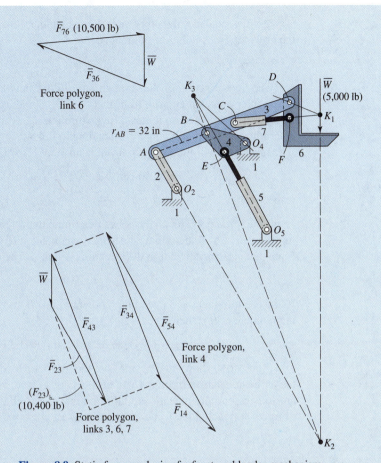

**Figure 8.9** Static force analysis of a front-end loader mechanism.

Because of the equal and opposite directions of interactive forces, we can easily determine the forces acting on link 3 at points $C$ and $D$. In addition, there are two other forces acting on link 3 through the turning pairs located at $A$ and $B$.

Link 3 can be referred to as a four-force member because it is connected to other links through four turning pairs. We are not able to solve for the forces applied to this link using the cases presented in Sections 8.2 and 8.3. Instead, it is necessary to consider a subsystem consisting of a combination of links 3, 6, and 7. This is permissible because under static conditions there is no relative motion between links.

The subsystem is subjected to three external forces. Therefore the properties of a three-force member may be applied. The line of action of the external load and that of the two-force member of link 2 intersect at point $K_2$. The line of action of the force through the turning pair at $B$ must also pass through the same intersection point. Having determined the three lines of action, we construct a force polygon corresponding to the subsystem. Included in this polygon is force $\overline{F}_{43}$.

We now consider the other three-force member, link 4. We employ the result from the previous force polygon and realize that

$$\overline{F}_{34} = -\overline{F}_{43} \tag{8.6-7}$$

The line of action of the force through $B$ has now been determined. This line of action intersects that coming from link 5 at $K_3$. The third line of action through point $O_4$ is determined because it must also go through $K_3$. From the three lines of action, we determine the force polygon corresponding to link 4.
The magnitudes of the forces in links 5 and 7 are

$$|\overline{F}_{54}| = 20,400\,\text{lb} \qquad \text{and} \qquad |\overline{F}_{76}| = 10,500\,\text{lb}$$

respectively.

(b) From the force polygon analysis, we determine the force $\overline{F}_{23}$, which acts through the turning pair at $A$. To determine the bending moment at $B$, we take the component of this force that is perpendicular to the direction of link 3. The result is

$$(F_{23})_{\perp} = 10,400\,\text{lb}$$

By inspection, there is no bending moment in link 3 at $A$. The magnitude of the bending moment at $B$ is then

$$M_B = r_{AB} \times (F_{23})_{\perp} = 32 \times 10,400 = 3.33 \times 10^5\,\text{in-lb}$$

Figure 8.10 shows two positions of the front-end loader mechanism. The speeds of the animation shown through [Model 8.10] are so slow that for all positions the mechanism is essentially in static equilibrium, that is, for all configurations, lines of action of forces for each three-force member intersect at a common point.

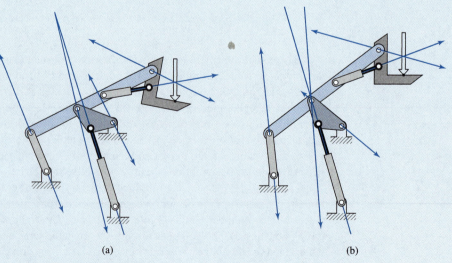

(a)                           (b)

**Figure 8.10** Front-end loader mechanism [Model 8.10].

## 8.7 PRINCIPLE OF SUPERPOSITION

When a mechanism is subjected to more than one force simultaneously, it is not possible to immediately determine their combined effect in one analysis. Instead, we may perform multiple analyses where each has only one of the forces being applied. The combined result may be found by superimposing individual results. This procedure is referred to as the *Principle of Superposition*. This principle is illustrated in Figure 8.11 and in the following example.

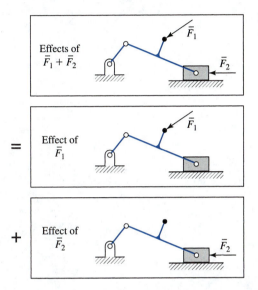

**Figure 8.11** Example of Principle of Superposition.

## EXAMPLE 8.4 Static Force Analysis of a Four-Bar Mechanism

Consider the four-bar mechanism shown in Figure 8.12(a). The mechanism is subjected to two forces, one on link 3, and the other on link 4. Determine the torque required to be supplied from the base link onto link 2 in order to keep the mechanism in equilibrium. Neglect the effect of gravity.

$$r_{O_2B} = 10 \, \text{in}$$

### SOLUTION

We begin by considering only load $\bar{P}$ on link 3 and neglecting load $\bar{Q}$ on link 4, as shown in Figure 8.12(b). In this instance, link 3 may be considered as a three-force member, whereas link 4 is a two-force member. Through the analysis shown, the force $(\bar{F}_{32})_P$ is determined.

Next, we remove load $\bar{P}$ from link 3 and apply load $\bar{Q}$ to link 4, as shown in Figure 8.12(c). In this instance, link 3 may be considered as a two-force member, and link 4 is now a three-force member. Through the analysis shown, the force $(\bar{F}_{32})_Q$ is determined.

The total load between links is found by superimposing the above results, that is,

$$\bar{F}_{32} = (\bar{F}_{32})_P + (\bar{F}_{32})_Q \tag{8.7-1}$$

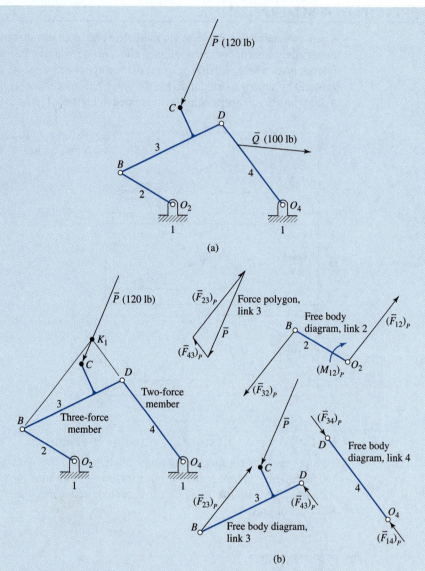

**Figure 8.12** Static force analysis of a four-bar mechanism: (a) mechanism, (b) load applied to link 3.

The result is shown in Figure 8.12(d), from which

$$|\bar{F}_{32}| = 63 \, \text{lb}$$

To maintain static equilibrium of link 2, we require

$$\bar{F}_{12} = -\bar{F}_{32} \qquad\qquad (8.7\text{-}2)$$

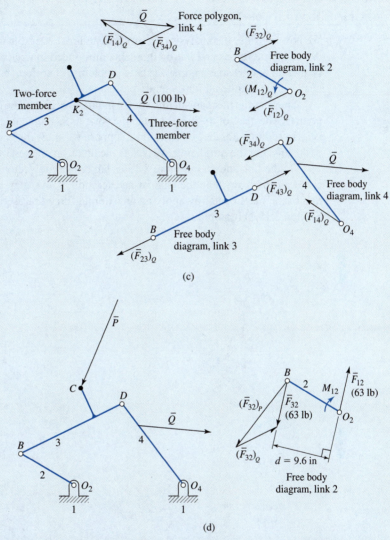

**Figure 8.12** (*Continued*) Static force analysis of a four-bar mechanism: (c) load applied to link 4, (d) superimposed loads.

In addition, a moment is supplied from the base link onto link 2. Its magnitude and direction are

$$M_{12} = |\bar{F}_{32}|d = 63 \times 9.6 = 605 \text{ in-lb CW}$$

where $d = 9.6$ in is the perpendicular distance between the lines of action of the forces acting on link 2. This moment must be supplied in the clockwise direction to counteract the counterclockwise moment created by the force couple.

## 8.8 GRAPHICAL DYNAMIC FORCE ANALYSIS OF A MECHANISM LINK—INERTIA CIRCLE

We now consider the analysis of a mechanism link under dynamic conditions. Through the use of inertia forces and inertia moments introduced in Chapter 2, dynamic problems can be treated in the same manner as static problems. Therefore, concepts introduced for static analyses are employed for dynamic problems.

Consider the mechanism link shown in Figure 8.13(a). The center of mass is indicated as point $G$. The mass of the link is $m$, and the polar mass moment of inertia about its mass center is $I_G$. In addition, for the given position, suppose we know the linear and angular accelerations of the link. For the time being, we isolate the link from the rest of the mechanism.

On the link shown in Figure 8.13(a) are drawn the vector of linear acceleration of the center of mass, $\bar{a}_G$, and the angular acceleration, $\ddot{\theta}$. We may generate the linear acceleration of the mass center by applying an external force that is parallel to $\bar{a}_G$. The force (see Section 2.10.1) required is

$$\bar{F} = m\bar{a}_G \qquad (8.8\text{-}1)$$

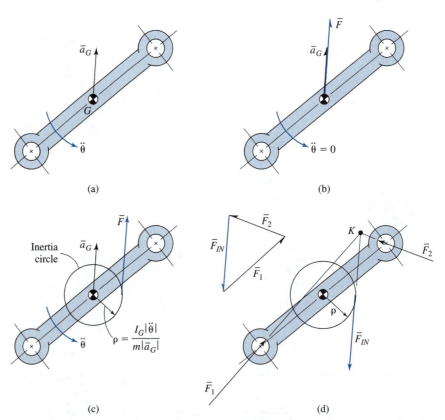

**Figure 8.13** Dynamic force analysis of a mechanism link: (a) given motion, (b) force generating linear acceleration, (c) force generating linear and angular accelerations, (d) inertia force and force polygon.

Equation (8.8-1) is independent of the location of application of $\overline{F}$. Therefore, regardless of where the force is applied, we always obtain the same acceleration of $G$. However, different angular accelerations of the link will be generated by varying the location of $\overline{F}$. If $\overline{F}$ is applied at the center of mass (Figure 8.13(b)), then it will not provide a moment about $G$; and based on Equation (2.10-4), the link will have zero angular acceleration. However, by applying the force at a position other than the mass center, the force will cause a moment about $G$. Using Equation (2.10-4), the applied moment will cause the link to undergo an angular acceleration. By applying the force as shown in Figure 8.13(c), $\overline{F}$ also causes a moment, $M_G$, about the mass center in the counterclockwise direction and as a result causes the link to accelerate in the same direction. We recognize that

$$|M_G| = |\overline{F}|\rho \qquad (8.8\text{-}2)$$

where $\rho$ is the perpendicular distance of the line of action of the force from $G$. Combining Equations (2.10-4), (8.8-1), and (8.8-2),

$$|M_G| = |\overline{F}|\rho = m|\overline{a}_G|\rho = I_G|\ddot{\theta}| \qquad (8.8\text{-}3)$$

Combining Equations (2.12-6), (2.12-9), and (8.8-3),

$$\rho = \frac{|M_G|}{|\overline{F}|} = \frac{I_G|\ddot{\theta}|}{m|\overline{a}_G|} = \frac{|M_{IN}|}{|\overline{F}_{IN}|} \qquad (8.8\text{-}4)$$

Substituting given motions in Equation (8.8-4), we may calculate $\rho$, which indicates how far distant $\overline{F}$ must be placed from $G$ in order to produce both the given linear and angular accelerations. As a guide, we draw an *inertia circle* of radius $\rho$, centered at $G$, as shown in Figure 8.13(c). Placing $\overline{F}$ tangent to the inertia circle ensures that the perpendicular distance is $\rho$.

It must be recognized, however, that acceleration of the link is not caused by $\overline{F}$. Rather, for the current illustration, forces are actually applied through turning pairs of adjacent links. Figure 8.13(d) illustrates forces $\overline{F}_1$ and $\overline{F}_2$ at the two turning pairs of the link, for which it is required that

$$\overline{F} = \overline{F}_1 + \overline{F}_2 \qquad (8.8\text{-}5)$$

Therefore, instead of $\overline{F}$, we introduce an inertia force that has the same line of action as $\overline{F}$ but points in the opposite direction, as shown in Figure 8.13(d). This inertia force when combined with forces $\overline{F}_1$ and $\overline{F}_2$ must sum to zero, that is,

$$\overline{F} + \overline{F}_{IN} = \overline{0} \qquad (8.8\text{-}6)$$

Also, the inertia force produces an inertia moment in a direction that is opposite to that produced by the actual forces, such that

$$M_{IN} = -M_G \qquad (8.8\text{-}7)$$

or

$$M_G + M_{IN} = 0 \qquad (8.8\text{-}8)$$

The form of Equations (8.8-6) and (8.8-8) is similar to that of a static problem when we consider one of the forces as an inertia force and one of the moments as an inertia moment. This link may be considered as a three-force member in which one of the forces is the inertia force.

The Principle of Superposition, presented in Section 8.7, may be used for dynamic analyses, where inertia forces would be considered one at a time, and the results could then be superimposed to find their combined effect.

## 8.9 DYNAMIC FORCE ANALYSIS USING FORCE POLYGON METHOD

To perform a dynamic force analysis of a mechanism, the following general steps are required:

1. Draw a diagram of the mechanism to scale, in the configuration for which the analysis is required.
2. Using prescribed input motion(s), construct an acceleration polygon and determine the linear accelerations of link mass centers, the link angular accelerations, the inertia forces, and the inertia moments.
3. Indicate on the diagram of the mechanism all externally applied and inertia forces (offset each inertia force from the mass center to provide the required inertia moment).
4. Treat each inertia force and externally applied load individually, draw force polygons for each, and determine the effect of each throughout the mechanism.
5. If the combined effect of more than one force is to be determined, employ the Principle of Superposition.

## EXAMPLE 8.5  Dynamic Force Analysis of a Slider Crank Mechanism

For the slider crank mechanism shown in Figure 8.14(a), the link lengths are

$$r_{O_2 B} = r_2 = 3.5 \text{ in}; \qquad r_{BD} = r_3 = 9.0 \text{ in}; \qquad r_{BG_3} = 4.5 \text{ in}$$

and the constant input rotational speed is

$$\dot{\theta}_2 = 500 \text{ rpm CCW} = 52.4 \text{ rad/sec CCW}$$

An outline of the acceleration polygon is given in Figure 8.14(b).

The coupler, link 3, has the shape of a uniform slender rod (see Appendix D, Section D.9) and has the following properties:

$$m_3^* = 3.0 \text{ lbm}; \qquad I_{G_3}^* = 47.0 \text{ lbm-in}^2$$

For the position shown, determine the forces on link 3 at the turning pairs due to the inertia of link 3.

### SOLUTION

Since the coupler has the shape of a uniform slender link, we locate the position of the center of mass as midway between $B$ and $D$ on the acceleration polygon, as shown in

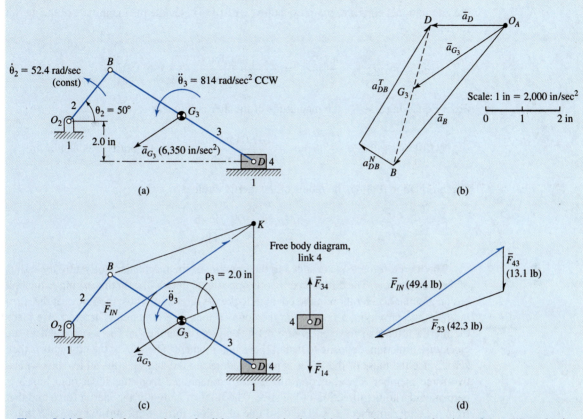

**Figure 8.14** Dynamic force analysis of a slider crank mechanism: (a) mechanism, (b) acceleration polygon, (c) mechanism with inertia circle, (d) force polygon.

Figure 8.14(b). The magnitude of the acceleration of the center of mass is

$$|\bar{a}_{G_3}| = 6,350 \, \text{in/sec}^2$$

and the tangential acceleration component between points $B$ and $D$ is

$$a_{DB}^T = 7,330 \, \text{in/sec}^2$$

The angular acceleration of link 3 is then

$$\ddot{\theta}_3 = \frac{a_{DB}^T}{r_{DB}} = \frac{7,330}{9.0} = 814 \, \text{rad/sec}^2 \, \text{CCW}$$

The units of the quantities employed in this example correspond to the third set listed in Table 2.3. In the calculation of the inertia force, we substitute the corresponding value of the gravitational constant, $g_c$, in Equation (2.12-6) to calculate the magnitude of the inertia force:

$$|\overline{F}_{IN}| = m_3|\overline{a}_{G_3}| = \frac{m_3^*}{g_c}|\overline{a}_{G_3}| = \frac{3.0\,\text{lbm}}{(386\,\text{lbm-in/lb-sec}^2)} \times 6{,}350\,\text{in/sec}^2 = 49.4\,\text{lb}$$

From Equation (2.12-9), the magnitude of the inertia moment is

$$|M_{IN}| = I_{G_3}|\ddot{\theta}_3| = \frac{I_{G_3}^*}{g_c}|\ddot{\theta}_3| = \frac{47.0\,\text{lbm-in}^2}{(386\,\text{lbm-in/lb-sec}^2)} \times 814\,\text{rad/sec}^2 = 99.1\,\text{in-lb}$$

Using Equation (8.8-4), the radius of the inertia circle is

$$\rho_3 = \frac{|M_{IN}|_3}{|\overline{F}_{IN}|_3} = \frac{99.1\,\text{in-lb}}{49.4\,\text{lb}} = 2.0\,\text{in}$$

The inertia force, $\overline{F}_{IN}$, is added to the drawing of the mechanism, offset by the radius of the inertia circle. The direction of $\overline{F}_{IN}$ is opposite to that of $\overline{a}_{G_3}$. In this example, the link has an angular acceleration in the counterclockwise direction. Thus, the forces at the turning pairs are providing a moment in the counterclockwise direction. Therefore, we place the inertia force as shown in Figure 8.14(c) so that it applies an inertia moment in the clockwise direction, opposite to that of the actual moment. The sum of the moments is then zero. Since the mass of the slider is not considered in the analysis, the slider (link 4) is a two-force member. The direction of the force transmitted from the slider onto link 3 is perpendicular to the direction of the slide. The lines of action of the inertia force and that from the slider intersect at $K$. The third line of action, through the turning pair at $B$, must pass through the same intersection point. Using the inertia force and the other two lines of action, the force polygon is completed, as shown in Figure 8.14(d). The magnitudes of the forces located at $B$ and $D$ as a result of the inertia in link 3 are

$$|\overline{F}_{43}| = 13.1\,\text{lb}; \qquad |\overline{F}_{23}| = 42.3\,\text{lb}$$

The directions of these forces are indicated in the force polygon.

## EXAMPLE 8.6  Dynamic Force Analysis of a Four-Bar Mechanism

Figure 8.15(a) shows a four-bar mechanism. Link 2 rotates at a constant rotational speed of 120 rad/sec in the counterclockwise direction. Also indicated in the figure are the results of the acceleration analysis. Pertinent properties of links 3 and 4 are listed in Table 8.1.

Determine the magnitude and sense of the moment to be applied to link 2 from base link 1 to overcome the inertias of links 3 and 4.

$$r_1 = 9.0\,\text{cm}; \qquad r_2 = 2.0\,\text{cm}; \qquad r_3 = 13.0\,\text{cm}; \qquad r_4 = 7.0\,\text{cm}$$

SOLUTION

We will solve this problem using the Principle of Superposition. For the first part of the solution, only the inertia of link 3 will be considered. In the second part, only the effects of the inertia of link 4 will be evaluated. Finally, the results will be superimposed. The units of the quantities employed in this example correspond to the second set listed in Table 2.2, for which $g_c = 1$.

For the first part of the problem, using the acceleration parameters given in Figure 8.15(a), the magnitude of the inertia force of link 3 is

$$|\overline{F}_{IN}|_3 = m_3|\overline{a}_{G_3}|$$

$$= \frac{m_3^*}{g_c}|\overline{a}_{G_3}| = \frac{0.50\,\text{kg}}{1} \times 236\,\text{m/sec}^2 = 118\,\text{N}$$

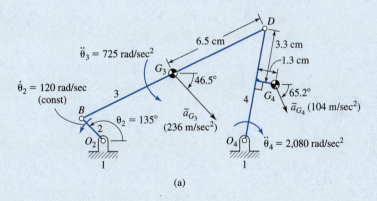

(a)

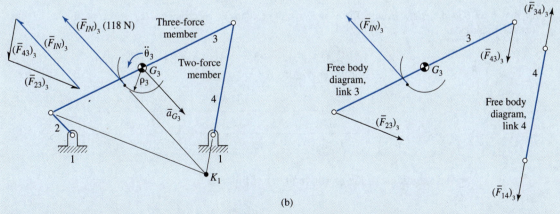

(b)

**Figure 8.15** Dynamic force analysis of a four-bar mechanism: (a) input parameters and results of acceleration analysis, (b) effect of inertia of link 3.

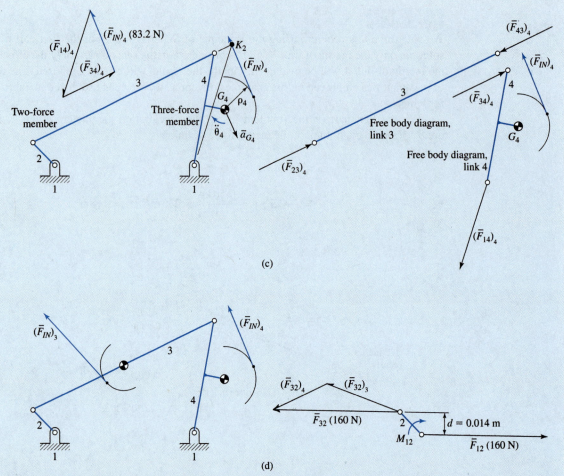

**Figure 8.15** (*Continued*) Dynamic force analysis of a four-bar mechanism: (c) effect of inertia of link 4, (d) superimposed result.

**TABLE 8.1** Parameters of the Four-Bar Mechanism Shown in Figure 8.15

|  | Link 3 | Link 4 |
|---|---|---|
| $m^\bullet$ (kg) | 0.50 | 0.80 |
| $I_G^\bullet$ (kg-m$^2$) | $2.52 \times 10^{-3}$ | $8.00 \times 10^{-4}$ |

The magnitude of the inertia moment of link 3 is

$$|M_{IN}|_3 = I_{G_3}\ddot{\theta}_3 = \frac{I_{G_3}^*}{g_c}\ddot{\theta}_3 = \frac{2.52 \times 10^{-3}\,\text{kg-m}^2}{1} \times 725\,\text{rad/sec}^2 = 1.83\,\text{N-m}$$

The radius of the inertia circle of link 3 is

$$r_3 = \frac{|M_{IN}|_3}{|\overline{F}_{IN}|_3} = \frac{1.83\,\text{N-m}}{118\,\text{N}} = 0.0155\,\text{m} = 1.55\,\text{cm}$$

Figure 8.15(b) shows the corresponding analysis in which only the inertia associated with link 3 is considered. This link is treated as a three-force member. Link 4 is assumed to have no mass and is treated as a two-force member. From this analysis, we determine

$$(\overline{F}_{32})_3 = -(\overline{F}_{23})_3$$

For the second part of the problem, only the inertia of link 4 is considered. In this instance, link 4 is treated as a three-force member, and link 3 is treated as a two-force member. Carrying out analyses similar to those completed for link 3 gives

$$|\overline{F}_{IN}|_4 = 83.2\,\text{N}; \qquad |M_{IN}|_4 = 1.66\,\text{N-m}; \qquad \rho_4 = 2.00\,\text{cm}$$

Figure 8.15(c) shows the force analysis for this part of the problem. The analysis provides

$$(\overline{F}_{32})_4 = -(\overline{F}_{23})_4$$

The superimposed results are illustrated in Figure 8.15(d). They include

$$|\overline{F}_{32}| = |(\overline{F}_{32})_3 + (\overline{F}_{32})_4| = 160\,\text{N}; \qquad d = 0.014\,\text{m}$$

where $d$ is the perpendicular distance between the lines of action of forces $\overline{F}_{32}$ and $\overline{F}_{12}$. The moment required from the base link on link 2 is then

$$M_{12} = |\overline{F}_{32}|d = 160 \times 0.014 = 2.24\,\text{N-m CW}$$

Results of this example show that the moment to be supplied from the base link onto link 2 is in the clockwise direction, opposite to its direction of rotational speed. A dynamic force analysis of the same mechanism throughout a complete cycle of motion is provided through [Model 8.16]. During the cycle, the driving moment takes on positive and negative values. Positive values correspond to the counterclockwise direction. With this model it is possible to adjust values of the masses and, consequently, values of polar mass moments of inertia about the mass centers. Figure 8.16(a) shows the torque curve when both links 3 and 4 have mass. Figure 8.16(b) shows the case where the mass of link 3 has been set to zero, and only the mass of link 4 remains. In this instance, link 3 acts as a two-force member, and as a result

**Model 8.16**
Dynamic Force
Analysis

$$|\overline{F}_{23}| = |\overline{F}_{34}|$$

Figure 8.16(c) shows the case where the mass of link 4 has been set to zero, and only the mass of link 3 remains. In this instance

$$|\overline{F}_{34}| = |\overline{F}_{14}|$$

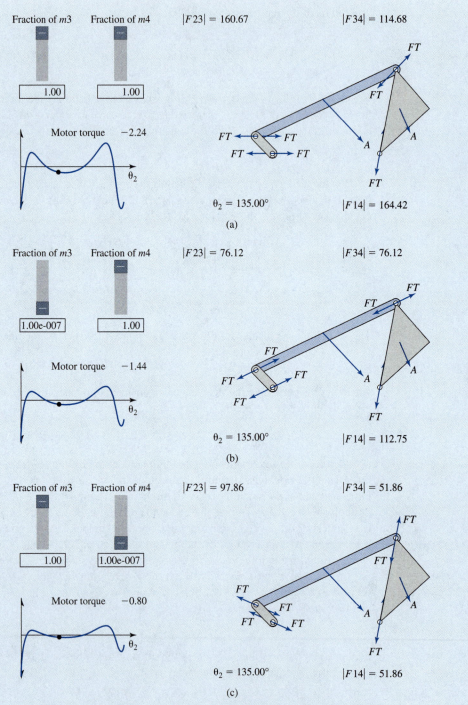

**Figure 8.16** Dynamic analysis of a four-bar mechanism [Model 8.16]: (a) links 3 and 4 have mass, (b) only link 4 has mass, (c) only link 3 has mass.

# PROBLEMS

**P8.1** Assuming the mechanism shown in Figure P8.1 is in static equilibrium, determine forces $\bar{Q}$ and $\bar{F}_{12}$.

$r_{O_2B} = 4.0\,\text{cm}$

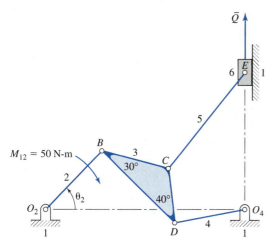

**Figure P8.1**

**P8.2** For the vise grip mechanism shown in Figure P8.2, use force polygons to determine

(a) the applied force $\bar{P}$ necessary to produce a force $\bar{Q}$ of magnitude 120 lb

(b) the force $\bar{Q}$ that can be produced by an applied force $\bar{P}$ of magnitude 15N

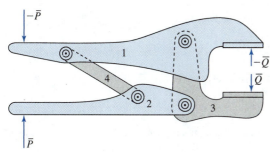

**Figure P8.2**

**P8.3** For the quick return mechanism shown in Figure P8.3, the cutting force $\bar{F}$ has a magnitude of 30 N. Neglecting inertia, determine

(a) the driving torque required on link 2

(b) the bearing forces at $O_2$ and $O_3$

$r_{O_2B_2} = 4.0\,\text{cm}$

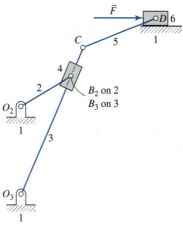

**Figure P8.3**

**P8.4** Construct a complete set of force polygons for the mechanism in static equilibrium shown in Figure P8.4, and if $\bar{Q} = 150$ lb, determine

(a) $\bar{P}$

(b) $\bar{F}_{41}$

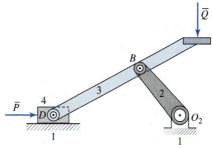

**Figure P8.4**

**P8.5** Figure P8.5 shows a scaled diagram of a hand-operated toggle clamp. In the position shown, draw the static force polygons. Determine the clamping force, $\bar{F}$, when the hand force, $\bar{P}$, of 25 lb is applied as shown.

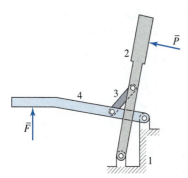

**Figure P8.5**

**P8.6** Determine the torque required from the base link on link 2 for static equilibrium of the mechanism shown in Figure P8.6. Force $\bar{F}$ has a magnitude of 20 lb.

$$r_{O_2 B_2} = 2.0 \text{ in}$$

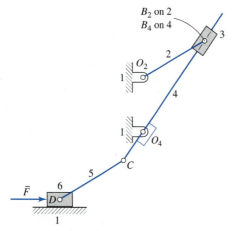

**Figure P8.6**

**P8.7** Determine the required cylinder pressure for static equilibrium of the mechanism shown in Figure P8.7. Torque $M_{14}$ has a magnitude of 25 N-m. The diameter of the piston is 1.0 cm.

$$r_{O_4 D} = 6.0 \text{ cm}$$

**P8.8** Determine the required input torque $M_{12}$ for static equilibrium of the mechanism shown in Figure P8.8. Force $\bar{F}$ has a magnitude of 1,000 N.

$$r_{O_2 B} = 3.0 \text{ cm}$$

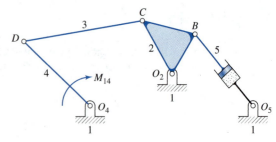

**Figure P8.7**

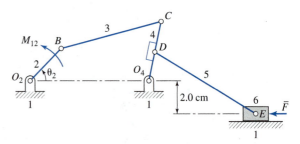

**Figure P8.8**

**P8.9** Determine the pressure required in the 2.0 cm diameter cylinder shown in Figure P8.9 to maintain static equilibrium. The image shown is to scale.

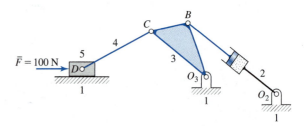

**Figure P8.9**

**P8.10** For the given position of the mechanism shown in Figure P8.10, determine the magnitude and sense of the torque required to be applied to crank $O_6 E$ by the base link, to maintain static equilibrium.

$$r_{O_2 B} = 7.0 \text{ cm}; \qquad r_{O_6 E} = 3.0 \text{ cm}$$

$$M_{12} = 5.0 \text{ N-m CW}$$

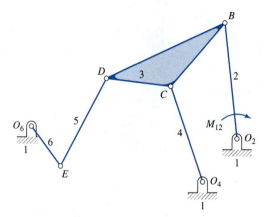

Figure P8.10

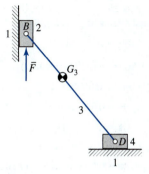

Figure P8.12

P8.11 For the mechanism shown in Figure P8.11, determine the magnitude and sense of the torque to be applied to link 2 from the base link to overcome the inertia of link 3.

$$r_{O_2B} = 5.0 \, \text{cm}; \qquad r_{BD} = 10.0 \, \text{cm}$$

$$r_{BG_3} = 5.0 \, \text{cm}; \qquad m_3 = 800 \, \text{gr}$$

$$I_{G_3} = 6.4 \times 10^{-3} \, \text{kg-m}^2$$

$$\dot{\theta}_2 = 600 \, \text{rpm CCW (constant)}$$

Determine the instantaneous force $\bar{F}$ required to produce this motion, assuming that the slider blocks are massless.

$$r_{BD} = 10.0 \, \text{in}; \qquad r_{BG_3} = 4.0 \, \text{in}$$

$$\bar{a}_{G_3} = 2,360 \, \text{in/sec}^2, \leftarrow$$

$$\ddot{\theta}_3 = 392 \, \text{rad/sec}^2 \, \text{CW} +$$

P8.13 For the mechanism shown in Figure P8.13, a gas force of 5,000 lb acts to the left on link 4. Determine the magnitude and sense of the torque to be applied to link 2 from base link 1 to overcome

(a) the gas force
(b) the inertia of link 3

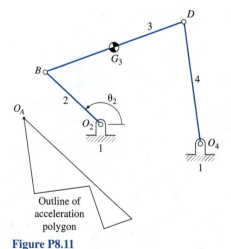

Figure P8.11

P8.12 For the mechanism shown in Figure P8.12, link 3 has a mass of 3 lbm and polar mass moment of inertia of 0.04 lb-in-sec$^2$ about its mass center $G_3$. Sliding block 2 has a constant velocity of 10 ft/sec upward.

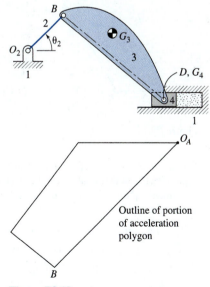

Figure P8.13

(c) the inertia of link 4

(d) the combined effects of parts (a), (b), and (c)

$$r_{O_2B} = 1.5\,\text{in}$$

$$\dot{\theta}_2 = 1{,}800\,\text{rpm CCW (constant)}$$

$$m_3 = 0.90\,\text{lbm}$$

$$I_{G_3} = 0.010\,\text{lb-in-sec}^2$$

$$m_4 = 0.60\,\text{lbm}$$

$$I_{G_4} = 5.0 \times 10^{-3}\,\text{lb-in-sec}^2$$

**P8.14** For the given position of the mechanism shown in Figure P8.14, determine the magnitude and sense of the torque required to be applied to link 2 by the base link, to overcome the inertia of link 3.

$$r_{O_2O_4} = 10.0\,\text{cm}; \qquad r_{O_2B} = 4.0\,\text{cm}$$

$$r_{BD} = 9.0\,\text{cm}; \qquad r_{BG_3} = 6.0\,\text{cm}$$

$$r_{O_4D} = 7.8\,\text{cm}; \qquad m_3 = 600\,\text{gr}$$

$$I_{G_3} = 1.5 \times 10^{-3}\,\text{kg-m}^2$$

$$\dot{\theta}_2 = 200\,\text{rpm CCW}; \qquad \ddot{\theta}_2 = 100\,\text{rad/sec}^2\,\text{CCW}$$

$$\bar{a}_{G_3} = 19.4\,\text{m/sec}^2,\ 30^\circ$$

$$\ddot{\theta}_3 = 122\,\text{rad/sec}^2\,\text{CW}$$

0.32 N-m CW.

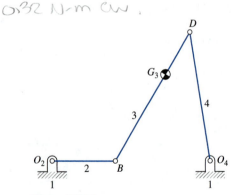

**Figure P8.14**

**P8.15** For the given position of the mechanism shown in Figure P8.15, determine the magnitude and sense of the torque required to be applied to link 2 by the base link, to overcome the inertia of link 5.

$$r_{O_2B} = 3.0\,\text{cm}; \qquad r_{CE} = 6.0\,\text{cm}$$

$$r_{CG_5} = 3.0\,\text{cm}; \qquad m_5 = 5.0\,\text{gr}$$

$$I_{G_5} = 5.0 \times 10^{-6}\,\text{kg-m}^2$$

$$\dot{\theta}_2 = 3{,}000\,\text{rpm CW (constant)}$$

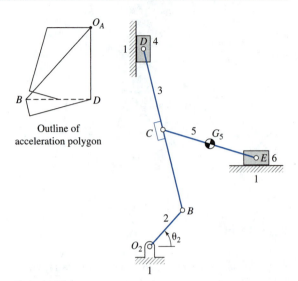

**Figure P8.15**

**P8.16** For the given position of the mechanism shown in Figure P8.16, determine the magnitude and sense of the torque required to be applied to link 2 by the base link, to overcome

(a) the inertia of link 3

(b) the inertia of link 4

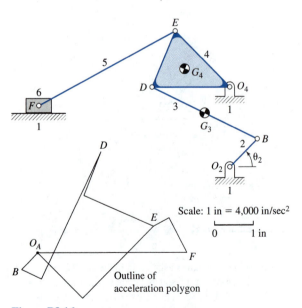

Scale: 1 in = 4,000 in/sec$^2$

**Figure P8.16**

(c)  the inertia of links 3 and 4

$$r_{O_2B} = 1.0\,\text{in}$$

$$\dot{\theta}_2 = 475\,\text{rpm CW (constant)}$$

$$m_3 = 9.0\,\text{lbm};\qquad I_{G_3} = 1.5 \times 10^{-2}\,\text{lb-in-sec}^2$$

$$m_4 = 11.5\,\text{lbm};\qquad I_{G_4} = 8.0 \times 10^{-2}\,\text{lb-in-sec}^2$$

**P8.17** For the given position of the mechanism shown in Figure P8.17, determine the magnitude and sense of the torque required to be applied to link 2 by the base link, to overcome

(a)  the inertia of link 3
(b)  the inertia of link 4
(c)  the inertia of links 3 and 4

$$r_{O_2B} = 5.0\,\text{cm};\qquad r_{BD} = 10.0\,\text{cm}$$

$$r_{BG_3} = 3.0\,\text{cm};\qquad m_3 = 400\,\text{gr}$$

$$m_4 = 600\,\text{gr};\qquad I_{G_3} = 6.3 \times 10^{-4}\,\text{kg-m}^2$$

$$\theta_2 = 40°;\qquad \dot{\theta}_2 = 200\,\text{rpm CW}$$

$$\ddot{\theta}_2 = 200\,\text{rad/sec}^2\,\text{CCW}$$

$$\bar{a}_{G_3} = 25.2\,\text{m/sec}^2,\,12°\,\nearrow$$

$$\ddot{\theta}_3 = 122\,\text{rad/sec}^2\,\text{CW}$$

$$\bar{a}_{G_4} = 27.8\,\text{m/sec}^2\,\longleftarrow$$

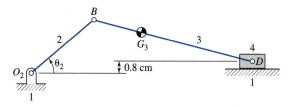

**Figure P8.17**

**P8.18** For the given position of the mechanism shown in Figure P8.18, determine the magnitude and sense of the torque required to be applied to link 2 by the base link, to overcome the inertia of link 4.

$$r_{O_2B} = 5.0\,\text{cm};\qquad r_{O_4D} = 8.0\,\text{cm}$$

$$r_{O_4G_4} = 4.0\,\text{cm};\qquad m_4 = 0.50\,\text{kg}$$

$$I_{G_4} = 5.0 \times 10^{-4}\,\text{kg-m}^2$$

$$\dot{\theta}_2 = 1,200\,\text{rpm CW (constant)}$$

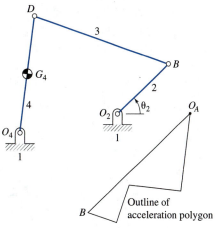

**Figure P8.18**

**P8.19** For the given position of the mechanism shown in Figure P8.19, determine the magnitude and sense of the torque required to be applied to link 2 by the base link, to overcome the inertia of link 6.

$$r_{O_2B} = 5.0\,\text{cm};\qquad m_6 = 50\,\text{gr}$$

$$I_{G_6} = 4.0 \times 10^{-5}\,\text{kg-m}^2;\qquad \dot{\theta}_2 = 100\,\text{rpm CW}$$

$$\ddot{\theta}_2 = 80\,\text{rad/sec}^2\,\text{CW}$$

$$\bar{a}_{G_6} = 380\,\text{cm/sec}^2,\,73°\,\nearrow$$

$$\ddot{\theta}_6 = 90\,\text{rad/sec}^2\,\text{CCW}$$

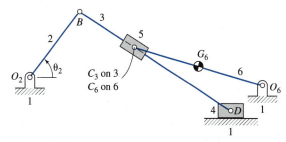

**Figure P8.19**

**P8.20** For the given position of the mechanism shown in Figure P8.20, determine the magnitude and sense of the torque required to be applied to crank $O_2B$ by the base link, to overcome the inertia of link 5.

$$\theta_2 = 135°;\qquad \dot{\theta}_2 = 150\,\text{rpm CCW (constant)}$$

$$r_{O_2B} = 4.0\,\text{cm};\qquad m_5 = 49.2\,\text{gr}$$

$$I_{G_5} = 6.0 \times 10^{-3} \, \text{kg-m}^2$$

$$\bar{a}_{G_5} = 145 \, \text{m/sec}^2, 42° \ \searrow$$

$$\ddot{\theta}_5 = 29.7 \, \text{rad/sec}^2 \, \text{CCW}$$

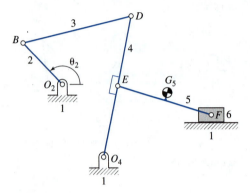

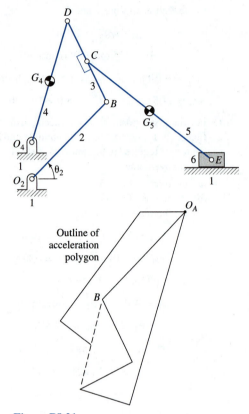

**Figure P8.20**

**P8.21** For the given position of the mechanism shown in Figure P8.21, determine the magnitude and sense of the torque required to be applied to crank $O_2 B$ by the base link, to overcome

(a)  the inertia of link 4
(b)  the inertia of link 5
(c)  the inertia of links 4 and 5

$$r_{O_2 B} = 6.0 \, \text{cm}$$

$$\dot{\theta}_2 = 12.0 \, \text{rad/sec CCW (constant)}$$

$$m_4 = 30 \, \text{gr}; \qquad I_{G_4} = 6.0 \times 10^{-5} \, \text{kg-m}^2$$

$$m_5 = 50 \, \text{gr}; \qquad I_{G_5} = 6.5 \times 10^{-5} \, \text{kg-m}^2$$

**Figure P8.21**

# 9 Analytical Force Analysis and Balancing of Planar Mechanisms

## 9.1 INTRODUCTION

In this chapter, analytical equations are presented for use in completing kinetic analyses of four-bar mechanisms and slider crank mechanisms. They are based on the governing equations of motion (Section 2.10) and the analytical expressions of kinematics of the links (Chapter 4). With use of a computer, multiple analyses throughout cycles of motion may be readily completed. Also presented are methods for either reducing or eliminating the net force imparted to the base link, called the *shaking force*, created by the unbalanced inertias of the moving links.

In all of these analyses, we restrict ourselves to cases where the rotational speed of input link 2 is constant. We will prescribe

$$\dot{\theta}_2 = \omega = \text{constant} \tag{9.1-1}$$

Therefore, we can set

$$\theta_2 = \omega t \tag{9.1-2}$$

Unless otherwise noted, we will assume that motions of all links of planar mechanisms take place in a single plane.

## 9.2 FORCE ANALYSIS OF A FOUR-BAR MECHANISM

Consider the four-bar mechanism illustrated in Figure 9.1(a). Locations of the centers of mass of links 2, 3, and 4 with respect to a turning pair are specified by distances $b_i$ ($i = 2, 3, 4$) and angles $\phi_i$ ($i = 2, 3, 4$). Free body diagrams of individual links are shown in Figure 9.1(b). The driving torque, $M_{12}$, is one of the quantities that must be determined.

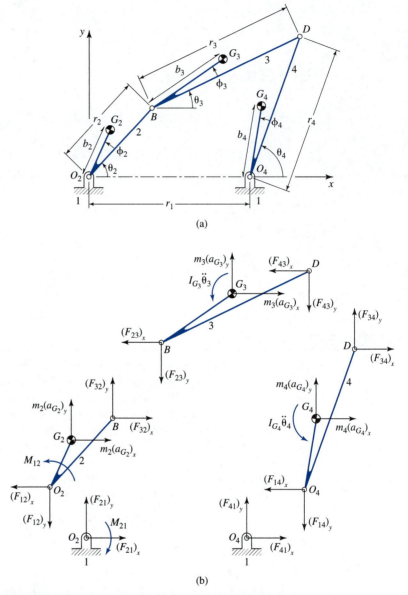

**Figure 9.1** Dynamic analysis of a four-bar mechanism.

To complete a kinetic analysis, we must determine accelerations of the centers of mass of the moving links. Referring to Figure 9.1(a), and implementing Equation (9.1-1), the results for link 2 are

$$(r_{G_2})_x = b_2 \cos(\theta_2 + \phi_2)$$
$$(a_{G_2})_x = (\ddot{r}_{G_2})_x = -b_2\omega^2 \cos(\theta_2 + \phi_2)$$
$$(r_{G_2})_y = b_2 \sin(\theta_2 + \phi_2)$$
$$(a_{G_2})_y = (\ddot{r}_{G_2})_y = -b_2\omega^2 \sin(\theta_2 + \phi_2)$$

(9.2-1)

The results for link 3 are

$$(r_{G_3})_x = r_2 \cos \theta_2 + b_3 \cos(\theta_3 + \phi_3)$$

$$(a_{G_3})_x = -r_2 \omega^2 \cos \theta_2 - b_3 \left[ \dot{\theta}_3^2 \cos(\theta_3 + \phi_3) + \ddot{\theta}_3 \sin(\theta_3 + \phi_3) \right]$$

$$(r_{G_3})_y = r_2 \sin \theta_2 + b_3 \sin(\theta_3 + \phi_3)$$ 
(9.2-2)

$$(a_{G_3})_y = -r_2 \omega^2 \sin \theta_2 - b_3 \left[ \dot{\theta}_3^2 \sin(\theta_3 + \phi_3) - \ddot{\theta}_3 \cos(\theta_3 + \phi_3) \right]$$

The results for link 4 are

$$(r_{G_4})_x = r_1 + b_4 \cos(\theta_4 + \phi_4)$$

$$(a_{G_4})_x = -b_4 \left[ \dot{\theta}_4^2 \cos(\theta_4 + \phi_4) + \ddot{\theta}_4 \sin(\theta_4 + \phi_4) \right]$$

$$(r_{G_4})_y = b_4 \sin(\theta_4 + \phi_4)$$ 
(9.2-3)

$$(a_{G_4})_y = -b_4 \left[ \dot{\theta}_4^2 \sin(\theta_4 + \phi_4) - \ddot{\theta}_4 \cos(\theta_4 + \phi_4) \right]$$

The governing equations of motion for link 2 are determined by considering the $x$ and $y$ components of Equation (2.10-1), as well as Equation (2.10-6). Referring to Figure 9.1(b),

$$(F_{32})_x - (F_{12})_x = m_2 (a_{G_2})_x \tag{9.2-4}$$

$$(F_{32})_y - (F_{12})_y = m_2 (a_{G_2})_y \tag{9.2-5}$$

$$-(F_{32})_x r_2 \sin \theta_2 + (F_{32})_y r_2 \cos \theta_2 + M_{12} = 0 \tag{9.2-6}$$

Similar equations may be generated for links 3 and 4. The governing equations of motion of all moving links may then be combined in matrix form and expressed as

$$[A]\{x\} = \{B\} \tag{9.2-7}$$

where

$$[A] = \begin{bmatrix} -1 & 0 & 1 & 0 & 0 & 0 & 0 & 0 & 0 \\ 0 & -1 & 0 & 1 & 0 & 0 & 0 & 0 & 0 \\ 0 & 0 & A_{3,3} & A_{3,4} & 0 & 0 & 0 & 0 & 1 \\ 0 & 0 & -1 & 0 & -1 & 0 & 0 & 0 & 0 \\ 0 & 0 & 0 & -1 & 0 & -1 & 0 & 0 & 0 \\ 0 & 0 & A_{6,3} & A_{6,4} & A_{6,5} & A_{6,6} & 0 & 0 & 0 \\ 0 & 0 & 0 & 0 & 1 & 0 & -1 & 0 & 0 \\ 0 & 0 & 0 & 0 & 0 & 1 & 0 & -1 & 0 \\ 0 & 0 & 0 & 0 & A_{9,5} & A_{9,6} & 0 & 0 & 0 \end{bmatrix} \tag{9.2-8a}$$

$$A_{3,3} = -r_2 \sin \theta_2; \qquad A_{3,4} = r_2 \cos \theta_2$$

$$A_{6,3} = -b_3 \sin(\theta_3 + \phi_3); \qquad A_{6,4} = b_3 \cos(\theta_3 + \phi_3)$$

$$A_{6,5} = r_3 \sin \theta_3 - b_3 \sin(\theta_3 + \phi_3); \qquad A_{6,6} = -r_3 \cos \theta_3 + b_3 \cos(\theta_3 + \phi_3)$$

$$A_{9,5} = -r_4 \sin \theta_4; \qquad A_{9,6} = r_4 \cos \theta_4$$

$$\{x\} = \begin{Bmatrix} (F_{12})_x \\ (F_{12})_y \\ (F_{23})_x \\ (F_{23})_y \\ (F_{34})_x \\ (F_{34})_y \\ (F_{14})_x \\ (F_{14})_y \\ M_{12} \end{Bmatrix}; \qquad \{B\} = \begin{Bmatrix} m_2(a_{G_2})_x \\ m_2(a_{G_2})_y \\ 0 \\ m_3(a_{G_3})_x \\ m_3(a_{G_3})_y \\ I_{G_3}\ddot{\theta}_3 \\ m_4(a_{G_4})_x \\ m_4(a_{G_4})_y \\ \left(I_{G_4} + m_4 b_4^2\right)\ddot{\theta}_4 \end{Bmatrix} \qquad (9.2\text{-}8b; \ 9.2\text{-}8c)$$

The above equations are functions of the angular displacements, angular velocities, and angular accelerations of links 3 and 4. These quantities may be determined using equations provided in Section 4.3.3.

Solving Equation (9.2-7) permits other results to be determined. The magnitude of the force transmitted through a turning pair between links $i$ and $j$ is

$$\left| \overline{F}_{ij} \right| = \left[ (F_{ij})_x^2 + (F_{ij})_y^2 \right]^{1/2} \qquad (9.2\text{-}9)$$

The direction of the force is

$$\alpha_{ij} = \tan 2^{-1}[(F_{ij})_x, (F_{ij})_y], \qquad -\pi \le \alpha_{ij} \le \pi \qquad (9.2\text{-}10)$$

The *shaking force* is transmitted through the base pivots onto the base link and is expressed as

$$\overline{F}_S = \overline{F}_{21} + \overline{F}_{41} \qquad (9.2\text{-}11)$$

The magnitude of the shaking force is

$$\left| \overline{F}_S \right| = \left\{ [(F_{21})_x + (F_{41})_x]^2 + [(F_{21})_y + (F_{41})_y]^2 \right\}^{1/2} \qquad (9.2\text{-}12)$$

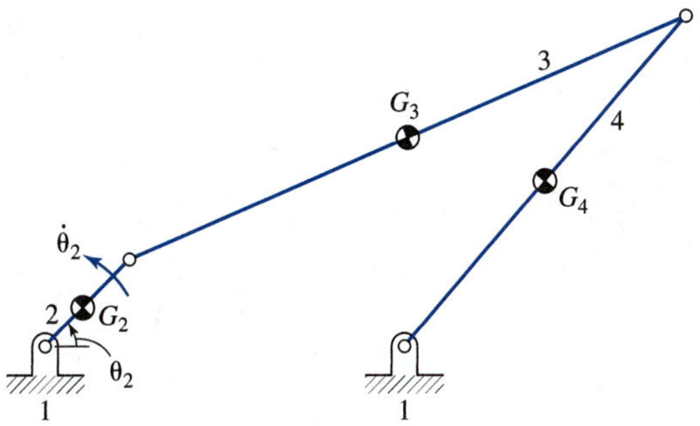

**Figure 9.2** Four-bar mechanism.

and its direction is

$$\alpha_S = \tan 2^{-1}\{[(F_{21})_x + (F_{41})_x], [(F_{21})_y + (F_{41})_y]\}, \qquad -\pi \le \alpha_S \le \pi \qquad (9.2\text{-}13)$$

The *shaking moment* about the base pivot $O_2$ is due to the reaction of the driving torque onto the base link and the pin force at base pivot $O_4$, and it is expressed as

$$M_S = -M_{21} + (F_{41})_y r_1 \qquad (9.2\text{-}14)$$

As an illustration of the application of the above equations, consider the four-bar mechanism shown in Figure 9.2. Link 2 is driven at $\omega = 50$ rad/sec CCW. The parameters of the links are given in Table 9.1.

Employing Equations (9.2-7) and (9.2-8), a plot of the driving torque is shown in Figure 9.3. Employing Equations (9.2-12) and (9.2-13), Figure 9.4(a) shows a polar plot of the shaking force. Each point on the curve defines the magnitude and direction of the force for each value of $\theta_2$. A typical shaking force vector is depicted for $\theta_2 = 30°$. The related calculated values are

$$(F_{12})_x = 7.47 \text{ N}; \qquad (F_{12})_y = 2.60 \text{ N}$$

$$(F_{14})_x = -3.93 \text{ N}; \qquad (F_{14})_y = -4.76 \text{ N}$$

**TABLE 9.1** Parameters of the Four-Bar Mechanism Shown in Figure 9.2

|  | Link 1 | Link 2 | Link 3 | Link 4 |
|---|---|---|---|---|
| $r_i$ (cm) | 6.00 | 2.00 | 10.0 | 7.00 |
| $b_i$ (cm) | — | 1.00 | 5.00 | 3.50 |
| $m$ (gr) | — | 6.24 | 31.2 | 21.8 |
| $\phi_i$ (degrees) | — | 0 | 0 | 0 |
| $I_{G_i}$ (gr-cm²) | — | — | 260 | 89.0 |

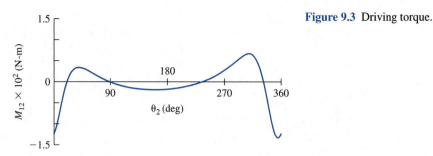

**Figure 9.3** Driving torque.

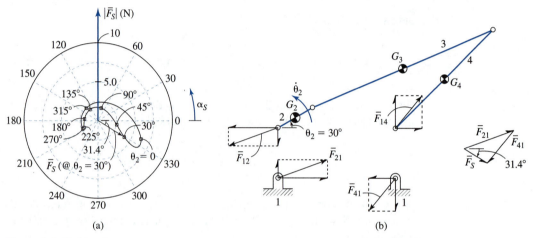

(a)

(b)

**Figure 9.4** Shaking force.

$$\overline{F}_S = [(F_{21})_x + (F_{41})_x]\overline{i} + [(F_{21})_y + (F_{41})_y]\overline{j}$$

$$= (7.47 - 3.93)\overline{i} + (2.60 - 4.76)\overline{j}$$

$$= 3.54\overline{i} - 2.16\overline{j} \text{ N}$$

$$\alpha_S = \tan 2^{-1}\{[(F_{21})_x + (F_{41})_x], [(F_{21})_y + (F_{41})_y]\}$$

$$= \tan 2^{-1}(3.54, -2.16) = -31.4°$$

Figure 9.4(b) shows the three moving links and the interactive forces acting on the base link. The shaking force for this configuration is also illustrated.

Employing Equations (9.2-9) and (9.2-10), Figure 9.5 is a polar plot of the force at the turning pair between links 2 and 3. Employing Equation (9.2-14), Figure 9.6 is a plot of the shaking moment.

Undue shaking force transmitted to the base link generally creates an undesirable condition. In Section 9.6, a procedure is presented whereby the net shaking force transmitted through the base pivots can be eliminated.

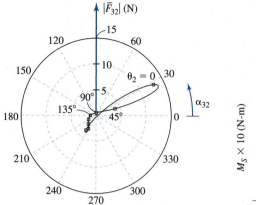

Figure 9.5  Force at turning pair.

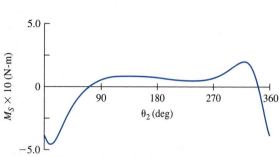

Figure 9.6  Shaking moment.

## 9.3  FORCE ANALYSIS OF A SLIDER CRANK MECHANISM

Figure 9.7(a) shows a slider crank mechanism. The rotational speed, $\omega$, of the crank is constant. Locations of the centers of mass of links 2 and 3 with respect to a turning pair are specified by distances $b_i$ ($i = 2, 3$) and angles $\phi_i$ ($i = 2, 3$). Free body diagrams of the links are shown in Figure 9.7(b). Based on these diagrams, we can generate a set of governing equations of motion expressed as

$$[A]\{x\} = \{B\} \tag{9.3-1}$$

where

$$[A] = \begin{bmatrix} -1 & 0 & 1 & 0 & 0 & 0 & 0 & 0 \\ 0 & -1 & 0 & 1 & 0 & 0 & 0 & 0 \\ 0 & 0 & A_{3,3} & A_{3,4} & 0 & 0 & 0 & 1 \\ 0 & 0 & -1 & 0 & -1 & 0 & 0 & 0 \\ 0 & 0 & 0 & -1 & 0 & -1 & 0 & 0 \\ 0 & 0 & A_{6,3} & A_{6,4} & A_{6,5} & A_{6,6} & 0 & 0 \\ 0 & 0 & 0 & 0 & 1 & 0 & 0 & 0 \\ 0 & 0 & 0 & 0 & 0 & 1 & -1 & 0 \end{bmatrix} \tag{9.3-2a}$$

$$A_{3,3} = -r_2 \sin\theta_2; \qquad A_{3,4} = r_2 \cos\theta_2$$

$$A_{6,3} = r_3 \sin\theta_3 - b_3 \sin(\theta_3 + \phi_3); \qquad A_{6,4} = -r_3 \cos\theta_3 + b_3 \cos(\theta_3 + \phi_3)$$

$$A_{6,5} = -b_3 \sin(\theta_3 + \phi_3); \qquad A_{6,6} = b_3 \cos(\theta_3 + \phi_3)$$

$$\{x\} = \begin{Bmatrix} (F_{12})_x \\ (F_{12})_y \\ (F_{23})_x \\ (F_{23})_y \\ (F_{34})_x \\ (F_{34})_y \\ (F_{14})_y \\ M_{12} \end{Bmatrix} ; \qquad \{B\} = \begin{Bmatrix} m_2(a_{G_2})_x \\ m_2(a_{G_2})_y \\ 0 \\ m_3(a_{G_3})_x \\ m_3(a_{G_3})_y \\ I_{G_3}\ddot{\theta}_3 \\ m_4(a_{G_4})_x \\ 0 \end{Bmatrix} \qquad \text{(9.3-2b; 9.3-2c)}$$

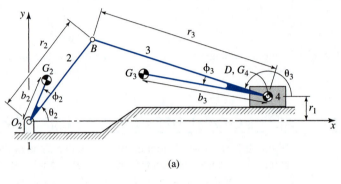

(a)

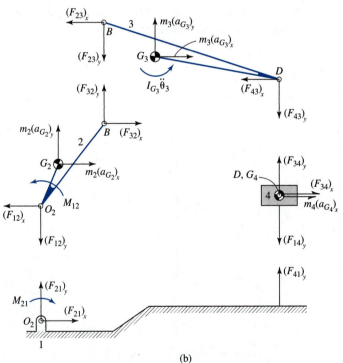

(b)

**Figure 9.7** Dynamic analysis of a slider crank mechanism.

Equations (9.3-2) contain expressions for the angular displacement, angular velocity, and angular acceleration of link 3. They may be determined using Equations (4.3-77), (4.3-82), and (4.3-84). Equation (9.3-2c) is a function of the linear accelerations of the centers of mass of links 2, 3, and 4. Expressions for these accelerations may be determined using the procedure presented in Section 4.4. The results for link 2 are

$$(a_{G_2})_x = -b_2\omega^2 \cos(\theta_2 + \phi_2); \qquad (a_{G_2})_y = -b_2\omega^2 \sin(\theta_2 + \phi_2) \qquad (9.3\text{-}3)$$

The results for link 3 are

$$(a_{G_3})_x = -r_2\omega^2 \cos\theta_2 + \dot{\theta}_3^2 [r_3 \cos\theta_3 - b_3 \cos(\theta_3 + \phi_3)]$$
$$+ \ddot{\theta}_3 [r_3 \sin\theta_3 - b_3 \sin(\theta_3 + \phi_3)]$$
$$(a_{G_3})_y = -r_2\omega^2 \sin\theta_2 + \dot{\theta}_3^2 [r_3 \sin\theta_3 - b_3 \sin(\theta_3 + \phi_3)]$$
$$- \ddot{\theta}_3 [r_3 \cos\theta_3 - b_3 \cos(\theta_3 + \phi_3)]$$

$$(9.3\text{-}4)$$

The result for link 4 is

$$(a_{G_4})_x = \frac{-r_2\omega^2 \cos(\theta_3 - \theta_2) + r_3\dot{\theta}_3^2}{\cos\theta_3} \qquad (9.3\text{-}5)$$

The shaking force is

$$\boxed{\overline{F}_S = \overline{F}_{21} + \overline{F}_{41}} \qquad (9.3\text{-}6)$$

The magnitude and direction of the shaking force are

$$\boxed{\begin{aligned} |\overline{F}_S| &= \left\{ (F_{21})_x^2 + [(F_{21})_y + (F_{41})_y]^2 \right\}^{1/2} \\ \alpha_S &= \tan 2^{-1}\{(F_{21})_x, [(F_{21})_y + (F_{41})_y]\}, \qquad -\pi \le \alpha_S \le \pi \end{aligned}} \qquad (9.3\text{-}7)$$

The magnitude and direction of a force transmitted through a turning pair may be found using Equations (9.2-12) and (9.2-13). The shaking moment is

$$\boxed{M_S = -M_{21} + (F_{41})_y(r_2 \cos\theta_2 - r_3 \cos\theta_3)} \qquad (9.3\text{-}8)$$

## EXAMPLE 9.1  Dynamic Force Analysis of a Slider Crank Mechanism

Figure 9.8 illustrates a slider crank mechanism. Link 2 is driven at $\omega = 55$ rad/sec CCW. The associated parameters are listed in Table 9.2. For one revolution of the crank, determine

(a) the plot of the driving torque
(b) the polar plot of the shaking force

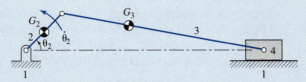

**Figure 9.8** Slider crank mechanism.

**TABLE 9.2** Parameters of the Slider Crank Mechanism Shown in Figure 9.8

|  | Link 1 | Link 2 | Link 3 | Link 4 |
|---|---|---|---|---|
| $r_i$ (cm) | 0 | 2.50 | 10.0 | — |
| $b_i$ (cm) | — | 1.25 | 6.67 | — |
| $m$ (gr) | — | 2.46 | 9.84 | 10.0 |
| $\phi_i$ (degrees) | — | 0 | 0 | — |
| $I_{G_i}$ (gr-cm$^2$) | — | — | 82.0 | — |

(c)  the polar plot of the force at the turning pair between links 2 and 3
(d)  the plot of the shaking moment

## SOLUTION

(a)  Employing Equations (9.3-1) and (9.3-2), the driving torque is shown in Figure 9.9.
(b)  Employing Equations (9.3-6) and (9.3-7), a polar plot of the shaking force is shown in Figure 9.10.
(c)  Employing Equations (9.2-12) and (9.2-13), a polar plot of the force between links 2 and 3 is shown in Figure 9.11.
(d)  Employing Equation (9.3-8), the shaking moment is shown in Figure 9.12.

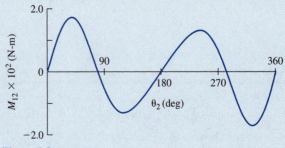

**Figure 9.9** Driving torque.

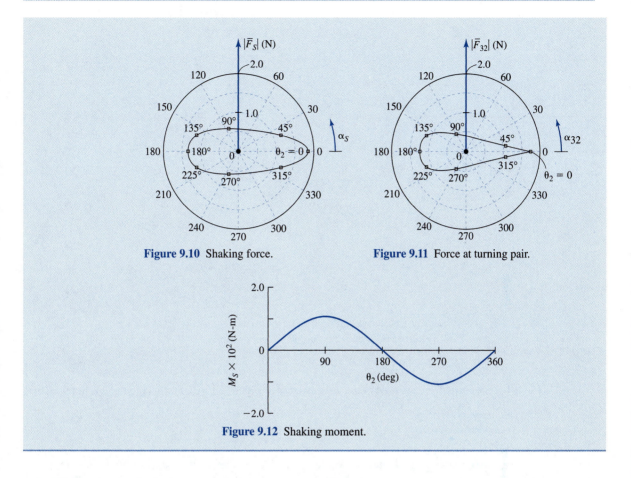

**Figure 9.10** Shaking force.

**Figure 9.11** Force at turning pair.

**Figure 9.12** Shaking moment.

## 9.4  UNBALANCE AND BALANCING

Virtually all machines contain rotating components. If the center of mass of a single rotating component does not coincide with its axis of rotation, then there is an *unbalance*. A rotating component having unbalance will produce inertia forces that are transmitted to connected links. If the component is connected to the base link, a shaking force is created.

In most instances, it is desirable to have machines operate smoothly with minimum shaking force. *Balancing* is the process of designing or modifying machinery in order to reduce unbalance to an acceptable level. The most common approach to achieve balancing is by selectively adding mass to, or removing mass from, links of the machine.

Figure 9.13(a) shows a link turning at rotational velocity $\omega$ about base pivot $O$. It is composed of two lumped masses, $m_A$ and $m_B$. The center of mass, $G$, is also illustrated, located distance $e$ from the base pivot. The system shown in Figure 9.13(a) is equivalent to that given in Figure 9.13(b), where $M = m_A + m_B$. This system will produce a shaking force with magnitude $Me\omega^2$, as shown in Figure 9.13(c). The direction of this force rotates at rate $\omega$ and is transmitted to the base link. To remove this shaking force, a *balance mass*, $m_c$, is added as shown, at distance $r_c$ from the base pivot. This mass is added diagonally

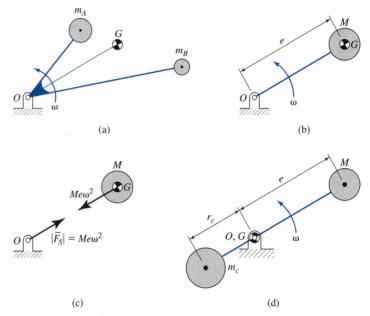

(a)

(b)

(c)

(d)

**Figure 9.13**  Force balancing of a rotating link.

opposite from the base pivot, as illustrated in Figure 9.13(d). For balancing, it is required that

$$m_c r_c = Me \tag{9.4-1}$$

After balancing, the revised center of mass coincides with the axis of rotation.

The above is an illustration of *rotating unbalance*. It can occur as a result of pure rotational motion of a body about a fixed axis. Besides rotating unbalance, many mechanisms simultaneously generate *reciprocating unbalance*. This is caused by the inertial forces associated with a translating mass. Reciprocating unbalance may be generated by the translational motion of the slider of a slider crank mechanism. The coupler of a four-bar mechanism may also produce reciprocating unbalance.

In the following sections, we will investigate conventional means of eliminating or reducing the shaking forces produced by four-bar and slider crank mechanisms.

## 9.5  FORCE BALANCING OF A FOUR-BAR MECHANISM

This section presents a method of achieving a four-bar mechanism that is completely force balanced [19]. This method involves determining an expression for the center of mass of the entire mechanism in terms of the link properties and angular displacement of the input link. This expression will be used to determine values of the link properties under which the center of mass of the *entire mechanism* remains stationary for all configurations of the mechanism. As a result, there will be no acceleration of the center of mass of the

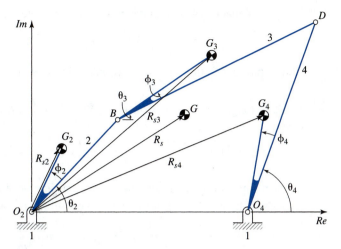

**Figure 9.14** Location of center of mass of a four-bar mechanism.

mechanism, and thus the net inertia force transmitted through the two base pivots to the base link will be zero.

Figure 9.14 illustrates a four-bar mechanism. The real axis of a complex plane is drawn through the base pivots. Vectors $R_{sj}$ ($j = 2, 3, 4$) point to the centers of mass of the moving links. As illustrated in Figure 9.1(a), parameters $b_i$ and $\phi_i$ define the positions of the mass centers of the links with respect to the appropriate turning pair.

The position of the center of mass, $G$, of the entire mechanism is

$$R_s = \frac{1}{M}(m_2 R_{s2} + m_3 R_{s3} + m_4 R_{s4}) \tag{9.5-1}$$

where

$$M = m_2 + m_3 + m_4 \tag{9.5-2}$$

From Figures 9.1(a) and 9.14

$$R_{s2} = b_2 e^{i(\theta_2 + \phi_2)}$$
$$R_{s3} = r_2 e^{i\theta_2} + b_3 e^{i(\theta_3 + \phi_3)} \tag{9.5-3}$$
$$R_{s4} = r_1 + b_4 e^{i(\theta_4 + \phi_4)}$$

Substituting Equations (9.5-3) in Equation (9.5-1), and rearranging,

$$MR_s = (m_2 b_2 e^{i\phi_2} + m_3 r_2) e^{i\theta_2} + (m_3 b_3 e^{i\phi_3}) e^{i\theta_3} + (m_4 b_4 e^{i\phi_4}) e^{i\theta_4} + m_4 r_1 \tag{9.5-4}$$

Equation (9.5-4) is a function of the angular displacement of the three moving links. One of the three angular displacements can be removed by substituting the loop closure equation (refer to Section 4.2), which in this case is

$$r_2 e^{i\theta_2} + r_3 e^{i\theta_3} - r_4 e^{i\theta_4} - r_1 = 0 \tag{9.5-5}$$

Solving Equation (9.5-5) for the unit vector associated with link 3,

$$e^{i\theta_3} = \frac{1}{r_3}(r_1 - r_2 e^{i\theta_2} + r_4 e^{i\theta_4}) \tag{9.5-6}$$

Substituting Equation (9.5-6) in Equation (9.5-4), and rearranging,

$$MR_s = \left( m_2 b_2 e^{i\phi_2} + m_3 r_2 - m_3 b_3 \frac{r_2}{r_3} e^{i\phi_3} \right) e^{i\theta_2} + \left( m_4 b_4 e^{i\phi_4} + m_3 b_3 \frac{r_4}{r_3} e^{i\phi_3} \right) e^{i\theta_4}$$
$$+ m_4 r_1 + m_3 b_3 \frac{r_1}{r_3} e^{i\phi_3} \tag{9.5-7}$$

In Equation (9.5-7), only $\theta_2$ and $\theta_4$ are time dependent. Therefore, the equation is independent of time if the coefficients that multiply $e^{i\theta_2}$ and $e^{i\theta_4}$ are zero. That is, if

$$m_2 b_2 e^{i\phi_2} + m_3 r_2 - m_3 b_3 \frac{r_2}{r_3} e^{i\phi_3} = 0 \tag{9.5-8}$$

and

$$m_4 b_4 e^{i\phi_4} + m_3 b_3 \frac{r_4}{r_3} e^{i\phi_3} = 0 \tag{9.5-9}$$

then vector $R_s$ must be constant. We can employ Equations (9.5-8) and (9.5-9) to determine the properties of the moving links that will provide force balance. These equations involve parameters associated with the three moving links. They may be satisfied by first specifying the parameters of any one of the moving links, say, link 3, and then by solving for the parameters of links 2 and 4. In this case, we rearrange Equations (9.5-8) and (9.5-9) and isolate the parameters of link 3 onto the right-hand sides as

$$m_2 b_2 e^{i\phi_2} = m_3 \left( b_3 \frac{r_2}{r_3} e^{i\phi_3} - r_2 \right) \tag{9.5-10}$$

$$m_4 b_4 e^{i\phi_4} = -m_3 b_3 \frac{r_4}{r_3} e^{i\phi_3} \tag{9.5-11}$$

To solve for the parameters of links 2 and 4, it is convenient to employ coordinate systems relative to each link (see Figure 9.15). The components of Equations (9.5-10) and (9.5-11) are

$$\boxed{\begin{array}{l} m_2 (b_2)_{\xi_2} = m_3 \left( b_3 \frac{r_2}{r_3} \cos \phi_3 - r_2 \right) \\[2mm] m_2 (b_2)_{\eta_2} = m_3 b_3 \frac{r_2}{r_3} \sin \phi_3 \end{array}} \tag{9.5-12}$$

$$\boxed{\begin{array}{l} m_4 (b_4)_{\xi_4} = -m_3 b_3 \frac{r_4}{r_3} \cos \phi_3 \\[2mm] m_4 (b_4)_{\eta_4} = -m_3 b_3 \frac{r_4}{r_3} \sin \phi_3 \end{array}} \tag{9.5-13}$$

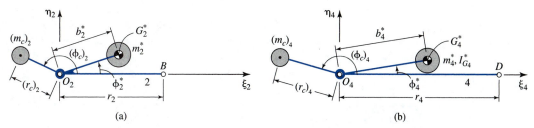

Figure 9.15 Local coordinates for placement of balance masses on a four-bar mechanism: (a) link 2, (b) link 4.

Equations (9.5-12) and (9.5-13) give the net required values to provide force balance. If a mechanism already exists, the magnitude and locations of additional balance masses required may be determined to provide force balance. This is illustrated in the example below.

## EXAMPLE 9.2 Force Balancing of a Four-Bar Mechanism

Figure 9.16(a) illustrates a four-bar mechanism having the following parameters:

$$r_1 = 6.0 \text{ cm}$$

$$r_2 = 2.0 \text{ cm}; \quad m_2^* = 250 \text{ gr}; \quad b_2^* = 0.70 \text{ cm}; \quad \phi_2^* = 5°$$
$$r_3 = 10.0 \text{ cm}; \quad m_3 = 500 \text{ gr}; \quad b_3 = 3.0 \text{ cm}; \quad \phi_3 = 10°$$
$$r_4 = 7.0 \text{ cm}; \quad m_4^* = 250 \text{ gr}; \quad b_4^* = 2.5 \text{ cm}; \quad \phi_4^* = 15°$$
$$I_{G_3} = 250 \text{ gr-cm}^2; \quad I_{O_4}^* = 500 \text{ gr-cm}^2$$

The constant rotational speed of link 2 is $\omega = 120$ rad/sec CCW. Figure 9.16(b) shows a polar plot of the shaking force.

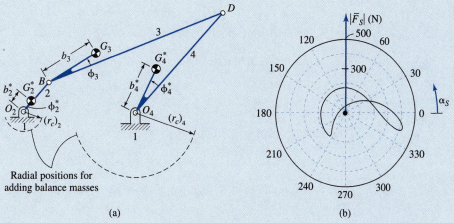

Figure 9.16 Four-bar mechanism prior to force balancing: (a) mechanism, (b) shaking force.

Determine the magnitudes and locations of balance masses to be added to links 2 and 4 that will provide complete force balance. Add the masses at the following radial distances from the base pivots:

$$(r_c)_2 = 1.0 \text{ cm}; \qquad (r_c)_4 = 3.5 \text{ cm}$$

## SOLUTION

From Equations (9.5-12) and (9.5-13)

$$m_2(b_2)_{\xi_2} = -705 \text{ gr-cm}; \qquad m_2(b_2)_{\eta_2} = 52.1 \text{ gr-cm}$$
$$m_4(b_4)_{\xi_4} = -1{,}000 \text{ gr-cm}; \qquad m_4(b_4)_{\eta_4} = -182 \text{ gr-cm}$$

To provide force balance of the mechanism, balance masses $(m_c)_2$ and $(m_c)_4$ are added to links 2 and 4, respectively. The combined effect caused by the given links, and that of the added balance masses, must provide the above calculated values. That is, for link 4

$$m_4(b_4)_{\xi_4} = m_4^*(b_4^*)_{\xi_4} + (m_c)_4 ((r_c)_4)_{\xi_4} \qquad (9.5\text{-}14a)$$

$$m_4(b_4)_{\eta_4} = m_4^*(b_4^*)_{\eta_4} + (m_c)_4 ((r_c)_4)_{\eta_4} \qquad (9.5\text{-}14b)$$

where

$$(b_4^*)_{\xi_4} = b_4^* \cos \phi_4^* = 2.42 \text{ cm}; \qquad (b_4^*)_{\eta_4} = b_4^* \sin \phi_4^* = 0.65 \text{ cm}$$

Using Equations (9.5-14a) and (9.5-14b), and substituting known values,

$$
\begin{aligned}
(\phi_c)_4 &= \tan 2^{-1}[(m_c)_4((r_c)_4)_{\xi_4}, \ (m_c)_4((r_c)_4)_{\eta_4}] \\
&= \tan 2^{-1}[m_4(b_4)_{\xi_4} - m_4^*(b_4^*)_{\xi_4}, \ m_4(b_4)_{\eta_4} - m_4^*(b_4^*)_{\eta_4}] \\
&= -168.1°
\end{aligned}
$$

The coordinates of the balance mass for link 4 are

$$((r_c)_4)_{\xi_4} = (r_c)_4 \cos(\phi_c)_4 = -3.4 \text{ cm}$$
$$((r_c)_4)_{\eta_4} = (r_c)_4 \sin(\phi_c)_4 = -0.72 \text{ cm}$$

From Equation (9.5-14a), the balance mass required on link 4 is

$$(m_c)_4 = \frac{m_4(b_4)_{\xi_4} - m_4^*(b_4^*)_{\xi_4}}{((r_c)_4)_{\xi_4}} = 478 \text{ gr}$$

The mass of link 4 after adding the balance mass is

$$m_4 = m_4^* + (m_c)_4 = 728 \text{ gr}$$

The polar mass moment of inertia of link 4 about its base pivot after addition of the balance mass is

$$I_{O_4} = I_{O_4}^* + (m_c)_4(r_c)_4^2 = 6{,}360 \text{ gr-cm}^2$$

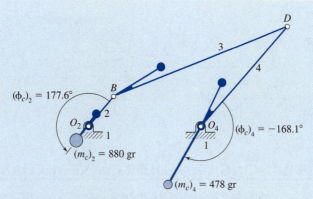

Figure 9.17 Force balanced mechanism.

A similar analysis for link 2 provides

$$(\phi_c)_2 = 177.6°; \qquad (m_c)_2 = 880 \text{ gr}; \qquad m_2 = 1,130 \text{ gr}$$

The force-balanced mechanism is shown in Figure 9.17. After force balancing, the polar plot of the shaking force is shown as a dot at the origin.

Force balancing has an effect on the forces at the turning pairs as well as the shaking moment of the mechanism. Considering Example 9.2, Figure 9.18 shows a polar plot of the force between links 2 and 3. The solid line indicates the result using the mechanism prior to force balancing, and the dashed line corresponds to the force-balanced mechanism. The maximum magnitude of the force in the turning pair is greater in the force-balanced mechanism. Figure 9.19 shows a plot of the shaking moment before and after force balancing. The maximum magnitude is greater after force balancing.

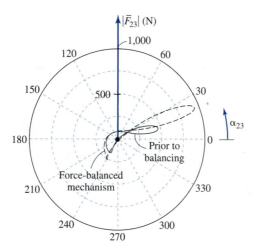

Figure 9.18 Force at turning pair, before and after force balancing.

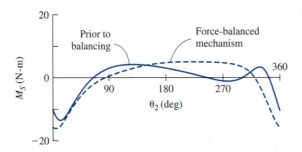

**Figure 9.19** Shaking moment, before and after force balancing.

## 9.6  FORCE BALANCING OF A SLIDER CRANK MECHANISM

Figure 9.20 illustrates a typical slider crank mechanism that will be considered in this section. The mechanism has zero offset, and the centers of mass of the crank and coupler are on lines connecting the turning pairs associated with each link. The location of the center of mass of link 3 is specified by dimensions $r_B$ and $r_D$. Comparing Figures 9.7 and 9.20, we have

$$r_1 = 0; \qquad \phi_2 = 0 \text{ (or } 180°\text{)}; \qquad \phi_3 = 0; \qquad b_3 = r_D; \qquad r_B + r_D = r_3 \qquad (9.6\text{-}1)$$

In addition, we consider geometries for which

$$\frac{r_2}{r_3} < 0.25 \qquad (9.6\text{-}2)$$

The conditions described by Equations (9.6-1) and (9.6-2) are commonly used in piston engines.

Based on the above, a simplified model will be developed for the kinetic analysis of a slider crank mechanism. This model includes development of a simplified expression of the linear acceleration of link 4. Also, we will replace the distributed mass of link 3 with a *lumped-mass model*. These modifications greatly reduce the computational effort required to complete an analysis, while still providing acceptably accurate results.

Figure 9.21 shows a mechanism having its center of mass of link 2 coinciding with the base pivot $O_2$, that is,

$$b_2 = 0 \qquad (9.6\text{-}3)$$

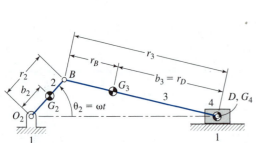

**Figure 9.20**  Slider crank mechanism.

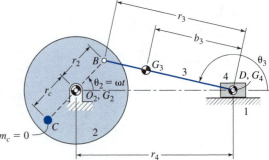

**Figure 9.21**  Reference mechanism.

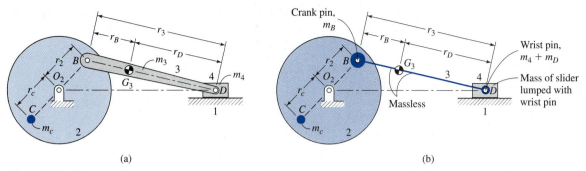

(a)                                                                    (b)

**Figure 9.22** Slider crank mechanism employing alternative models of coupler: (a) distributed mass model, (b) lumped-mass model.

This is regarded as the reference mechanism. With respect to this mechanism, the shaking force can be adjusted by adding a balance mass, $m_c$. As shown in Figure 9.21, a balance mass is added to link 2 at distance $r_c$ from the base pivot, and at angular position $\theta_2 + 180°$.

## 9.6.1 Lumped-Mass Model of the Coupler

A simplified model of the coupler is obtained by replacing the distributed mass of the coupler shown in Figure 9.22(a) with two lumped masses, $m_B$ and $m_D$. The lumped masses are located at the *crank pin* and the *wrist pin* and are connected by a massless rigid rod, as shown in Figure 9.22(b).

The two couplers shown in Figure 9.22 would generate equivalent results in a kinetic analysis provided they have the same total mass, the same location of the center of mass, and the same polar mass moment of inertia with respect to the center of mass. These conditions are expressed as

$$m_B + m_D = m_3 \tag{9.6-4}$$

$$m_B r_B = m_D r_D \tag{9.6-5}$$

$$m_B r_B^2 + m_D r_D^2 = I_{G_3} \tag{9.6-6}$$

Since there are only two adjustable quantities ($m_B$ and $m_D$) in Equations (9.6-4)–(9.6-6), not all three conditions can be satisfied. Thus, the lumped-mass coupler is an approximate representation of the distributed mass coupler. Solving Equations (9.6-4) and (9.6-5) for $m_B$ and $m_D$, and implementing Equation (9.6-1),

$$m_B = \left(\frac{r_D}{r_B + r_D}\right) m_3 = \frac{r_D}{r_3} m_3; \qquad m_D = \frac{r_B}{r_3} m_3 \tag{9.6-7}$$

Although this lumped-mass model does not satisfy Equation (9.6-6), it will be shown that related values of calculated shaking force are acceptably accurate.

## 9.6.2 Accelerations of Lumped Masses

In this subsection we will determine the accelerations at the crank pin (point $B$), the wrist pin (point $D$), and the location at which the balance mass is added (point $C$). Points $B$ and $C$

execute circular motion. Expressions of their accelerations can be readily determined to be

$$\bar{a}_B = -r_2\omega^2(\cos\omega t\,\bar{i} + \sin\omega t\,\bar{j})$$
(9.6-8)

and

$$\bar{a}_C = r_c\omega^2(\cos\omega t\,\bar{i} + \sin\omega t\,\bar{j})$$
(9.6-9)

The wrist pin undergoes rectilinear motion. An expression of its acceleration is given in Equation (9.3-5), being a function of the lengths, angular displacements, and angular velocities of links 2 and 3. To develop a simplified expression for this acceleration, we begin by employing Equation (4.3-77) for the angular displacement of link 3. When the offset, $r_1$, is zero, and substituting Equation (9.1-2),

$$\theta_3 = 180° - \sin^{-1}\left(\frac{r_2}{r_3}\sin\omega t\right)$$
(9.6-10)

Employing the identity

$$\sin^{-1} u = \cos^{-1}[1 - u^2]^{1/2}$$
(9.6-11)

for which in this case

$$u = \frac{r_2}{r_3}\sin\omega t$$
(9.6-12)

then Equation (9.6-10) becomes

$$\theta_3 = 180° - \cos^{-1}\left[1 - \left(\frac{r_2}{r_3}\sin\omega t\right)^2\right]^{1/2}$$
(9.6-13)

Now using the infinite series

$$[1 - u^2]^{1/2} = 1 - \tfrac{1}{2}u^2 - \tfrac{1}{8}u^4 - \cdots$$

Equation (9.6-13) is expressed as

$$\theta_3 = 180° - \cos^{-1}\left[1 - \frac{1}{2}\left(\frac{r_2}{r_3}\right)^2\sin^2\omega t - \frac{1}{8}\left(\frac{r_2}{r_3}\right)^4\sin^4\omega t - \cdots\right]$$
(9.6-14)

For the condition described in Equation (9.6-2), the maximum value of the third term inside the square brackets of Equation (9.6-14) is less than $1/64$ that of the second term. It is therefore reasonable to drop the third term and all other higher-order terms of the infinite series. We also employ the identity

$$\sin^2\omega t = \tfrac{1}{2} - \tfrac{1}{2}\cos 2\omega t$$

and Equation (9.6-14) becomes

$$\theta_3 \approx 180° - \cos^{-1}\left[1 - \frac{1}{4}\left(\frac{r_2}{r_3}\right)^2(1 - \cos 2\omega t)\right]$$
(9.6-15)

Employing Figure 9.21, the position of the link 4 is

$$r_4 = r_2 \cos \theta_2 - r_3 \cos \theta_3 \qquad (9.6\text{-}16)$$

Substituting Equations (9.1-2) and (9.6-15) in Equation (9.6-16),

$$r_4 = r_2 \cos \omega t - r_3 \cos \left\{ 180° - \cos^{-1} \left[ 1 - \frac{1}{4} \left( \frac{r_2}{r_3} \right)^2 (1 - \cos 2\omega t) \right] \right\} \qquad (9.6\text{-}17)$$

Employing the identity

$$\cos(a - b) = \cos a \cos b + \sin a \sin b$$

in Equation (9.6-17) for which

$$a = 180°; \qquad b = \cos^{-1} \left[ 1 - \frac{1}{4} \left( \frac{r_2}{r_3} \right)^2 (1 - \cos 2\omega t) \right] \qquad (9.6\text{-}18)$$

and rearranging and simplifying,

$$r_4 = r_2 \cos \omega t + r_3 \left[ 1 - \frac{1}{4} \left( \frac{r_2}{r_3} \right)^2 (1 - \cos 2\omega t) \right] \qquad (9.6\text{-}19)$$

Differentiating Equation (9.6-19) twice with respect to time,

$$\ddot{r}_4 = -r_2 \omega^2 \left( \cos \omega t + \frac{r_2}{r_3} \cos 2\omega t \right) \qquad (9.6\text{-}20)$$

Therefore, the acceleration of link 4 is

$$\boxed{\bar{a}_D = -r_2 \omega^2 \left( \cos \omega t + \frac{r_2}{r_3} \cos 2\omega t \right) \bar{i}} \qquad (9.6\text{-}21)$$

Equation (9.6-21) is an explicit function of only the link dimensions and the rotational speed of link 2.

## 9.6.3 Force Analysis and Balancing

We now implement the lumped-mass model of the coupler introduced in Section 9.6.1, and expressions for accelerations presented in Section 9.6.2, as the basis of completing a force analysis. Figure 9.23 illustrates a slider crank mechanism along with the accelerations of points $B$, $C$, and $D$. Also shown are the lumped masses at each of these locations. At the crank pin is the lumped mass $m_B$ of the coupler as given by Equation (9.6-7). Point $C$ is the location where a balancing mass may be added. At the wrist pin we combine the mass of the slider, $m_4$, with lumped mass $m_D$ (Equation (9.6-7)).

The shaking force is imparted to the base link as a result of the inertia of the links. For this model, it is the negative of the sum of the masses at $B$, $C$, and $D$ multiplied by their respective accelerations. It is expressed as

$$\boxed{\bar{F}_S = (F_S)_x \bar{i} + (F_S)_y \bar{j}} \qquad (9.6\text{-}22)$$

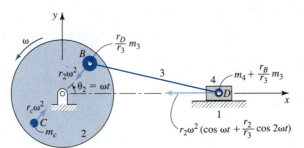

**Figure 9.23** Accelerations in a lumped mass model.

where after simplification

$$(F_S)_x = (F_S)_{x,\text{primary}} + (F_S)_{x,\text{secondary}}$$ (9.6-23a)

$$(F_S)_{x,\text{primary}} = [(m_3 + m_4)r_2 - m_c r_c]\,\omega^2 \cos \omega t$$ (9.6-23b)

$$(F_S)_{x,\text{secondary}} = \left(m_4 + \frac{r_B}{r_3}m_3\right)\frac{r_2^2\omega^2}{r_3} \cos 2\omega t$$ (9.6-23c)

and

$$(F_S)_y = -\left(\frac{r_D}{r_3}m_3 r_2 - m_c r_c\right)\omega^2 \sin \omega t$$ (9.6-24)

The first term on the right-hand side of Equation (9.6-23a) is called the *primary component*. By inspection of Equation (9.6-23b), its frequency of oscillation equals the rotational speed of the crank. The second term on the right-hand side of Equation (9.6-23a) is the *secondary component*. From Equation (9.6-23c), we see that its frequency of oscillation is twice the rotational speed of the crank.

An examination of Equations (9.6-23) and (9.6-24) reveals that the value of $m_c r_c$ has no influence on the secondary component. Also, it is impossible to find a value of $m_c r_c$ that will simultaneously eliminate the $y$ component of the shaking force and the primary component in the $x$ direction. It is thus impossible to achieve complete force balance by adding a balance mass to the crank. However, by properly sizing the balance mass, the magnitude of the shaking force can be considerably reduced from that generated by the reference mechanism.

Consider the slider crank mechanism illustrated in Figure 9.24 with

$$r_2 = 2.5 \text{ cm} = r_c; \qquad r_3 = 10.0 \text{ cm}; \qquad m_3 = 8.0 \text{ gr}; \qquad m_4 = 12.0 \text{ gr}$$

$$r_B = 4.0 \text{ cm}; \qquad r_D = 6.0 \text{ cm}; \qquad \omega = 55.0 \text{ rad/sec CCW}$$

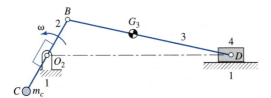

**Figure 9.24** Slider crank mechanism.

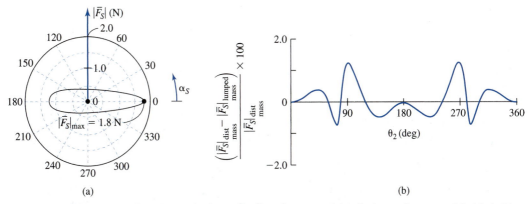

**Figure 9.25** Comparison of results employing a distributed mass model and a lumped mass model: (a) shaking forces, (b) percentage differences.

Figure 9.25(a) shows a polar plot of the shaking force of the reference mechanism (i.e., $m_c = 0$). Results are illustrated using both the lumped-mass model introduced in this section and the distributed-mass model presented in Section 9.3. There is no discernable difference between the two. The percentage differences obtained using the two models are shown in Figure 9.25(b). The maximum absolute difference is less than 1.5 percent. This result justifies utilization of a lumped-mass model to obtain acceptable results. The maximum magnitude of the shaking force is 1.80 N.

Figure 9.26 shows the shaking force for three levels of balance mass, $m_c$. All have a balance mass added according to the relation

$$m_c = \frac{\dfrac{r_D}{r_3}m_3 r_2 + \gamma\left(\dfrac{r_B}{r_3}m_3 + m_4\right)r_2}{r_c}, \qquad 0 \le \gamma \le 1.0 \tag{9.6-25}$$

Figure 9.26(a) corresponds to case for which $\gamma = 0$, and from Equation (9.6-25)

$$m_c = \frac{r_D}{r_3}\frac{r_2}{r_c}m_3 = 4.8 \text{ gr} \tag{9.6-26}$$

In this case, using Equation (9.6-24), the component of shaking force in the $y$ direction is eliminated. The dashed line in Figure 9.26(a) corresponds to the shaking force when

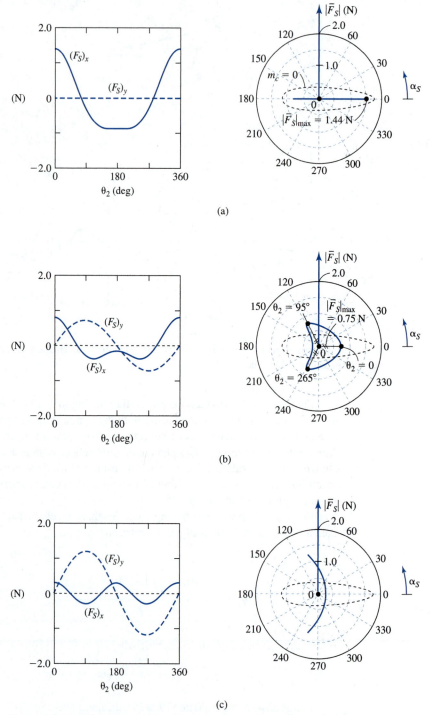

**Figure 9.26** Illustrative example of employing different levels of balance mass in a slider crank mechanism: (a) $m_c = 4.8$ gr, (b) $m_c = 13.8$ gr, (c) $m_c = 20.0$ gr.

$m_c = 0$. The maximum magnitude of the shaking force is now 1.44 N. Its value has been reduced 20 percent compared to that produced by the reference mechanism.

We now consider increasing the balance mass, using values of $\gamma$ in the range $0 < \gamma < 1.0$. By introducing various values of $\gamma$, we obtain the result illustrated in Figure 9.26(b) showing the maximum magnitude of the shaking force during a cycle has been minimized. This is the condition of optimum balancing.

Under this condition, the same maximum magnitude of shaking force occurs three times during each cycle. For this illustration, they occur when $\theta_2 = 0°$, 95°, and 265°. The maximum shaking force has magnitude of 0.754 N, which is a 58 percent reduction, compared to that generated by the reference mechanism.

For this illustration, the condition of optimum balance is achieved using $\gamma = 0.595$, which corresponds to $m_c = 13.8$ gr. In general, the value of $\gamma$ that produces optimum balancing depends on the relative masses of the coupler and slider, and the relative lengths of links 2 and 3. Optimum force balancing is typically obtained by searching in the range $0.50 < \gamma < 0.75$.

Figure 9.26(c) corresponds to case for which $\gamma = 1.0$. Using Equations (9.6-1) and (9.6-25),

$$m_c = \frac{r_2}{r_c}(m_3 + m_4) = 20.0 \text{ gr} \tag{9.6-27}$$

Comparing Equations (9.6-23b), (9.6-23c), and (9.6-27), it can be shown in this instance that the primary component of the shaking force in the $x$ direction has been counterbalanced, and only the secondary component remains. Although the shaking force in the $x$ direction now has a relatively small value, the maximum magnitude of the shaking force is greater than that produced for the condition of optimum balancing. Employing $\gamma = 1.0$ in Equation (9.6-25) yields a calculated value of balance mass that, if used, will cause a shaking force higher than the optimum condition.

# PROBLEMS

P9.1  For the parameters of the four-bar mechanism (see Figure 9.1) given in Table P9.1, determine

(a) the shaking force throughout a cycle of motion as a result of the inertia of links 3 and 4

(b) the shaking moment throughout a cycle of motion as a result of the inertia of links 3 and 4

(Mathcad program: fourbarforce)

$\omega = 50.0$ rad/sec CCW

TABLE P9.1

|  | Link 1 | Link 2 | Link 3 | Link 4 |
|---|---|---|---|---|
| $r_i$ (cm) | 6.0 | 1.5 | 9.0 | 8.0 |
| $b_i$ (cm) | — | 1.0 | 4.0 | 3.0 |
| $m$ (gr) | — | 2.0 | 10.0 | 7.0 |
| $\phi_i$ (degrees) | — | 0 | 5.0 | 0 |
| $I_{G_i}$ (gr-cm²) | — | — | 80.0 | 30.0 |

**TABLE P9.2**

|  | Link 1 | Link 2 | Link 3 | Link 4 |
|---|---|---|---|---|
| $r_i$ (cm) | 0.5 | 2.0 | 9.0 | — |
| $b_i$ (cm) | — | 1.0 | 7.0 | — |
| $m$ (gr) | — | 2.5 | 10.0 | 15.0 |
| $\phi_i$ (degrees) | — | 0 | 5.0 | — |
| $I_{G_i}$ (gr-cm²) | — | — | 80.0 | — |

**P9.2** For the parameters of the slider crank mechanism (see Figure 9.7) given in Table P9.2, determine

    (a) the shaking force throughout a cycle of motion as a result of the inertia of links 3 and 4

    (b) the shaking moment throughout a cycle of motion as a result of the inertia of links 3 and 4

**(Mathcad program: slidercrankforce)**

$$\omega = 50.0 \text{ rad/sec CCW}$$

**P9.3** For the given parameters of a four-bar mechanism (see Figure 9.1), determine the masses and the locations of the centers of mass of links 2 and 4 so that the mechanism is completely force balanced. **(Mathcad program: fourbarbalance)**

$r_1 = 6.0$ cm;    $r_2 = 2.0$ cm;    $r_4 = 7.0$ cm

$r_3 = 10.0$ cm;    $m_3 = 600$ gr;    $b_3 = 6.0$ cm

$$\phi_3 = 10°; \qquad I_{G_3} = 200 \text{ gr-cm}^2$$

$$(r_c)_2 = 2.0 \text{ cm}; \qquad (r_c)_4 = 3.5 \text{ cm}$$

**P9.4** For the given parameters of a four-bar mechanism (see Figure 9.16), determine the magnitudes and locations of the balance masses to be added to links 2 and 4 so that the mechanism is completely force balanced. **(Mathcad program: fourbarbalance)**

$r_1 = 6.0$ cm;    $r_2 = 2.0$ cm;    $m_2^* = 200$ gr

$b_2^* = 0.70$ cm;    $\phi_2^* = -5°$;    $r_3 = 10.0$ cm

$m_3 = 450$ gr;    $b_3 = 3.0$ cm;    $\phi_3 = -10°$

$r_4 = 7.0$ cm;    $m_4^* = 250$ gr;    $b_4^* = 2.5$ cm

$\phi_4^* = -15°$;    $(r_c)_2 = 2.0$ cm;    $(r_c)_4 = 2.5$ cm

$$I_{G_3} = 300 \text{ gr-cm}^2; \qquad I_{O_4}^* = 800 \text{ gr-cm}^2$$

**P9.5** For the given parameters of a slider crank mechanism (see Figure 9.22), determine the balance mass to be added to link 2 so that the mechanism has optimum force balancing. **(Mathcad program: slidercrankbalance)**

$r_2 = 2.0$ cm $= r_c$;    $r_3 = 9.0$ cm

$m_3 = 10.0$ gr;    $m_4 = 12.0$ gr

$r_B = 4.0$ cm;    $r_D = 5.0$ cm

$$\omega = 50.0 \text{ rad/sec CCW}$$

**P9.6** For the given parameters of a slider crank mechanism (see Figure 9.22), determine the balance mass to be added to link 2 so that the mechanism has:

    (a) balanced shaking force in the $y$ direction

    (b) optimum balancing

    (c) balanced primary shaking force component in the $x$ direction

**(Mathcad program: slidercrankbalance)**

$r_2 = 2.0$ cm $= r_c$;    $r_3 = 10.0$ cm

$m_3 = 6.0$ gr;    $m_4 = 15.0$ gr

$r_B = 4.0$ cm;    $r_D = 6.0$ cm

$$\omega = 55.0 \text{ rad/sec CCW}$$

# 10  Flywheels

## 10.1 INTRODUCTION

A *flywheel* is a mechanical component designed to store and release kinetic energy. It is mounted on a rotating shaft of a machine, and its mass is symmetrical about the axis of rotation so that shaking forces are reduced. Some flywheels can be simply a solid metal disc.

Flywheels are commonly used for reducing periodic speed fluctuations that occur in internal combustion engines under steady-state operation. Figure 10.1(a) illustrates a single-cylinder engine (see Appendix B, Section B.8), with a flywheel attached to the crankshaft. A free body diagram of the flywheel is illustrated in Figure 10.1(b), indicating the directions of the applied torques. The drive torque is supplied by the engine, and the load torque is the reaction on the flywheel from the externally driven system.

A plot of the drive torque generated by a single-cylinder engine is shown in Figure 10.1(c). This torque is a superposition of the effects of combustion of the mixture of fuel and air during the power stroke and the inertia of the links. Under steady-state operating conditions, this torque repeats itself every two revolutions ($4\pi$ radians). The load torque may also fluctuate depending on the application. When the torques vary, there will be fluctuations in the rotational speed, $\omega$. When the magnitude of the drive torque is greater than the magnitude of the load torque, the net torque on the flywheel will cause an angular acceleration of the crankshaft and an increase in rotational speed. Alternatively, when the magnitude of the load torque is greater than the magnitude of the drive torque, the rotational speed will decrease. Adding a flywheel to the crankshaft increases the polar mass moment of inertia of the rotating elements. For a given set of drive and load torques, the addition of a flywheel will reduce the speed fluctuations. The greater the polar mass moment of inertia of the flywheel, the lower these speed fluctuations.

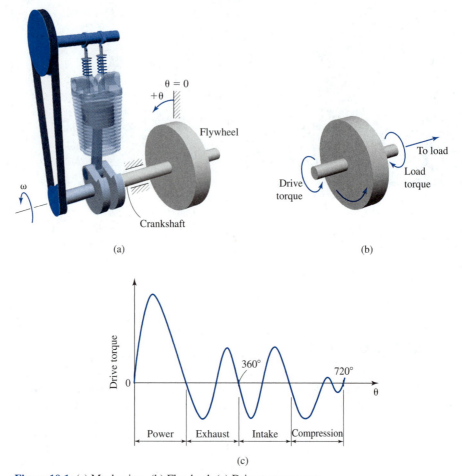

(a)

(b)

(c)

**Figure 10.1**  (a) Mechanism. (b) Flywheel. (c) Drive torque curve.

A power lawn mower is an example of the system illustrated in Figure 10.1. The blade of the lawn mower is rigidly attached to the crankshaft and provides a significant portion of the polar mass moment of inertia about the axis of rotation, and therefore it also acts as a form of flywheel. A properly designed system prevents excessive fluctuations of speed when there are modest variations in the load torque requirements.

A single-cylinder engine produces a greater variation in drive torque compared to that produced by a multi-cylinder engine. Flywheels, in some form, are used in the design of virtually all internal combustion engines.

Flywheels are also integrated in systems for which the drive torque can be considered constant, as when incorporating electric motors, when load torque is variable. A punch press is a typical example. The punching operation is performed during only that portion of the cycle when the load torque on the flywheel is larger than the drive torque. The electric motor drives the flywheel to increase its rotational speed and thus provide the source of energy for the sudden surge of power needed in the punching operation. The power rating of an electric motor used in a punch press can be significantly reduced by use of a flywheel.

Flywheels have recently become the subject of extensive research as energy storage devices. A typical system consists of a flywheel suspended by nearly frictionless magnetic bearings inside a vacuum chamber and can be connected to either an electric motor or an electric generator. These flywheels are made from composite materials that permit greater rotational speed and increased capacity for energy storage. Here, the flywheel stores kinetic energy by driving it with an electric motor. The energy may later be retrieved while engaging the spinning flywheel with an electric generator.

This chapter will be restricted to providing a mathematical model for determining the required size of a flywheel to keep the cyclical rotational speed within stated limits, based on applied torques under cyclical steady-state operating conditions. Systems will be considered for which all components attached to a rotating shaft have the same rotational speed, and then those systems that involve more than one rotational speed will be examined.

## 10.2 MATHEMATICAL FORMULATION

The governing equation for rotational motion of a body (see Equation (2.10-4)) is

$$M_G = I_G \ddot{\theta} \tag{10.2-1}$$

A free body diagram of a flywheel is shown in Figure 10.2, indicating positive directions of rotational speed, $\omega$, the *drive torque*, $T_D$, and *load torque*, $T_L$, applied to the flywheel. Employing Equation (10.2-1) and dropping subscript $G$,

$$M = T_D + T_L = I \frac{d^2\theta}{dt^2} \tag{10.2-2}$$

or

$$M = I \frac{d}{dt}\left(\frac{d\theta}{dt}\right) \tag{10.2-3}$$

We substitute the following for the rotational speed:

$$\omega = \frac{d\theta}{dt} \tag{10.2-4}$$

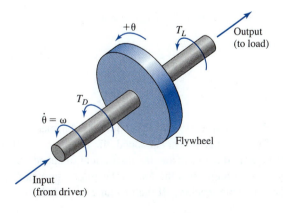

**Figure 10.2** Flywheel—free body diagram.

and Equation (10.2-3) becomes

$$M = I\frac{d\theta}{dt}\frac{d}{d\theta}\left(\frac{d\theta}{dt}\right) = I\omega\frac{d\omega}{d\theta} \tag{10.2-5}$$

or

$$M\,d\theta = I\omega\,d\omega \tag{10.2-6}$$

Integrating both sides of Equation (10.2-6) produces

$$\int_{\theta=\theta_1}^{\theta=\theta_2} M\,d\theta = \int_{\omega=\omega_1}^{\omega=\omega_2} I\omega\,d\omega = \tfrac{1}{2}I\left(\omega_2^2 - \omega_1^2\right) \tag{10.2-7}$$

The left-hand side of Equation (10.2-7) represents the work done on the flywheel by the torques and will be designated as $\Delta E$. The right-hand side represents the corresponding change of kinetic energy stored in the flywheel (see Appendix D, Section D.10.1).

We are interested in the maximum speed fluctuation during each cycle of motion. This requires that we find those values of $\theta$ that generate the maximum value of $\Delta E$, designated as $\Delta E_{max}$. Under these circumstances, from Equation (10.2-7)

$$\Delta E_{max} = \tfrac{1}{2}I\left(\omega_{max}^2 - \omega_{min}^2\right) = \tfrac{1}{2}I(\omega_{max} + \omega_{min})(\omega_{max} - \omega_{min}) \tag{10.2-8}$$

We introduce the average rotational speed:

$$\omega_0 = \frac{(\omega_{max} + \omega_{min})}{2} \tag{10.2-9}$$

and Equation (10.2-8) becomes

$$\Delta E_{max} = I\omega_0(\omega_{max} - \omega_{min}) \tag{10.2-10}$$

Further, we introduce the *coefficient of speed fluctuation*:

$$\boxed{C_S = \frac{\omega_{max} - \omega_{min}}{\omega_0}} \tag{10.2-11}$$

and Equation (10.2-10) becomes

$$\Delta E_{max} = I\omega_0^2 C_S \tag{10.2-12}$$

or

$$\boxed{I = \frac{\Delta E_{max}}{\omega_0^2 C_S}} \tag{10.2-13}$$

Equation (10.2-13) is an expression for the required polar mass moment of inertia of a flywheel in terms of the average rotational speed, the maximum change of kinetic energy in the flywheel throughout one cycle, and the coefficient of speed fluctuation.

Equation (10.2-13) represents the total polar mass moment of inertia of not only the flywheel but for the whole machine. If the machine has a polar mass moment of inertia of

**TABLE 10.1** Typical Values of the Coefficient of Speed Fluctuation

| Application | Typical Value of $C_S$ |
|---|---|
| Alternators, generators | 0.005 |
| Punch press | 0.10 |
| Rock crusher | 0.20 |

$I_{\text{machine}}$, then the polar mass moment of inertia required for the flywheel is

$$I_{\text{flywheel}} = I - I_{\text{machine}} \qquad (10.2\text{-}14)$$

The coefficient of speed fluctuation is a dimensionless quantity. In the analysis presented, we will consider situations in which there are small speed fluctuations about the average value $\omega_0$, that is, the coefficient of speed fluctuation will be much less than unity. Typical allowable values of the coefficient of speed fluctuation are given in Table 10.1.

If steady-state operating conditions are assumed, the system will repeat itself after an integer number of revolutions. For instance, speed fluctuation in a reciprocating engine will repeat itself every two revolutions ($4\pi$ radians) of the crankshaft.

A convenient method of analyzing the requirements of a flywheel is to plot the drive torque and load torque curves on the same graph. Using the notation presented in this section, it is recommended that the *positive* of the drive torque and the *negative* of the load torque be drawn onto the graph. An illustrative graph is shown in Figure 10.3.

Under steady-state conditions, during each cycle, the energy input to the flywheel provided by the drive torque must equal the energy removed from the flywheel by the load torque. Therefore, throughout a cycle, the average absolute values of the drive and load torques must be equal.

The net torque on the flywheel, $M$, is the sum of the drive and load torques (Equation (10.2-2)). For the graph in Figure 10.3(a), since the negative of the load torque is plotted, the net torque on the flywheel for a given rotation is the *difference* between the two curves. A plot of the net torque on the flywheel is given in Figure 10.3(b). Points of intersection of the curves in Figure 10.3(a) correspond to there being no net torque applied to the flywheel. At these locations, there is no acceleration or deceleration, and rotational speed is at a local extremum. A plot of the rotational energy throughout a cycle is shown in Figure 10.3(c). At each point of intersection, the energy is also at a local extremum.

When the driving torque has a magnitude greater than the load torque, there is a net torque in the direction of rotation, and the rotational speed (and energy in the flywheel) increases. Alternatively, when the magnitude of the load torque is greater than that of the drive torque, the net torque is in a direction opposite to that of the direction of rotation, and the rotational speed decreases.

Since these graphs are in terms of torque versus rotation, areas between the drive torque and load torque curves represent changes of energy in the flywheel. Also, since intersection points of the curves correspond to local extrema of energy, it is possible to compare the relative levels of energy at all intersection points. The value of $\Delta E_{\max}$ can

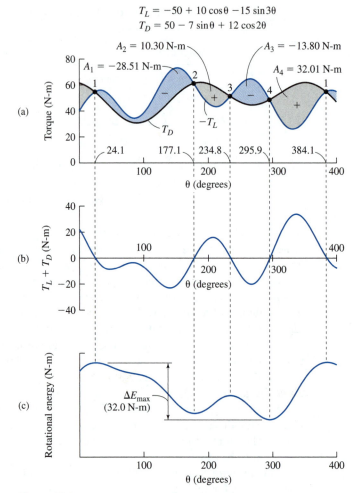

$$T_L = -50 + 10\cos\theta - 15\sin3\theta$$
$$T_D = 50 - 7\sin\theta + 12\cos2\theta$$

**Figure 10.3** Typical load and drive torque curves.

then be determined by comparing all of the energies, selecting the maximum and minimum values, and taking the difference.

As another demonstration, Figure 10.4 shows a model of a single-cylinder engine operating under steady-state conditions. The drive and load torque curves are also given. The load torque is constant, and the negative of its value is plotted. The drive torque curve is highly simplified and only deviates from a constant value during the power and compression strokes. In order for this system to operate under steady-state conditions, average values of the drive torque and of the negative of the load torque must be equal.

Each cycle in this demonstration lasts two revolutions ($4\pi$ radians) of crank rotation. In any given cycle, there are only two points of intersection between the torque curves; each point corresponds to a local extremum of rotational energy.

Intersection point 1A is set to coincide with zero rotation. It corresponds to a local minimum. During the interval of rotation between intersection points 1A and 2A, since the drive

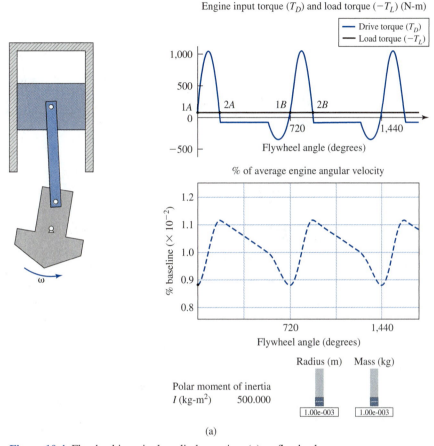

(a)

**Figure 10.4** Flywheel in a single-cylinder engine: (a) no flywheel.

torque is greater in magnitude than the load torque, both energy and rotational speed increase. This increase will continue until intersection point 2A, at which time both energy and rotational speed will have reached maximums. Beyond point 2A, and up until point 1B, both energy and rotational speed decrease. Between points 1A and 2A, the area bounded by torque curves equals that bounded by the torque curves between points 2A and 1B. In either case, the area is

$$\Delta E_{max} = 1{,}905 \text{ N-m}$$

The average rotational speed is

$$\omega_0 = 4.0 \text{ rad/sec}$$

The amount of fluctuation from this average depends on the polar mass moment of inertia of the rotating components. The polar mass moment of inertia of the machine without a flywheel is assumed to have a constant value of

$$I = I_{machine} = 500 \text{ kg-m}^2$$

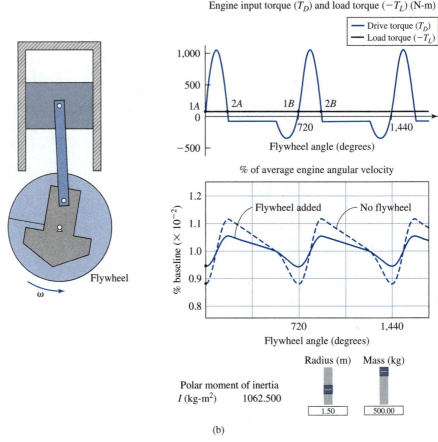

(b)

**Figure 10.4** (*Continued*) Flywheel in a single-cylinder engine: (b) flywheel added [Model 10.4].

The coefficient of speed fluctuation is

$$C_S = \frac{\Delta E_{max}}{\omega_0^2 I} = \frac{1,905}{4.0^2 \times 500} = 0.238$$

The corresponding speed curve in this case is provided in Figure 10.4(a).

Using **[Model 10.4]**, it is possible to adjust both the mass and radius of the solid disc flywheel. For instance, Figure 10.4(b) shows the case where

$$m = 500 \text{ kg}; \qquad r = 1.5 \text{ m}$$

and the corresponding polar mass moment of inertia of the flywheel is

$$I_{flywheel} = \tfrac{1}{2}mr^2 = 560 \text{ kg-m}^2$$

Thus

$$I = I_{machine} + I_{flywheel} = 500 + 560 = 1,060 \text{ kg-m}^2$$

and the coefficient of speed fluctuation has reduced to

$$C_S = \frac{\Delta E_{max}}{\omega_0^2 I} = \frac{1,905}{4.0^2 \times 1,060} = 0.112$$

A corresponding speed curve is plotted in Figure 10.4(b). For purposes of comparison, the case where there is no flywheel is also drawn.

## EXAMPLE 10.1

The load and drive torques are given in Figure 10.5. In this example, the load torque is constant, and the drive torque is variable.

The energies at the intersection points between the torque curves are listed in Table 10.2. Intersection point 1 is given an arbitrary energy level of $E$.

Also given is

$$\omega_0 = 800 \text{ rpm}; \qquad C_S = 0.02$$

Determine the required polar mass moment of inertia.

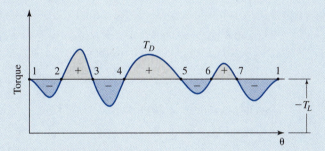

**Figure 10.5** Torque curves.

**TABLE 10.2** Energy Values at the Intersection Points in Figure 10.5

| Point | Energy Level (N-m) |
| --- | --- |
| 1 | $E$ |
| 2 | $E - 1,600$ |
| 3 | $E + 3,200$ |
| 4 | $E - 800$ |
| 5 | $E + 4,800$ |
| 6 | $E + 1,500$ |
| 7 | $E + 3,400$ |

SOLUTION

Examination of Table 10.2 shows that the maximum energy level corresponds to point 5, and the minimum energy level is at point 2. Thus

$$\Delta E_{max} = E_5 - E_2 = (E + 4{,}800) - (E - 1{,}600) = 6{,}400 \text{ N-m}$$

Substituting in Equation (10.2-13),

$$I = \frac{\Delta E_{max}}{\omega_0^2 C_S}$$

$$= \frac{6{,}400 \text{ N-m}}{\left(800 \times \dfrac{2\pi}{60}\right)^2 \text{rad}^2/\sec^2 \times 0.02}$$

$$= 45.5 \text{ N-m-sec}^2$$

$$= 45.5 \text{ kg-m}^2$$

## EXAMPLE 10.2

Figure 10.6 illustrates a graph that has a variable load torque and constant drive torque. Each cycle lasts three revolutions ($6\pi$ radians). In addition

$$\omega_0 = 180 \text{ rpm} = 18.85 \text{ rad/sec}; \qquad I_{machine} = 125 \text{ kg-m}^2$$

Assuming steady-state conditions, determine

(a) the average power required
(b) the maximum and minimum rotational speeds throughout a cycle
(c) the mass of a 0.6-meter-diameter solid disc flywheel to produce $C_S = 0.025$

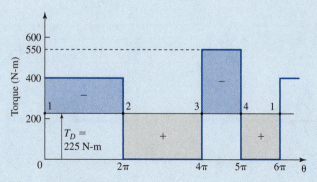

**Figure 10.6** Torque curves.

SOLUTION

Since the averages of the absolute values of the drive and load torques must be equal, the constant drive torque is

$$T_D = -\frac{\text{area under load torque curve}}{\text{cycle of operation}}$$

$$= \frac{400 \times 2\pi + 550 \times \pi}{6\pi} = 225 \text{ N-m}$$

Energies at the intersection points between the load and drive torque curves are listed in Table 10.3.

TABLE 10.3 Energy Values at the Intersection Points in Figure 10.6

| Point | Energy Level (N-m) |
|-------|--------------------|
| 1 | $E_1 = E$ |
| 2 | $E_2 = E_1 - (400 - 225) \times 2\pi = E_1 - 350\pi$ |
| 3 | $E_3 = E_2 + 225 \times 2\pi = E_1 + 100\pi$ |
| 4 | $E_4 = E_3 - (550 - 225) \times \pi = E_1 - 225\pi$ |

Examination of Table 10.3 shows that the maximum energy level corresponds to point 3, and the minimum energy level is at point 2. Thus, $\Delta E_{max}$ is calculated as

$$\Delta E_{max} = E_3 - E_2 = 450\pi \text{ N-m}$$

Based on the above, the solutions are

(a)

$$T_D\omega_0 = 225 \text{ N-m} \times 18.85 \text{ rad/sec} = 4,240 \text{ N-m/sec} = 4.24 \text{ kW}$$

(b)

$$I = \frac{\Delta E_{max}}{\omega_0^2 C_S}; \qquad I = I_{machine}$$

and therefore

$$C_S = \frac{\Delta E_{max}}{\omega_0^2 I} = \frac{450\pi}{18.85^2 \times 125} = 0.0318$$

However

$$C_S = \frac{\omega_{max} - \omega_{min}}{\omega_0}$$

Therefore

$$\omega_{max} - \omega_{min} = C_S \times \omega_0 = 0.0318 \times 18.85 = 0.600 \text{ rad/sec}$$

and

$$\omega_{min} = \omega_0 - \frac{\omega_{max} - \omega_{min}}{2} = 18.85 - 0.300 = 18.55 \text{ rad/sec}$$

$$\omega_{max} = 18.85 + 0.300 = 19.15 \text{ rad/sec}$$

(c)

$$I = \frac{\Delta E_{\text{max}}}{\omega_0^2 C_S} = \frac{450\pi}{18.85^2 \times 0.025} = 159 \text{ kg-m}^2$$

Therefore

$$I_{\text{flywheel}} = I - I_{\text{machine}} = 159 - 125 = 34 \text{ kg-m}^2$$

But

$$I_{\text{flywheel}} = \tfrac{1}{2} \text{mass } r^2 = 34 \text{ kg-m}^2$$

$$= \tfrac{1}{2} \text{mass } (0.30 \text{ m})^2$$

$$\text{mass} = 756 \text{ kg}$$

## EXAMPLE 10.3

The driving torque and load torque are

$$T_D = 100 + 20 \sin 3\theta \text{ N-m}$$

$$T_L = -100 - 80 \sin \theta \text{ N-m}$$

The torque curves are plotted in Figure 10.7. Also

$$I = 4.0 \text{ kg-m}^2$$

Determine

(a) $\Delta E_{\text{max}}$
(b) the maximum angular acceleration during a cycle of motion

As required for steady-state conditions, the absolute average values of the drive and load torques are equal.

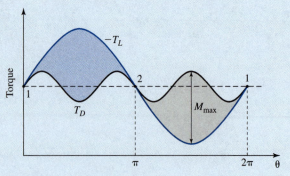

**Figure 10.7** Torque curves.

## SOLUTION

(a) By inspection, rotational energy is gained in the range from $\theta = \pi$ to $\theta = 2\pi$. The minimum energy level is at point 2, and the maximum energy level is at point 1. Therefore

$$\Delta E_{max} = \int_{\theta=\pi}^{\theta=2\pi} (T_D + T_L)\, d\theta = \int_{\theta=\pi}^{\theta=2\pi} (20 \sin 3\theta - 80 \sin \theta)\, d\theta = 147 \text{ N-m}$$

(b) The net moment, $M$, on the flywheel corresponds to the difference between the two torque curves. By inspection of Figure 10.7, the maximum difference occurs when $\theta = 3\pi/2$, and the maximum net moment is

$$M_{max} = (T_D + T_L)|_{\theta=3\pi/2} = 20 + 80 = 100 \text{ N-m}$$

Using Equation (10.2-1),

$$\ddot{\theta}_{max} = \frac{M_{max}}{I} = \frac{100 \text{ N-m}}{4.0 \text{ kg-m}^2} = 25 \text{ rad/sec}^2$$

## 10.3 BRANCHED SYSTEMS

In the operation of a machine, components usually rotate at different speeds. Consider the geared system shown in Figure 10.8. Gear $A$ is rotating at $\omega_2$. Then, the magnitude of the rotational speed of gear $B$ is

$$|\omega_3| = \frac{r_A}{r_B}|\omega_2| \tag{10.3-1}$$

Drive and load torques are applied to the shaft on which gear $A$ is mounted. A flywheel is mounted on the same shaft as gear $B$. If the polar mass moments of inertia of the two gears and flywheel are $I_A$, $I_B$, and $I_{\text{flywheel}}$, respectively, then the total rotational energy may be expressed as

$$\text{energy} = \tfrac{1}{2}I_A\omega_2^2 + \tfrac{1}{2}(I_B + I_{\text{flywheel}})\omega_3^2$$

$$= \tfrac{1}{2}I_A\omega_2^2 + \tfrac{1}{2}(I_B + I_{\text{flywheel}})\left(\frac{r_A}{r_B}\omega_2\right)^2 = \tfrac{1}{2}I_{\text{eff}}\,\omega_2^2 \tag{10.3-2}$$

where

$$I_{\text{eff}} = I_A + (I_B + I_{\text{flywheel}})\left(\frac{r_A}{r_B}\right)^2 \tag{10.3-3}$$

can be considered to be the effective polar mass moment of inertia with respect to shaft $A$.

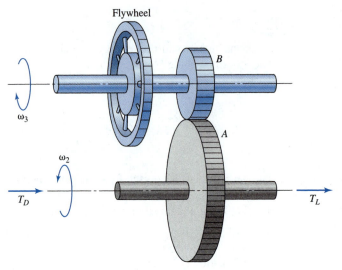

Figure 10.8  Branched system.

# EXAMPLE 10.4  Flywheel Analysis of a Branched System

For the branched system of Figure 10.8, the drive torque is illustrated in Figure 10.9. The load torque is constant. Gear $A$ has an average speed of 300 rpm, and its diameter is two times that of gear $B$. The polar mass moments of inertia of the gears are

$$I_A = 8.0 \text{ kg-m}^2; \qquad I_B = 1.0 \text{ kg-m}^2$$

Determine the polar mass moment of inertia of a flywheel on the same shaft as gear $B$ that is required to have a coefficient of speed fluctuation of 0.005.

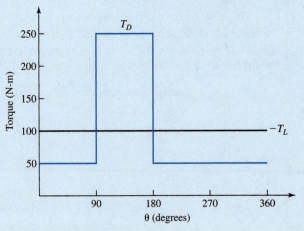

Figure 10.9  Torque curves.

SOLUTION

From the drive torque curve, it may be determined that

$$T_L = -100 \text{ N-m}; \qquad \Delta E_{max} = 150\frac{\pi}{2} \text{ N-m}$$

From Equation (10.2-13)

$$I = I_{eff} = \frac{\Delta E_{max}}{\omega_0^2 C_S} = \frac{150\frac{\pi}{2} \text{ N-m}}{\left(300 \times \frac{2\pi}{60}\right)^2 \text{ rad}^2/\text{sec}^2 \times 0.005} = 47.6 \text{ kg-m}^2$$

Rearranging Equation (10.3-3) gives

$$I_{flywheel} = (I_{eff} - I_A)\left(\frac{r_B}{r_A}\right)^2 - I_B$$

$$= (47.6 - 8.0)\left(\frac{1}{2}\right)^2 - 1.0 = 8.90 \text{ kg-m}^2$$

# PROBLEMS

**P10.1** Figure P10.1 shows the load torque diagram of an engine that operates at an average speed of 2,000 rpm. Determine the value of the constant driving torque necessary to move the crankshaft at this speed and hence determine the mass of a solid disc flywheel required to limit speed fluctuation to 0.8 percent of average speed, given that the outside diameter of the flywheel is 1.2 m. Compare this with the mass of a rim disc flywheel (i.e., assume all mass at the radius of the rim) of the same diameter that is required for the same speed fluctuation.

**P10.2** The crankshaft of a punch press rotates at a maximum speed of 100 rpm, which falls by 10 percent with each punching operation. The flywheel must be able to provide 10 kJ of energy during the 20° while the holes are punched. Determine

(a) the coefficient of speed fluctuation

(b) the mass of a rim disc flywheel of diameter 2.8 m necessary for the press

**P10.3** The maximum variation in stored energy of a flywheel is 3,000 N-m over each cycle of operation. For a coefficient of speed fluctuation of 0.05 at an average speed of 300 rpm, determine the radius of gyration of the 100 kg rim disc flywheel. Determine also the maximum and minimum angular velocities of the machine.

**P10.4** A steel (7,000 kg/m³) solid disc flywheel has a diameter of 1.2 m and a thickness of 20 mm. Determine

(a) the difference in stored kinetic energy of the flywheel for a speed increase from 200 to 210 rpm

(b) the coefficient of speed fluctuation

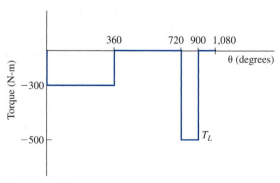

**Figure P10.1**

**P10.5**   A flywheel of mass 600 kg and radius of gyration 1.05 m rotates at 3,000 rpm. Determine the kinetic energy it delivers for a 2 percent drop in speed.

**P10.6**   Figure P10.6 shows the load torque variation for 1 cycle of a shaft driven by a constant torque.

   (a) Determine the energy delivered in each cycle for a flywheel mounted on the shaft so that the coefficient of speed fluctuation is 0.05 at an average speed of 200 rpm.
   (b) If the flywheel is a solid steel (7,000 kg/m$^3$) disc of thickness 10 mm, determine its diameter.

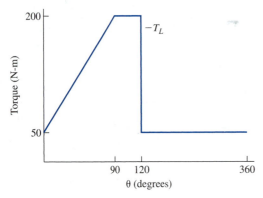

**Figure P10.6**

**P10.7**   A solid disc flywheel has a maximum variation in stored kinetic energy amounting to 1,000 J. Given that its diameter is 0.9 m and the velocity of the edge varies between 8 and 9 m/sec, determine its mass.

**P10.8**   A 1 m diameter solid disc flywheel of mass 100 kg is designed to handle 1,500 N-m of kinetic energy change at an average speed of 200 rpm. Determine the coefficient of speed fluctuation.

**P10.9**   A machine performs one complete cycle of operations in three revolutions, and the load torque curve is shown in Figure P10.1. The driving torque is constant, and the polar mass moment of inertia of the machine is 100 kg-m$^2$. The average speed is 200 rpm. Determine

   (a) the coefficient of speed fluctuation
   (b) the maximum acceleration and deceleration
   (c) the mass of a 60 cm diameter solid disc flywheel in order to reduce the coefficient of speed fluctuation to 0.020

**P10.10**   The main shaft of a rotating machine has a constant load torque and is driven by the torque shown in

Figure P10.10. The machine has a moment of inertia of 8.0 kg-m$^2$ and an average operating speed of 400 rpm. Determine

   (a) the minimum and maximum angular speeds in each cycle
   (b) the greatest acceleration and deceleration in each cycle
   (c) the thickness of a steel (7,000 kg/m$^3$) 30 cm diameter solid disc flywheel, which when attached to the shaft will prevent the maximum rotational speed from exceeding 410 rpm

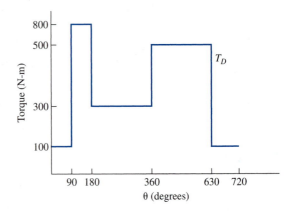

**Figure P10.10**

**P10.11**   A rotating engine with constant driving torque has a load of −5,000 N-m for half a turn, 250 N-m for the next 90°, and −500 N-m for the final 90° of its cycle. The minimum and maximum speeds are 98 and 104 rpm, respectively.

   (a) Determine the constant input torque, and sketch the input and load torque curves against shaft rotation.
   (b) Sketch the output speed curve below the torque curves, showing roughly the maximum, minimum, and average values.
   (c) Determine the polar mass moment of inertia from the given data.
   (d) Determine the thickness of a steel (7,000 kg/m$^3$) solid disc flywheel, 100 cm in diameter, which when mounted on the shaft reduces the coefficient of speed fluctuation to 0.052.

**P10.12**   Figure P10.12 is the negative of the load torque curve of an engine for one cycle of operation.

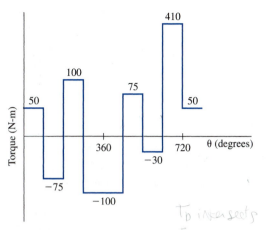

**Figure P10.12**

Given that the average operating speed of the engine is 400 rpm, determine

(a) the required constant drive torque
(b) the crank angles at which the angular velocity is a maximum or minimum in every cycle
(c) the coefficient of speed fluctuation of the engine given that its polar mass moment of inertia is 8 kg-m$^2$

**P10.13** An engine operates at an average speed of 3,500 rpm, with its speed fluctuating 20 rpm about this mean value. Given that the engine is fitted with a 2.5 m diameter rim disc flywheel of mass 35 kg, determine the kinetic energy consumed per cycle.

**P10.14** The torque exerted on the crankshaft of an engine is given by $20,000 + 1,500 \sin 2\theta - 1,320 \cos 2\theta$ N-m. For a constant load torque, determine

(a) the average power of the engine operating at 1,000 rpm
(b) the mass of a 1.2 m diameter solid disc flywheel so that the speed variation is less than 0.3 percent about the mean
(c) the rotational acceleration of the flywheel when $\theta = 45°$.

**P10.15** For an engine with a drive torque given by $12,000 + 2,366 \sin 3\theta$ N-m and a load torque given by $-12,000 - 1,200 \sin \theta$ N-m operating at 200 rpm, determine the coefficient of speed fluctuation, given that the polar mass moment of inertia of the flywheel is 120 kg-m$^2$.

**P10.16** In the system of Figure 10.8, the load torque of the system is $T_L = -100 \sin^2(\theta/2)$ N-m. Gear $A$ turns at an average speed of 300 rpm, driven by a constant torque, and its diameter is three times that of gear $B$. Determine the diameter of a rim disc flywheel of mass 100 kg that is needed to keep speed fluctuations within 0.9 percent of average speed. If the flywheel had been mounted on the main shaft, what would its diameter have been? Neglect the effect of the gears.

# 11 Synthesis of Mechanisms

## 11.1 INTRODUCTION

Most of the problems presented so far have involved the *analysis of mechanisms.* Some dealt with the kinematics of given mechanisms. An input motion was prescribed, and the task was to determine the resultant motions. Consider the mechanism given in Figure 11.1(a), where through analysis we obtain the function graph shown in Figure 11.1(b). Other analyses included the determination of loads between links for a given input motion, or determination of the torque required to maintain a given input motion.

*Synthesis of mechanisms* can be regarded as the reverse task of analysis. A typical problem starts with a desired motion of at least one of the links. It is then necessary to select the most suitable type of mechanism for that particular situation and determine its dimensions such that the motion comes as close as possible to that desired. Given the

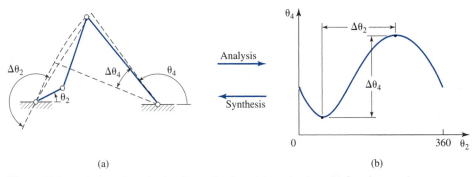

**Figure 11.1** Analysis and synthesis of a mechanism: (a) mechanism, (b) function graph.

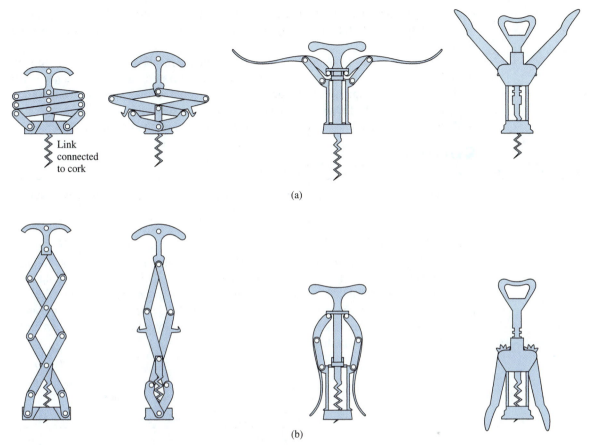

(a)

(b)

**Figure 11.2** Corkscrew mechanisms [Model 11.2].

function graph in Figure 11.1(b), the problem of synthesis would be to select the appropriate type of mechanism for the task and determine its dimensions. The study of cams, presented in Chapter 7, is another example involving synthesis of mechanisms. Examples started with the desired motion of the follower, as set out by a prescribed displacement diagram, and it was then required to determine the cam profile that would produce that motion.

More than one type of mechanism may be designed to provide similar output motion. Figure 11.2 illustrates four varieties of corkscrew mechanisms [9]. All produce vertical motion of the link connected to the cork. These mechanisms may be compared based on their cost, number of links, input motion required, etc.

*Type synthesis* involves selection of the kind of mechanism to be used for a particular application. Designers generally rely on their experience as a guide to carry out this task and therefore should be familiar with the capabilities and typical applications of a variety of mechanisms.

*Dimension synthesis* entails determination of the proportions of the selected type of mechanism. Determining the shape of a disc cam is an example of dimension synthesis.

**Model 11.2**
Corkscrew
Mechanisms

Numerous graphical and analytical methods have been developed for dimension synthesis of planar four-bar and slider crank mechanisms. Some common algorithms are presented in this chapter.

Still other studies of synthesis involve the *kinetics* of mechanisms. A typical problem is to design a particular type of mechanism that will operate in a desired manner when applying a specified input force or input torque. This could involve varying either the distribution of mass in the links or the sizing of flywheels attached to rotating cranks. Henceforth in this chapter, unless otherwise noted, *synthesis* will imply dimension synthesis.

## 11.2  CLASSIFICATION OF SYNTHESIS PROBLEMS

Problems of synthesis can generally be assigned to one of three categories. Each is described in a subsection.

### 11.2.1  Function Generation

A synthesis problem of *function generation* involves determining the dimensions of a mechanism so that it will coordinate either linear or angular motions of two links in a desired manner. For example, Figure 11.3 shows a disc cam mechanism for which the output motion of the follower is a function of the rotation of the cam. For this mechanism, it is often possible to design a cam for which the desired motion of the follower is obtained precisely for every cam rotation.

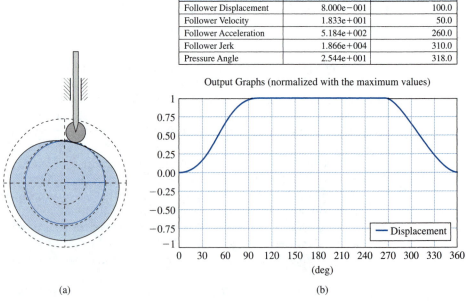

Absolute Maximum Values

| Function | Absolute Maximum | Maximum at (deg) |
|---|---|---|
| Follower Displacement | 8.000e−001 | 100.0 |
| Follower Velocity | 1.833e+001 | 50.0 |
| Follower Acceleration | 5.184e+002 | 260.0 |
| Follower Jerk | 1.866e+004 | 310.0 |
| Pressure Angle | 2.544e+001 | 318.0 |

Output Graphs (normalized with the maximum values)

(a)    (b)

**Figure 11.3** Cam mechanism.

A four-bar mechanism may also be employed for function generation. For the mechanism shown in Figure 11.1(a), we consider $\theta_4$ as a function of $\theta_2$. The form of the functional relation generated by a four-bar mechanism was covered in Chapter 4. Combining Equations (4.3-54) and (4.3-56)–(4.3-59), the relationship may be expressed as

$$\theta_4 = g(\theta_2, r_1, r_2, r_3, r_4) \tag{11.2-1}$$

As indicated, function $g$ depends on the input angular displacement, $\theta_2$, and the four link lengths.

Suppose that a four-bar mechanism is to be designed that will coordinate angular displacements according to the relation

$$\theta_4 = f_{\theta_2}(\theta_2) \tag{11.2-2}$$

In selecting a four-bar mechanism to generate a function, there are only a finite number of link dimensions that may be adjusted to achieve a desired result. Therefore, it is generally not possible to obtain the desired value of a function for every value of $\theta_2$. Instead, we must settle for the likelihood of a difference between the desired function, $f_{\theta_2}$, and that actually produced, $g$. This difference is known as *structural error, e*, which is expressed as

$$\boxed{e(\theta_2, r_1, r_2, r_3, r_4) = g(\theta_2, r_1, r_2, r_3, r_4) - f_{\theta_2}(\theta_2)} \tag{11.2-3}$$

Figure 11.4(b) shows typical plots of a desired function and an approximation of the function generated by a four-bar mechanism illustrated in Figure 11.4(a). The link lengths

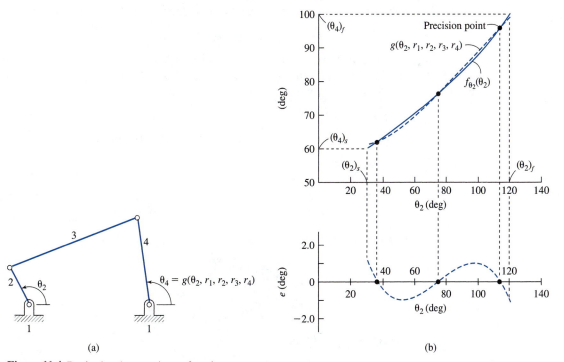

(a)

(b)

**Figure 11.4** Desired and approximate functions.

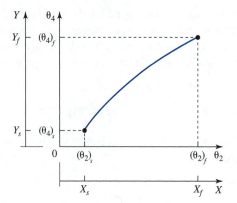

**Figure 11.5** Function generation curve.

were determined using a procedure presented in this chapter. For the plots shown, we do not consider full rotation of links 2 and 4. Rather, link 2 rotates in the range starting from $(\theta_2)_s$ and finishing at $(\theta_2)_f$, while link 4 moves between $(\theta_4)_s$ and $(\theta_4)_f$. The structural error is also shown in a separate plot. An intersection of the two functions is called a *precision point*. Procedures have been developed for the design of a four-bar mechanism where it is possible to specify four or five precision points [10]. However, they are quite complicated compared to an algorithm that requires only three precision points. A procedure requiring only specification of three precision points, which is sufficient for many design problems, is presented in this chapter.

It is often convenient to express a functional relation in the dimensionless form of

$$Y = f_X(X) \tag{11.2-4}$$

where $X$ is proportional to the input motion, and $Y$ is proportional to the output motion. As illustrated in Figure 11.5, we may relate the values of $X$ and $Y$ to the rotational movements of links 2 and 4 by recognizing that

$$
\begin{aligned}
X = X_s \quad &\text{when} \quad \theta_2 = (\theta_2)_s, \qquad X = X_f \quad \text{when} \quad \theta_2 = (\theta_2)_f \\
Y = Y_s \quad &\text{when} \quad \theta_4 = (\theta_4)_s, \qquad Y = Y_f \quad \text{when} \quad \theta_4 = (\theta_4)_f
\end{aligned}
\tag{11.2-5}
$$

Furthermore, if we define

$$
\begin{aligned}
\Delta X = X_f - X_s; \qquad \Delta\theta_2 = (\theta_2)_f - (\theta_2)_s \\
\Delta Y = Y_f - Y_s; \qquad \Delta\theta_4 = (\theta_4)_f - (\theta_4)_s
\end{aligned}
\tag{11.2-6}
$$

then for any value between the starting and finishing positions of link 2,

$$\frac{\theta_2 - (\theta_2)_s}{X - X_s} = \frac{\Delta\theta_2}{\Delta X} = r_X \tag{11.2-7}$$

or

$$\theta_2 = r_X(X - X_s) + (\theta_2)_s \tag{11.2-8}$$

Similarly, for the rotation of link 4

$$\theta_4 = f_{\theta_2}(\theta_2) = r_Y(Y - Y_s) + (\theta_4)_s \tag{11.2-9}$$

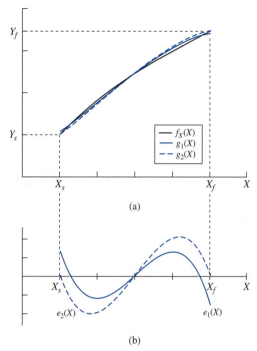

Figure 11.6

where

$$r_Y = \frac{\Delta\theta_4}{\Delta Y} \tag{11.2-10}$$

The positions of precision points play a role in the distribution of structural error. Consider function $f_X$ in Figure 11.6(a), where two different approximations of this function are shown. Each approximation has its distinct locations of the precision points. The structural error for each is shown in Figure 11.6(b).

Chebyshev spacing is commonly employed to locate the precision points for a given range of input motion [3]. Figure 11.7 illustrates the method of graphically locating three precision points. A semicircle is drawn on the $X$ axis with diameter $\Delta X$, and center at $(X_s + X_f)/2$. Half of a regular hexagon is then inscribed in the semicircle with two of its sides perpendicular to the $X$ axis. Lines drawn perpendicular to the $X$ axis from the vertices of the half polygon determine the locations of the precision points. The positions may be expressed as

$$X_1 = \frac{X_f + X_s}{2} - \frac{X_f - X_s}{2}\cos 30° = 0.933X_s + 0.0670X_f$$

$$X_2 = \frac{X_f + X_s}{2} = 0.500X_s + 0.500X_f \tag{11.2-11}$$

$$X_3 = \frac{X_f + X_s}{2} + \frac{X_f - X_s}{2}\cos 30° = 0.0670X_s + 0.933X_f$$

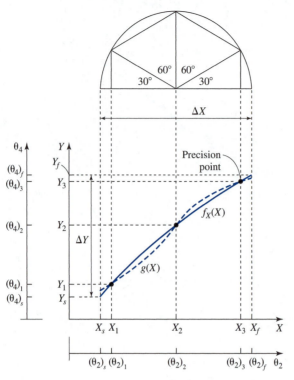

**Figure 11.7** Chebyshev spacing of three precision points.

Based on Equations (11.2-8) and (11.2-9), we can relate the positions of links 2 and 4 to the values of $X$ and $Y$ at the precision points using

$$(\theta_2)_j = r_X(X_j - X_s) + (\theta_2)_s, \qquad j = 1, 2, 3 \tag{11.2-12}$$

$$(\theta_4)_j = r_Y(Y_j - Y_s) + (\theta_4)_s, \qquad j = 1, 2, 3 \tag{11.2-13}$$

A slider crank mechanism may be designed to generate prescribed linear motion of the slider as a function of the rotational motion of the crank. For the mechanism shown in Figure 11.8(a), the corresponding function is shown in Figure 11.8(b). The function produced will depend on the offset and the lengths of the crank and coupler. Therefore, the generated function will likely be an approximation of that which is desired. Using the algorithm presented in this chapter, the desired function and that actually produced may be matched at three precision points.

We may relate positions of the slider to the values of $Y$ of a function using

$$s = r_Y(Y - Y_s) + s_s \tag{11.2-14}$$

where

$$r_Y = \frac{\Delta s}{\Delta Y} \tag{11.2-15}$$

Also, corresponding to the precision points,

$$s_j = r_Y(Y_j - Y_s) + s_s, \qquad j = 1, 2, 3 \tag{11.2-16}$$

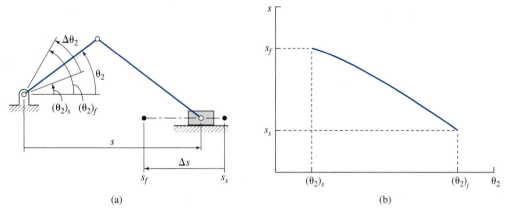

(a)                                                                          (b)

**Figure 11.8** Slider crank mechanism and function generation curve.

### 11.2.2 Path Generation

A synthesis problem of *path generation* involves determining dimensions of a mechanism that will guide a point on a link along a specified path.

Figure 1.16 shows an application of path generation. A point on the coupler is utilized in a film drive mechanism.

A wide variety of shapes of paths may be generated. Figure 1.14 illustrates a small sampling of the path of the trace point from the coupler of a four-bar mechanism. One of the curves illustrated includes a *cusp*: a point on the coupler path for which there are multiple tangents to the curve.

An extensive catalog of coupler curves that are generated by four-bar mechanisms was prepared by Hrones and Nelson [11]. Using a catalog of coupler curves, we may select mechanism dimensions to perform a specific function.

### 11.2.3 Rigid-Body Guidance

A synthesis problem of *rigid-body guidance* involves determining the dimensions of a mechanism so that during its motion a point on one of the links (i.e., the rigid body) passes through prescribed positions, while at the same time the link is constrained to undergo desired rotations. Figure 11.9(a) shows an example where it is required to design a mechanism in which one link starts from position and orientation 1 and moves to final position and orientation 3. An intermediate position and orientation 2 are also specified. For each position, point $A$ on the link has prescribed coordinates $(x_i, y_i)$, $i = 1, 2, 3$. Also, rotations of the link between positions 1 and 2 and between positions 1 and 3 are prescribed as $\theta_{12}$ and $\theta_{13}$, respectively. This synthesis problem is called *three-position rigid-body guidance*. Figure 11.9(b) shows a four-bar mechanism that provides the required motion. It was designed using a synthesis procedure presented in this chapter.

It is possible to design a four-bar mechanism to satisfy a finite number of positions of rigid-body guidance of the coupler. A procedure involving four-position rigid-body guidance is suited to the use of a computer [12], but impractical for hand calculation. However, synthesis procedures for two- or three-position rigid-body guidance are relatively easy to

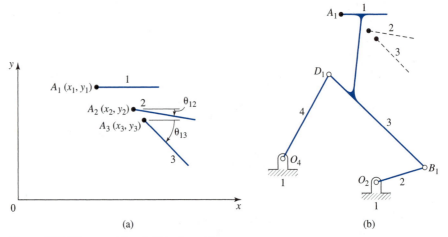

**Figure 11.9** Three-position rigid-body guidance.

complete either graphically or analytically, using a calculator. These methods are sufficient to solve a wide class of practical design problems and are presented in this chapter.

In many instances it will be possible to generate more than one solution to a particular synthesis problem. The solutions may be compared for suitability based on the dimensions of the links, the extent of motions, and the transmission angles.

## 11.3 ANALYTICAL DESIGN OF A FOUR-BAR MECHANISM AS A FUNCTION GENERATOR

In this section, a method is presented for the design of a four-bar mechanism to generate a function having three precision points. Employing Equation (4.3-53) along with the identity

$$\cos(\theta_i - \theta_j) = \cos \theta_i \cos \theta_j + \sin \theta_i \sin \theta_j$$

gives the *Freudenstein equation*:

$$h_1 \cos \theta_4 - h_3 \cos \theta_2 + h_5 = \cos(\theta_2 - \theta_4) \qquad (11.3\text{-}1)$$

Three distinct equations may be generated from Equation (11.3-1) by inserting three pairs of angles for links 2 and 4 corresponding to the precision points, that is,

$$h_1 \cos(\theta_4)_i - h_3 \cos(\theta_2)_i + h_5 = \cos((\theta_2)_i - (\theta_4)_i), \qquad i = 1, 2, 3 \qquad (11.3\text{-}2)$$

Quantities $h_1$, $h_3$, and $h_5$ can be determined from the solution of these three equations. If the length of the base link is selected, then the lengths of the three moving links may be determined. We employ the following equations, which were found by rearranging Equations (4.3-54):

$$r_2 = \frac{r_1}{h_1}; \qquad r_4 = \frac{r_1}{h_3}; \qquad r_3 = \left(r_1^2 + r_2^2 + r_4^2 - 2r_2 r_4 h_5\right)^{1/2} \qquad (11.3\text{-}3)$$

In determining configurations of the mechanism corresponding to a precision point, a negative value obtained from Equations (11.3-3) must be interpreted in a vector sense. That is, the link is drawn in the opposite direction to that defined in Figure 4.3(c).

## EXAMPLE 11.1 Synthesis of a Four-Bar Generating Mechanism with Three Precision Points

Design a four-bar mechanism that will approximately generate the function

$$Y = X^{1/2}, \qquad X_s = 1.0 \leq X \leq 5.0 = X_f \tag{11.3-4}$$

Also

$$r_1 = 1.00 \text{ cm}; \qquad \Delta\theta_2 = -90°; \qquad \Delta\theta_4 = -40° \tag{11.3-5}$$

Use Chebyshev spacing of three precision points.

### SOLUTION

This example specifies the motions of links 2 and 4. However, there is no stipulation of the starting angular configurations of these links. Therefore, we are free to select

$$(\theta_2)_s = 120°; \qquad (\theta_4)_s = 100° \tag{11.3-6}$$

Using Equations (11.2-11),

$$\begin{aligned}
X_1 &= 0.933X_s + 0.0670X_f \\
&= 0.933 \times 1.0 + 0.0670 \times 5.0 = 1.268 \\
X_2 &= 3.000; \qquad X_3 = 4.732
\end{aligned} \tag{11.3-7}$$

Also, from Equation (11.2-6)

$$\Delta X = X_f - X_s = 4.000 \tag{11.3-8}$$

Then from Equations (11.2-7), (11.2-12), (11.3-7), and (11.3-8)

$$\begin{aligned}
(\theta_2)_1 &= (\theta_2)_s + \frac{X_1 - X_s}{\Delta X}\Delta\theta_2 \\
&= 120° + \frac{1.268 - 1.000}{4.000}(-90°) = 113.97° \\
(\theta_2)_2 &= 75.00°; \qquad (\theta_2)_3 = 36.03°
\end{aligned} \tag{11.3-9}$$

Using Equation (11.3-4), the values of $Y$ at the precision points are

$$Y_j = X_j^{1/2}, \qquad j = 1, 2, 3$$

$$Y_1 = X_1^{1/2} = 1.126; \qquad Y_2 = 1.732; \qquad Y_3 = 2.175 \tag{11.3-10}$$

Also

$$\Delta Y = Y_f - Y_s = X_f^{1/2} - X_s^{1/2} = 1.236 \tag{11.3-11}$$

Combining Equations (11.2-13), (11.3-10), and (11.3-11),

$$(\theta_4)_1 = (\theta_4)_s + \frac{Y_1 - Y_s}{\Delta Y}\Delta\theta_4$$

$$= 100° + \frac{1.126 - 1.000}{1.236}(-40°) = 95.92° \tag{11.3-12}$$

$$(\theta_4)_2 = 76.31°; \qquad (\theta_4)_3 = 61.97°$$

Substituting the values from Equations (11.3-9) and (11.3-12) in Equation (11.3-2),

$$\begin{bmatrix} -0.103 & 0.406 & 1.000 \\ 0.237 & -0.259 & 1.000 \\ 0.470 & -0.809 & 1.000 \end{bmatrix} \begin{Bmatrix} h_1 \\ h_3 \\ h_5 \end{Bmatrix} = \begin{Bmatrix} 0.951 \\ 1.000 \\ 0.899 \end{Bmatrix} \tag{11.3-13}$$

Solving for the unknowns,

$$h_1 = 2.959; \qquad h_3 = 1.438; \qquad h_5 = 0.672 \tag{11.3-14}$$

Substituting Equations (11.3-14) in Equations (11.3-3), the lengths of the links are

$$r_2 = 0.338 \text{ cm}; \qquad r_3 = 1.132 \text{ cm}; \qquad r_4 = 0.695 \text{ cm}$$

Figure 11.10 shows the mechanism in three configurations corresponding to the precision points. Figure 11.11(a) shows plots of the desired function and that produced by the mechanism. A plot of the structural error is provided in Figure 11.11(b).

Other synthesized mechanisms can be produced by selecting alternative starting angular configurations of links 2 and 4. If we select

$$(\theta_2)_s = 160°; \qquad (\theta_4)_s = 90° \tag{11.3-15}$$

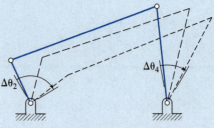

**Figure 11.10** Four-bar mechanism.

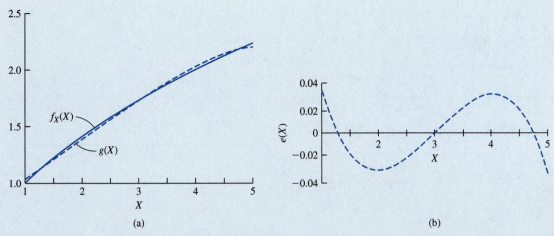

(a)

(b)

**Figure 11.11** Function-generating four-bar mechanism—first solution.

**Figure 11.12** Four-bar mechanism—second solution.

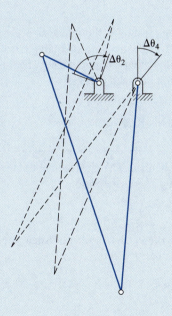

and reproduce the above procedure, the results are

$$r_2 = 1.685 \text{ cm}; \qquad r_3 = 6.375 \text{ cm}; \qquad r_4 = -5.279 \text{ cm}$$

The corresponding mechanism and function graph are shown in Figures 11.12 and 11.13. Note in this case that the value of structural error has changed. Although the structural error is reduced compared to the first solution, a designer may not wish to incorporate the second solution because the link lengths are greater, and the transmission angle is smaller.

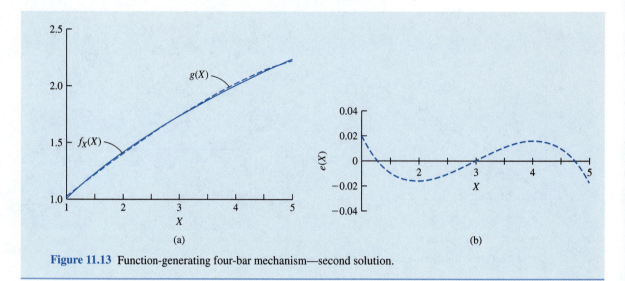

(a)

(b)

**Figure 11.13** Function-generating four-bar mechanism—second solution.

There is no guarantee the calculated results will produce an acceptable mechanism as far as link dimensions and transmission angle are concerned. In general, a solution must be checked relative to its suitability in completing a required task. The use of Working Model 2D software is very effective in carrying out such checks.

## 11.4 ANALYTICAL DESIGN OF A SLIDER CRANK MECHANISM AS A FUNCTION GENERATOR

Figure 11.14 shows a slider crank mechanism. The displacement of the slider is a function of the rotation of the crank. This function takes the form

$$s = f_{\theta_2}(\theta_2, r_1, r_2, r_3) \qquad (11.4\text{-}1)$$

As link 2 moves between positions $(\theta_2)_s$ and $(\theta_2)_f$, link 4 translates between $s_s$ and $s_f$. The $x$ and $y$ coordinates of turning pair $B$ are

$$x_B = r_2 \cos \theta_2; \qquad y_B = r_2 \sin \theta_2 \qquad (11.4\text{-}2)$$

**Figure 11.14** Slider crank mechanism.

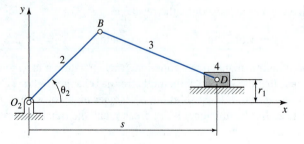

and those of turning pair $D$ are

$$x_D = s; \qquad y_D = r_1 \qquad\qquad (11.4\text{-}3)$$

The square of the distance between turning pairs $B$ and $D$ is

$$r_3^2 = (x_D - x_B)^2 + (y_D - y_B)^2 \qquad\qquad (11.4\text{-}4)$$

Substituting Equations (11.4-2) and (11.4-3) in Equation (11.4-4), rearranging, and simplifying,

$$s^2 = k_1 s \cos \theta_2 + k_2 \sin \theta_2 - k_3 \qquad\qquad (11.4\text{-}5)$$

where

$$k_1 = 2r_2; \qquad k_2 = 2r_1 r_2; \qquad k_3 = r_1^2 + r_2^2 - r_3^2 \qquad\qquad (11.4\text{-}6)$$

Three distinct equations may be generated from Equation (11.4-5) by inserting three pairs of angles of link 2 and the positions of the slider, that is,

$$\boxed{s_i^2 = k_1 s_i \cos(\theta_2)_i + k_2 \sin(\theta_2)_i - k_3, \qquad i = 1, 2, 3} \qquad\qquad (11.4\text{-}7)$$

Quantities $k_1$, $k_2$, and $k_3$ can be determined from the solution of these equations. Employing Equations (11.4-6), the mechanism geometry is determined using

$$\boxed{r_2 = \frac{k_1}{2}; \qquad r_1 = \frac{k_2}{2r_2}; \qquad r_3 = \left(r_1^2 + r_2^2 - k_3\right)^{1/2}} \qquad\qquad (11.4\text{-}8)$$

Chebyshev spacing of precision points (Equations (11.2-11)) may also be used for synthesizing a slider crank mechanism.

## EXAMPLE 11.2 Synthesis of a Slider Crank Generating Mechanism with Three Precision Points

Design a slider crank mechanism that will approximate the function

$$Y = X^{3/2}, \qquad X_s = 1.0 \le X \le 4.0 = X_f$$

by utilizing three precision points with Chebyshev spacing. Also

$$\Delta \theta_2 = -120°; \qquad \Delta s = 5.0 \text{ cm}$$

### SOLUTION

We make the selection

$$(\theta_2)_s = 150°; \qquad s_s = 2.000 \text{ cm}$$

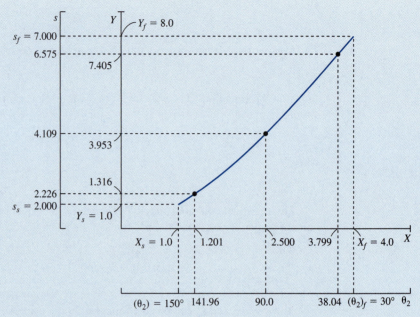

**Figure 11.15** Slider crank mechanism.

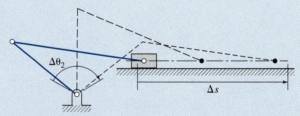

**Figure 11.16** Function-generating slider crank mechanism.

Then from the function graph shown in Figure 11.15

$$X_1 = 1.201; \qquad Y_1 = 1.316; \qquad (\theta_2)_1 = 141.96°; \qquad s_1 = 2.226 \text{ cm}$$

$$X_2 = 2.500; \qquad Y_2 = 3.953; \qquad (\theta_2)_2 = 90.00°; \qquad s_2 = 4.109 \text{ cm}$$

$$X_3 = 3.799; \qquad Y_3 = 7.405; \qquad (\theta_2)_3 = 38.04°; \qquad s_3 = 6.575 \text{ cm}$$

Substituting the above values in Equation (11.4-7), and solving for the unknowns,

$$k_1 = 5.522 \text{ cm}; \qquad k_2 = 5.865 \text{ cm}^2; \qquad k_3 = -11.02 \text{ cm}^2$$

Using Equations (11.4-8), the dimensions of the mechanism are

$$r_1 = 1.062 \text{ cm}; \qquad r_2 = 2.761 \text{ cm}; \qquad r_3 = 4.446 \text{ cm}$$

Figure 11.16 shows the mechanism in three configurations corresponding to the precision points. Figure 11.17 illustrates the ideal and approximately generated functions along with the structural error.

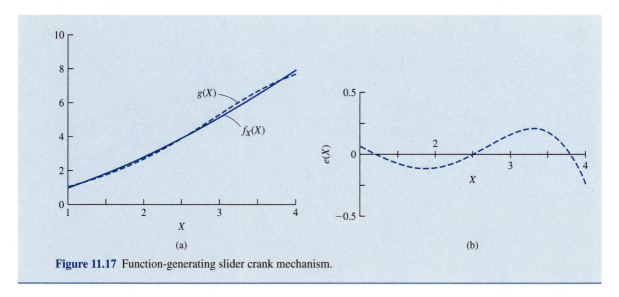

**Figure 11.17** Function-generating slider crank mechanism.

## 11.5 GRAPHICAL DESIGN OF MECHANISMS FOR TWO-POSITION RIGID-BODY GUIDANCE

Suppose that it is necessary to move the rigid body shown in Figure 11.18(a) from position 1 to position 2. Two points on the body are identified as $A$ and $B$, and subscripts indicate the position number. Figure 11.18(b) illustrates one method of achieving this, by connecting a slider at $A$ and $B$ through a turning pair and having each slider move along a straight and stationary slide. If an actuator is attached to the slider at $A$ and drives it from $A_1$ to $A_2$, then point $B$ will follow in the required manner. In this instance, points $A$ and $B$ move along straight lines. Alternatively, as shown in Figure 11.18(c), a point on the body may be driven along a circular path by implementing a crank arm that is pinned to the rigid body. Point $A$ on the body will move from $A_1$ to $A_2$ provided that the base pivot is located on the perpendicular bisector of line segment $A_1A_2$. Either or both points $A$ and $B$ may be connected to a crank arm through a turning pair. Figure 11.18(c) shows the case where one crank guides point $A$ from position 1 to position 2. For point $B$, we employ a straight slide and slider, as was used in Figure 11.18(b). This construction yields a slider crank mechanism. Note that there are an infinite number of locations where the base pivot could be located on the perpendicular bisector. Still another mechanism is shown in Figure 11.18(d). Here, we use one crank to guide point $A$ and another to guide point $B$. In this instance we have a four-bar mechanism.

In this section, three different mechanisms were identified to carry out the same task, where the motion between the positions is distinct. A designer must select the most appropriate type of mechanism for each particular application (i.e., carry out type synthesis).

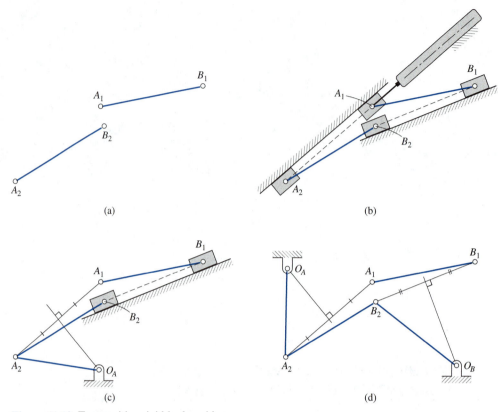

**Figure 11.18** Two-position rigid-body guidance.

## 11.6 GRAPHICAL DESIGN OF A FOUR-BAR MECHANISM FOR THREE-POSITION RIGID-BODY GUIDANCE

Consider the three positions of a rigid body containing points $A$ and $B$, as shown in Figure 11.19(a). The three positions of point $A$ are labeled $A_1$, $A_2$, and $A_3$. These points define a circle. The center of this circle is located at the intersection of the perpendicular bisectors of line segments $A_1A_2$ and $A_2A_3$ and is labeled $O_A$. A link pinned to the body at point $A$ and to the base link at $O_A$ can guide point $A$ through its three positions. Likewise, the three positions of point $B$, labeled $B_1$, $B_2$, and $B_3$, define a circle centered at $O_B$. A rigid link pinned to the body at $B$ and pinned to a base pivot at $O_B$ will guide point $B$ through its three positions. This construction has formed a four-bar mechanism $O_A A B O_B$, which guides the body through the three specified positions. Figure 11.19(b) illustrates the mechanism.

The procedure presented in this section started with selecting the location of the turning pairs on the coupler, and ended up by determining the base pivots. A trial and error procedure may be needed to determine a desirable location of the base pivots. In Section 11.7, an analytical procedure is presented whereby the base pivots may be specified, and locations of the moving pivots are determined.

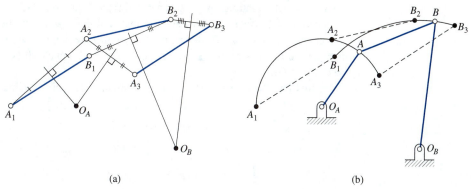

Figure 11.19 Three-position rigid-body guidance.

## 11.7 ANALYTICAL DESIGN OF A FOUR-BAR MECHANISM FOR THREE-POSITION RIGID-BODY GUIDANCE

An analytical procedure is presented for the synthesis of a four-bar mechanism to carry out three-position rigid-body guidance (see Figure 11.9). It employs the concept of displacement matrices presented in Appendix D. We consider the case where we have a rotation about a fixed origin followed by a translation. The coordinates of the point before the motion are $(x_{B_1}, y_{B_1})$. The coordinates after the motion are $(x_{B_n}, y_{B_n})$, $n = 2, 3$. Using Equations (D.4-1), (D.4-15), and (D.4-16), the relationships between the coordinates are

$$x_{B_n} = C_{1n}x_{B_1} - S_{1n}y_{B_1} + A_{13n}$$
$$y_{B_n} = S_{1n}x_{B_1} + C_{1n}y_{B_1} + A_{23n} \tag{11.7-1}$$

where

$$A_{13n} = x_{B_n} - x_{B_1}C_{1n} + y_{B_1}S_{1n}$$
$$A_{23n} = y_{B_n} - x_{B_1}S_{1n} - y_{B_1}C_{1n} \tag{11.7-2}$$
$$C_{1n} = \cos\theta_{1n}; \qquad S_{1n} = \sin\theta_{1n}$$

Employing a crank to guide the body through the positions, the coordinates of the base pivot are $(x_O, y_O)$, and the coordinates of the moving pivot in the starting position are $(x_{B_1}, y_{B_1})$. Since the length of the crank is constant, the distance between the base pivot $O_2$ and the moving turning pair is the same for all positions. Therefore

$$[(x_{B_1} - x_{O_2})^2 + (y_{B_1} - y_{O_2})^2]^{1/2} = [(x_{B_n} - x_{O_2})^2 + (y_{B_n} - y_{O_2})^2]^{1/2} \tag{11.7-3}$$

Squaring both sides of Equation (11.7-3), substituting Equations (11.7-1), and simplifying,

$$\begin{aligned} &x_{B_1}(A_{13n}C_{1n} + A_{23n}S_{1n} - x_{O_2}C_{1n} - y_{O_2}S_{1n} + x_{O_2}) \\ &\quad + y_{B_1}(A_{23n}C_{1n} - A_{13n}S_{1n} + x_{O_2}S_{1n} - y_{O_2}C_{1n} + y_{O_2}) \\ &= A_{13n}x_{O_2} + A_{23n}y_{O_2} - \tfrac{1}{2}(A_{13n}^2 + A_{13n}^2), \qquad n = 2, 3 \end{aligned} \tag{11.7-4}$$

Two distinct equations may be generated from Equation (11.7-4), one for $n = 2$, and the other for $n = 3$. If we specify the coordinates of the base pivot, then the equations may be solved for $x_{B_1}$ and $y_{B_1}$. To determine the coordinates of $D_1$ (see Figure 11.9), we start with Equation (11.7-4), replace $O_2$ with $O_4$, and then replace the coordinates of $B_1$ with those of point $D_1$.

The link lengths of the four-bar mechanism are determined using

$$
\begin{aligned}
r_1 &= [(x_{O_4} - x_{O_2})^2 + (y_{O_4} - y_{O_2})^2]^{1/2} \\
r_2 &= [(x_{B_1} - x_{O_2})^2 + (y_{B_1} - y_{O_2})^2]^{1/2} \\
r_3 &= [(x_{D_1} - x_{B_1})^2 + (y_{D_1} - y_{B_1})^2]^{1/2} \\
r_4 &= [(x_{D_1} - x_{O_4})^2 + (y_{D_1} - y_{O_4})^2]^{1/2}
\end{aligned}
\qquad (11.7\text{-}5)
$$

Care must be taken to ensure that the calculated values of link lengths produce a suitable mechanism.

## EXAMPLE 11.3 Design of a Four-Bar Mechanism for Three-Position Rigid-Body Guidance

It is required to move a headlight cover from the open position to the closed position as shown in Figure 11.20. Design a four-bar mechanism to perform this task by specifying three positions of rigid-body guidance. Employ the given locations of base pivots $O_2$ and $O_4$. During its motion, the headlight cover must not cross the fender of the automobile.

### SOLUTION

Corresponding to the given information provided in Figure 11.20, we have

$$
x_{O_2} = 0.0; \qquad y_{O_2} = 0.0; \qquad x_{O_4} = 5.0 \text{ cm}; \qquad y_{O_4} = -1.0 \text{ cm} \qquad (11.7\text{-}6)
$$

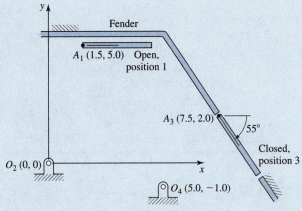

**Figure 11.20** Three-position rigid-body guidance.

The open and closed positions of the headlight cover are numbered 1 and 3, respectively. Thus

$$x_1 = 1.5 \text{ cm}; \quad y_1 = 5.0 \text{ cm}$$
$$x_3 = 7.5 \text{ cm}; \quad y_3 = 2.0 \text{ cm}; \quad \theta_{13} = -55° \tag{11.7-7}$$

For the second position, we have a choice for the position and orientation of the headlight cover. If we select the following intermediate values between positions 1 and 3:

$$x_2 = 4.5 \text{ cm}; \quad y_2 = 3.5 \text{ cm}; \quad \theta_{12} = -27.5° \tag{11.7-8}$$

Substituting values from Equations (11.7-7) and (11.7-8) in Equations (D.4-16),

$$C_{12} = \cos \theta_{12} = 0.887; \quad S_{12} = -0.462$$

$$A_{132} = x_2 - x_1 C_{12} + y_1 S_{12} = 0.861 \text{ cm}; \quad A_{232} = 0.361 \text{ cm}$$

$$C_{13} = 0.574; \quad S_{13} = -0.819; \quad A_{133} = 2.544 \text{ cm}; \quad A_{233} = 0.361 \text{ cm}$$

Substituting the above values in Equations (11.7-4), one for $n = 2$, and the other for $n = 3$, and placing in matrix form,

$$\begin{bmatrix} 0.875 & 0.182 \\ 1.164 & 2.291 \end{bmatrix} \begin{Bmatrix} x_{B_1} \\ y_{B_1} \end{Bmatrix} = \begin{Bmatrix} -0.400 \\ -3.301 \end{Bmatrix} \tag{11.7-9}$$

Solving Equation (11.7-9) gives

$$x_{B_1} = -0.175 \text{ cm}; \quad y_{B_1} = -1.352 \text{ cm} \tag{11.7-10}$$

Employing equations similar to (11.7-4) and solving for the coordinates of $D_1$ gives

$$x_{D_1} = 3.281 \text{ cm}; \quad y_{D_1} = -0.418 \text{ cm} \tag{11.7-11}$$

and using Equations (11.7-5) gives

$$r_1 = 5.099 \text{ cm}; \quad r_2 = 1.363 \text{ cm}$$
$$r_3 = 3.580 \text{ cm}; \quad r_4 = 1.815 \text{ cm} \tag{11.7-12}$$

**Model 11.21A**
Headlight Cover, Undesirable Design

**Model 11.21B**
Headlight Cover, Desirable Design

Figure 11.21(a) shows the resulting mechanism in its starting position. Although the mechanism can be assembled in each of the three positions, links 2 and 4 do not have continuous rotational motion in one direction in order to move through the three positions (see [Model 11.21A]). For this reason, it is not a desirable solution.

To obtain a desirable solution, one may try altering the location and orientation of the headlight cover for the second position. If we select

$$x_2 = 6.2 \text{ cm}; \quad y_2 = 2.8 \text{ cm}; \quad \theta_{12} = -44° \tag{11.7-13}$$

then the calculated link lengths are

$$r_1 = 5.099 \text{ cm}; \quad r_2 = 2.665 \text{ cm}$$
$$r_3 = 2.622 \text{ cm}; \quad r_4 = 1.777 \text{ cm} \tag{11.7-14}$$

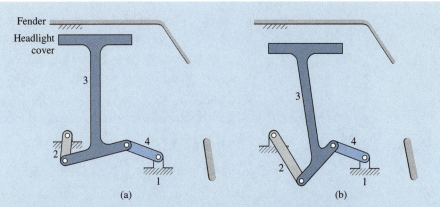

**Figure 11.21** Headlight cover: (a) undesirable design [Model 11.21A], (b) desirable design [Model 11.21B].

The result is illustrated in Figure 11.21(b), and in this instance link 4 moves continuously in one direction while the headlight cover moves from the open position to the closed position (see **[Model 11.21B]**).

# PROBLEMS

**P11.1** Design a four-bar mechanism that will approximately generate the function

$$Y = X^{1/2}, \qquad 1 \le X \le 4$$

Also

$$r_1 = 1.00 \text{ cm}; \qquad (\theta_2)_s = 120°$$

$$(\theta_2)_f = 30°; \qquad (\theta_4)_s = 100°$$

$$(\theta_4)_f = 50°$$

Use Chebyshev spacing of three precision points. **(Mathcad program: fourbarfuncsyn)**

**P11.2** Design a slider crank mechanism that will approximately generate the function

$$Y = X^{3/2}, \qquad 1 \le X \le 3$$

Also

$$(\theta_2)_s = 150°; \qquad (\theta_2)_f = 90°$$

$$s_s = 2.0 \text{ cm}; \qquad s_f = 4.0 \text{ cm}$$

Use Chebyshev spacing of three precision points. **(Mathcad program: slidercrankfuncsyn)**

**P11.3** Figure P11.3 illustrates a function graph. Determine a slider crank mechanism that will generate the function

with three precision points. Employ Chebyshev spacing of the precision points. **(Mathcad program: slidercrankfuncsyn)**

$$\Delta\theta_2 = -50°; \qquad \Delta s = 4.0 \text{ cm}$$

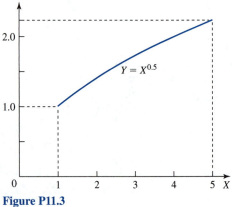

**Figure P11.3**

**P11.4** Figure P11.3 illustrates a function graph. Determine a four-bar mechanism that will generate the function with three precision points. Employ Chebyshev spacing of the precision points. (**Mathcad program: fourbarfuncsyn**)

$$r_1 = 1.00 \text{ cm}; \qquad \Delta\theta_2 = -30°$$

$$\Delta\theta_4 = -30°$$

**P11.5** Given the three positions of the rigid body shown in Figure P11.5, graphically design a four-bar mechanism that will guide the mechanism through the three positions.

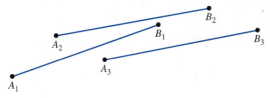

**Figure P11.5**

**P11.6** Design a four-bar mechanism, with the given locations of the base pivots, that will guide a rigid body through the specified three positions. (**Mathcad program: fourbarrbg**)

$$x_{O_2} = 0.0; \qquad y_{O_2} = 0.0$$

$$x_{O_4} = 5.0 \text{ cm}; \qquad y_{O_4} = -1.0 \text{ cm}$$

$$x_1 = 1.5 \text{ cm}; \qquad y_1 = 5.0 \text{ cm}$$

$$x_2 = 6.3 \text{ cm}; \qquad y_2 = 3.0 \text{ cm}$$

$$\theta_{12} = -40°; \qquad x_3 = 7.0 \text{ cm}$$

$$y_3 = 2.5 \text{ cm}; \qquad \theta_{13} = -50°$$

# Design Projects Using Working Model 2D

Twenty-six design projects are presented in this appendix. For each, a machine is described and a figure is given showing its links and kinematic pairs. The motion of the machine is to be modeled and animated using the Working Model 2D (Version 5.0) software. The specified requirements must be met while employing the given number of links and kinematic pairs.

Start by creating a Working Model 2D representation of the machine that closely resembles the one shown in the figure. You will find that this model will not allow proper motion. Then proceed by adjusting the geometry of the links, and the placement of the kinematic pairs, to meet the specified requirements. The computer file of the satisfactorily designed mechanism will be part of your submitted material.

The animation of the machine should be smooth and move at a reasonable speed. (The display rate of the animation may be controlled by using the dialogue box under the **World, Accuracy** menu of the Working Model 2D software.) The entire animation should not exceed 800 frames and will probably require at least 400. The computer file should be less than 1.5 Megabytes.

A written report is required to accompany the computer file. The report should not exceed three double-spaced printed pages, excluding the title page, abstract, and appendixes. The following format is suggested:

- Title page
- Abstract
- Introduction
- Data and Observations
- Discussion (explain the most important parameters for the machine and answer all questions specified for the particular project)
- Conclusions
- Appendixes

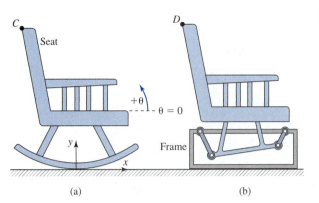

**Figure A.1** Rocking chair.

## A.1 ROCKING CHAIR

Figure A.1(a) shows a traditional rocking chair. For the first part of this project, create this chair using Working Model 2D and perform an analysis of its motion. As shown in Figure A.1(a), $\theta$ is the angle of the seat with respect to the horizontal. Create the ground surface directly on the $x$ axis, and line up the center of the chair with the $y$ axis. Create two graphs next to your chair model, the first showing how $x_C$ ($x$ position of point $C$) varies with $\theta$, and the second showing how $y_C$ varies with $\theta$, for the range $-30° \le \theta \le 30°$.

For the second part of this project, design a new rocking chair using a four-bar mechanism (Figure A.1(b)). Your new design must have a range of motion similar to the traditional rocking chair. In order to compare the motions of the two chairs, add $x_D$ versus $\theta$ to the first graph, and add $y_D$ versus $\theta$ to the second graph. Line up the new curves with the traditional curves as closely as possible by varying link lengths and joint positions. Employ an appropriate coordinate system for the new design. Specify the link lengths and joint positions.

Employ the same seat dimensions for both chairs, and place both designs within the viewing area of the screen.

## A.2 CLAMPING AND CUTTING MACHINE

The machine, illustrated in Figure A.2, is designed to clamp down on the material to be cut, and then to cut through it with two cutting edges.

By pressing down on the foot pedal, the clamp moves down, pressing upon the material to be cut. This action is completed before either of the cutting edges touch the material. Once the clamp is pressed against the material to be cut, the cutting edges continue to move toward the material. The bottom cutting edge and the top cutting edge should contact either side of the material simultaneously, thereby cutting through it. After passing through the material, the cutting edges will continue to move past each other for a short distance, then slow down, and then open again.

Create a computer simulation of this mechanism using Working Model 2D. Use the **Force** option to apply a force to the foot pedal to drive the simulation. Also include a graph showing the contact force between the clamp and the material.

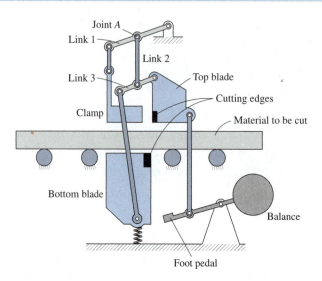

**Figure A.2** Clamping and cutting machine.

Once you have successfully simulated the mechanism, you may try to vary some of the parameters. Referring to the illustration of the clamping and cutting machine, explain the effect of varying the connection point of joint A along link 1. What is the effect of varying the length of link 2?

## A.3  SLICING MACHINE

The slicing motion of this machine is obtained from the rotation of two eccentric disks sliding within guides. The two eccentric disks are driven by an input shaft attached to a stand on the ground, as shown in Figure A.3. The two guides are welded together. Eccentric disk A slides only within guide A, and eccentric disk B slides only within guide B. In the position shown, eccentric disk B provides the horizontal cutting movement, and eccentric disk A provides the up-and-down movement. Create a simulation of this mechanism using Working Model 2D. Use the **Torque** button to apply a torque to the input shaft to drive the simulation. In addition, create a point on the tip of the blade. Using the **Tracking** feature, create a track for every fourth frame of motion. This will trace out the path of the blade in space. This path must be big enough to encompass a square object of 1.0 m by 1.0 m. The overall

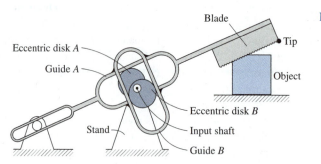

**Figure A.3** Slicing machine.

height of the mechanism must be less than 5.0 m, and the overall length must be less than 10.0 m.

Explain the significance of the size of the disks and the amount of eccentricity. Explain what would be the result if the diameter of disk *A* (and the width of guide *A*) were larger than disk *B*.

## A.4 STRAIGHT-LINE MECHANISM

The mechanism, illustrated in Figure A.4, lifts objects from the ground, up through a height in a straight line, as traced out by the tip of its fork. The mechanism begins the lift with the tip of the fork positioned at a height below the base of the hydraulic cylinder, as shown. Create a simulation of this described mechanism using Working Model 2D. The tip of the fork must travel at least 7.0 m in the *y* direction, during which it must not vary by more than 20 cm in the *x* direction. The overall height of the mechanism, in the position illustrated, must be less than 4.0 m, and the overall length less than 9.0 m. Use the **Actuator** button to create a simulated hydraulic cylinder to drive the simulation. Create a graph that shows the *x* position of the tip versus the *y* position. Specify the dimensions of links 1, 2, and 3. Also, add a point on the tip of the fork, and using the **Tracking** feature, track this point for every fourth frame of motion. This will trace out the path of the fork in space.

What is the effect of varying the dimensions of link 1? What is the effect of varying link 2?

## A.5 SINE FUNCTION MACHINE

The machine, shown in Figure A.5, consists of two independent four-bar mechanisms, which restrain a block but permit it to slide along the linkage rods. The block causes the L-shaped link to pivot, which, in turn, drives shaft *M* up and down. Create a simulation of this mechanism, such that the range of motion *y* varies by no more than 5.0 mm, while angle *a*1 varies through the range $45° \leq a1 \leq 135°$, and angle *a*2 varies through $45° \leq a2 \leq 135°$. Use the **Torque** button to apply a torque at point *A* to drive link 1 and a torque at point *E* to drive link 4. The base link length of both four-bar mechanisms should be at least 7.0 m. Also, create a graph beside the model showing the motion in the *y* direction versus angle *a*1. (Hint: In order to simplify your model, both four-bar kinematic chains should form parallelograms.)

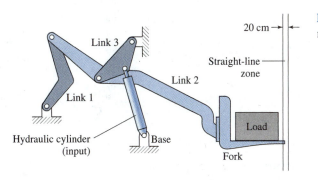

20 cm

Straight-line zone

Link 3

Link 2

Link 1

Load

Hydraulic cylinder (input)

Base

Fork

**Figure A.4** Straight-line mechanism.

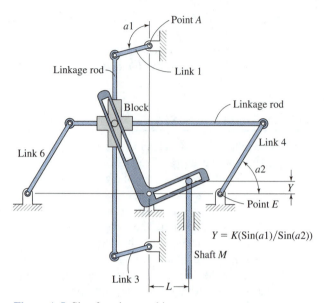

**Figure A.5** Sine function machine.

This machine is called a *sine function machine*, and the motion in the *y* direction can be described by the equation $y = K(\sin(a1)/\sin(a2))$, where $K$ is a constant. For the mechanism you have created, what is the value of $K$, and what is the ratio $a1{:}a2$? What is the effect of varying the lengths of link 1 and link 3 together?

## A.6 FLEXIBLE-FINGERED ROBOT GRIPPER

Create the robot gripper mechanism shown in Figure A.6. The gripper has fingers that flex as they close inwards. The starting position for the fingers is shown by the line *AB*. Links 1 and 2 should be parallel to the line *AB*. Links 1 and 3 have three joints each, and links 2 and 4 have two joints each. Link 1 should be 50 mm long. Distance $AA'$ should be 80 mm. Both fingers should be able to touch each other tip to tip when they meet.

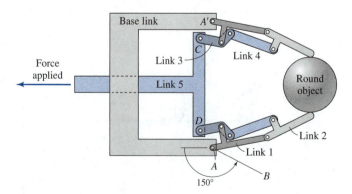

**Figure A.6** Flexible-fingered robot gripper.

Once your mechanism is complete, create a round object 30 mm in diameter. By applying a force of 10 N to link 5, as shown in the figure, determine the steady-state contact force exerted upon the round object. Create a graph of normal force with respect to time. What is the mechanical efficiency of your mechanism (i.e., contact output force divided by input force)? How can you increase the mechanical efficiency? What is the effect on finger motion and force, when distance $CD$ on link 5 is made shorter, but all other links remain the same?

## A.7 WINDSHIELD WIPER

Create the windshield wiper mechanism shown in Figure A.7. Wiper blade 1 and wiper blade 2 must be able to swing through exactly 120° of motion (60° to the left, and 60° to the right), as shown in the figure. Both wiper blades should be parallel to each other at all times. The length of the working part of each blade should be 50 cm, and link 3 should be 60 cm long. All links, except both wiper blades, should fit into a space less than 40 cm tall, while moving through the full range of motion.

Create a graph showing the rotation angle of wiper blades 1 and 2 with respect to time. Make the wiper blades oscillate at a frequency of 1.0 Hz. What motor speed is required to do this? Apply a damper with a value of 1.0 N-s/m at the end of wiper blade 1, as shown in the figure, to simulate water being cleared. Create a graph of motor torque versus time. What is the peak motor torque experienced during a cycle?

## A.8 PARALLEL-JAW PLIERS

Create the parallel-jaw pliers shown in Figure A.8. When force is applied to the handles of the pliers, the jaw link will close downward to meet the base link. As this occurs, line segment $AB$ on the jaw link must remain parallel to line segment $CD$ on the base link. Line segment $CD$ should be at an angle of 0° with the horizontal at all times. The overall size of the pliers must not exceed 20 cm in length and 10 cm in height when fully open.

Create a graph showing the rotation angle of line segment $AB$ and line segment $CD$ with respect to the $y$ displacement of the jaw link. A typical adult hand can squeeze with a grip force of 60 N. How much force will a round object 10 mm in diameter experience in

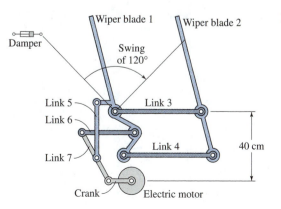

**Figure A.7** Windshield wiper.

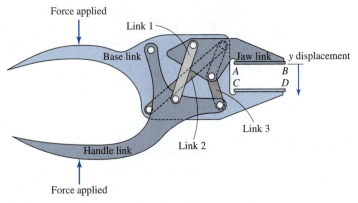

**Figure A.8** Parallel-jaw pliers.

the jaws of these pliers, with that force applied? What is the effect of moving the force applied on the handle farther back along the handle (away from the jaws)?

## A.9 OVERHEAD GARAGE DOOR

An overhead garage door, as illustrated in Figure A.9, must be created such that the clearance height of a vehicle passing below is at least 2.0 m. The garage door is 2.3 m tall. When the door is fully closed, it should be perpendicular to the ground and lined up at the top and bottom with the garage wall. While the door is in motion, no part of it must pass through the garage roof or the garage floor.

For the simulation, start the garage door in the open position. To give your mechanism motion, apply a downwards vertical force to the garage door at point $B$. Create a trace of

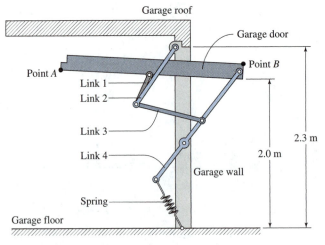

**Figure A.9** Overhead garage door.

point *A* and point *B* on the garage door while it is closing, using the tracking feature of Working Model 2D. What is the purpose of the spring in the mechanism? Create a graph of the spring length and spring tension with respect to the *x* position of point *B*. (Hint: Activate the **Collision** option between the garage roof, door, and floor, to assist you in designing the mechanism.)

## A.10 OVERHEAD LAMP

Create an overhead lamp, as shown in Figure A.10, that is capable of moving right to left, while keeping the bottom of the lamp head approximately parallel to the horizontal. Point *A* on the lamp head should be 5.0 cm lower than any other part of the lamp at all times. Design the lamp so that point *A* can extend out (i.e., to the right) as far as 100 cm from the wall and collapse inward as close as 40 cm to the wall. The overall mechanism height should be no more than 1.5 m when fully collapsed, and the overall mechanism length should be no more than 1.5 m when fully extended.

Start the mechanism in the fully extended position. To move the mechanism, apply a small horizontal force to the lamp head to make the mechanism close. Create a graph showing the *y* position of point *A* with respect to the *x* position of point *A*. The *y* position of point *A* should not deviate more than 5.0 cm throughout the full travel of the lamp head. Also create a graph showing the rotation angle of the lamp head with respect to the *x* position of point *A*.

## A.11 STEERING MECHANISM

Create the steering mechanism shown in Figure A.11(a). The tires used are model #P195/75R14. The center-to-center distance between the two tires, when they are both straight, is 142 cm. When the *pitman arm* is rotated, the two tires turn. They must turn in such a way that their motion closely approximates the *Ackerman steering* condition, which is illustrated in Figure A.11(b). The condition states that lines drawn perpendicular to each

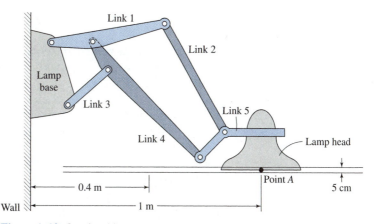

**Figure A.10** Overhead lamp.

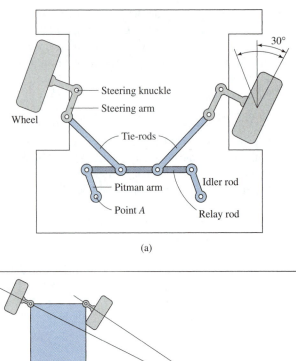

(a)

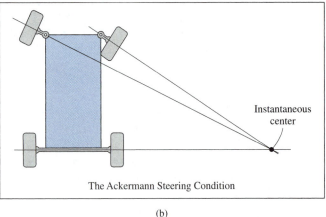

The Ackermann Steering Condition

(b)

**Figure A.11**  Steering mechanism.

of the four tires must have a common point of intersection for any radius of turn. This minimizes lateral sliding action of the tires and reduces tire wear. The center-to-center distance between the front and rear tires is 300 cm.

The steering mechanism can be driven by placing a motor at point $A$. Create a motor that produces a sinusoidal motion of the pitman arm. The right front tire should be able to turn to the right by up to 30°. The left front tire should be able to turn to the left by up to 30°.

When your mechanism is completed, create three lines perpendicular to the tires. This can be done by rigidly attaching a thin beam, 1.0 cm wide and 30 m long, to each of the front tires, and one through the rear tires. As the motor drives the front tires, the three lines should approximately have a common point of intersection, as shown in Figure A.11(b), through the full range of motion of the tires. The closer, the better.

Create a graph for each front tire, showing its turning angle with respect to time.

What is the purpose of the idler rod? Explain the function of the tie-rods. How does their length and their connection point on the relay rod effect the mechanism?

## A.12 SIDE-TIPPING ORE RAILCAR

Create the railcar shown in Figure A.12. The maximum size of the car must be no greater than 2.0 m wide by 1.5 m tall, from the car base to the top door. The tilting bed of the car must be able to pivot about point $A$, by 20°. As it tilts, three simultaneous motions must occur. (1) The latch must be released as link $B$ rotates up; (2) the top door must rotate upwards to 45° (with respect to the tilting bed); and (3) the side door must rotate downward 90° (with respect to the tilting bed).

In order to drive the mechanism, create the pump jack, using an actuator to exert a force from the car base onto the tilting bed. If necessary, create a stopper mechanism, to ensure the side door stops at 90°, so that it is parallel and in line with the tilting bed.

Create three randomly shaped objects, and place them in the car. When the simulation runs, the three objects should slide across the tilting bed, across the open side door, and off the car.

Create a graph showing the angle of the side door, and top door, with respect to the angle of the tilting bed.

What is the role of links $A$, $B$, and $C$? If the pump jack were required to tilt the car bed when it carries a specific load, a certain force would be required. What parameters would this force be a function of?

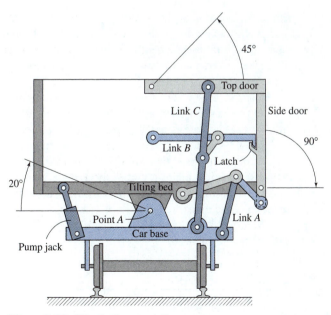

**Figure A.12**  Side-tipping ore railcar.

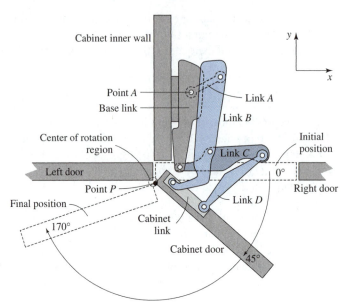

**Figure A.13**  170° cabinet hinge.

## A.13  170° CABINET HINGE

Create the cabinet hinge mechanism shown in Figure A.13. Include the two adjacent cabinet doors and an inner wall, indicated in gray, with your mechanism. All doors are 18 mm thick. The cabinet door is 300 mm long. The initial position of the cabinet door is shown as the dotted rectangle. In the initial position, there is a clearance of 2.0 mm between the door and the right door, and 2.0 mm between the door and the left door. The final position of the door is shown as a dotted rectangle, rotated 170° from its initial position.

When the hinge mechanism is fully collapsed (when the door is in the initial position), the mechanism should occupy a space of no more than 90 mm deep (depth measured parallel to the inner wall), and 70 mm wide (width measured parallel to the doors).

To create the motion of the mechanism, place a motor at point A to drive link A. Create a motor that causes a sinusoidal motion of the door from the 0° position to the 170° position.

For the simulation, create a point P on the door as indicated. During the door's rotation, point P should be approximately at the center of rotation. Point P should stay within a circle of diameter 10 mm through the full range of rotation of the door. Create a graph showing the y position of P with respect to the x position.

The door must not pass through the right door, left door, or inner wall, throughout its full range of motion.

What is the role of link A? What effect does the initial position and length of link A have on the mechanism?

## A.14  ELECTRIC GARDEN SHEAR

Create the shear mechanism shown in Figure A.14. The unit must not exceed 40 cm in length. The cutting shears are to be 10 cm deep and 15 cm wide. The rear of the unit must

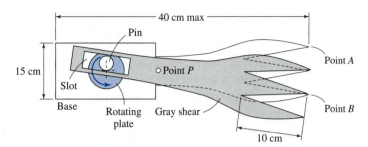

**Figure A.14** Electric garden shear.

be a maximum of 15 cm wide as indicated in the diagram. The center tip of the gray shear should have a range of travel from point *A* to point *B*. The center tip of the gray shear is illustrated at point *B* in the figure. The mechanism is to be driven by an off-center pin connected to a rotating plate. The rotating plate is connected to a motor at its center. The gray shear has a slot cut into it, which the pin is able to travel within. The motion of the slot, on the gray shear, must stay within the 15 cm band indicated in the diagram, throughout its full range of travel. The gray shear and pin are to be made of steel.

The motor to be used for this design must deliver 333 mN-m of torque. In order to satisfy the design requirements, the shears must be able exert a minimum force of 6.0 N on a 2.0 cm diameter object between any two cutting blades.

Create a graph showing the motor speed and motor torque with respect to time. Create a graph showing the acceleration of the center tip of the gray shear.

What effect does the location of point *P* have on this mechanism? If the gray shear and pin were made of wood, what effect would this have on the simulation? What effect does the location of the 2.0 cm diameter test object, within the shears, have on the force exerted on the object by the shears?

## A.15 AIRCRAFT LANDING GEAR

Create the aircraft landing gear mechanism as shown in Figure A.15. The tire diameter is 0.30 m. When the landing gear is fully extended, as illustrated in the figure, the distance from point *O* to point *A* must be 0.66 m. The length of the shock strut must be 0.53 m

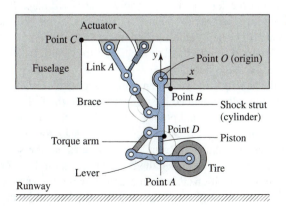

**Figure A.15** Aircraft landing gear.

measured from point $O$ to point $D$. When folded upward, the landing gear mechanism must fit within a space in the aircraft fuselage. The space within the fuselage can be defined as a rectangle, with the top left designated as point $C$, and the bottom right as point $B$. Point $O$ in the figure has $x$ and $y$ coordinates $(0, 0)$. Point $C$ is located at coordinates $(-0.83, 0.43)$, and point $B$ at coordinates $(0.06, -0.13)$. All coordinates are measured in meters.

Part I: The landing gear mechanism must be driven using an actuator. The approximate location of the actuator is shown in the diagram. Use a sinusoidal function of length as an input for the actuator. The mechanism should move from the fully extended position to the fully folded position in the simulation. No portion of the landing gear mechanism is allowed to penetrate the fuselage during any part of the simulation. The piston should be parallel to the shock strut at all times. Join the piston and the shock strut with a spring damper. Create a graph showing the $x$ and $y$ positions of the tire with respect to time. What is the purpose of the torque arm?

Part II: For the written report, simulate the landing of the landing mechanism. To do this, give the actuator a fixed value for length during the landing simulation. The landing can be simulated by creating a rectangular body (representing the runway) about 1.0 m below the tire. Constrain the runway using the anchor tool in Working Model 2D. Remove the constraint (anchor) from the fuselage, and restrict the motion of the fuselage to pure translation in the $y$ direction only. Apply a force on the fuselage so that it accelerates downward. Measure the maximum contact force between the tire and the runway during landing. Also measure the maximum force at joint $O$ (point $O$). Answer the following questions in your report: What is the effect of increasing or decreasing the spring stiffness of the shock strut? Should the spring be made rigid? What is the effect of the increasing the damping? Is this an accurate simulation of a landing? Why or why not?

Note: Do not include the landing simulation of Part II in your electronic submission. Submit only the mechanism described in Part I.

## A.16 FOLDOUT SOFA BED MECHANISM

Create the foldout sofa bed mechanism as shown in Figure A.16. The sofa bed is shown in the fully extended position. When fully folded, the sofa bed fits within a rectangular space defined by points $A$, $B$, and $C$. Point $O$ in the figure has $x$ and $y$ coordinates $(0, 0)$. Point $A$ is located at $(0.37, -0.12)$, at the lower right of the rectangular space within the sofa bed. Point $B$ defines the bottom left of the sofa backrest at coordinates $(0.04, 0.22)$. Point $C$ defines the top of the sofa front at coordinates $(-0.46, 0.17)$. The bed frame is 0.66 m long from point $L$ to point $H$. The bed frame and bed mattress together are 0.15 m thick. The bed mattress can be considered rigidly fixed to the top of the bed frame. The underside of the bed frame is 0.39 m above the floor.

The sofa bed mechanism must be driven using an actuator. For the simulation, the sofa bed should begin in the fully extended position and end in the fully folded position. Connect an actuator between point $P$ and point $Q$ in the diagram to create the folding motion. Use a sinusoidal function of length as an input for the actuator. No part of the sofa bed mechanism should penetrate the sofa front, the sofa backrest, the sofa back, or the floor during any part of the simulation. Further, the bed mattress should be horizontal at the fully extended and fully folded positions. Create a graph showing the angle of the bed mattress with respect to time. Create a graph of the angle of the bed leg with respect to the bed

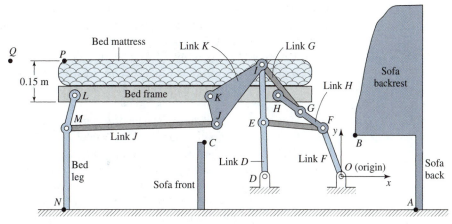

**Figure A.16** Foldout sofa bed mechanism.

frame. Will the bed frame be able to fold inward without link *J*? What is the purpose of link *J*? What is the maximum joint force exerted on link *K*, and on which joint?

## A.17 WINDOW MECHANISM

Create the window mechanism as shown in Figure A.17. The window mechanism is shown in the partially open position. In the initial position, the window is parallel to the windowsill and fits flush within the windowsill (i.e., point *D* is on top of point *E*). In the final

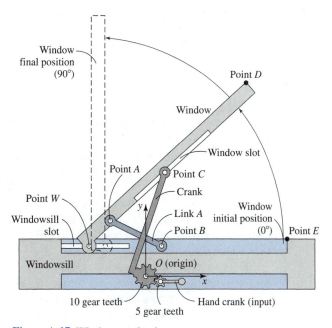

**Figure A.17** Window mechanism.

position, the window is perpendicular to the windowsill. Point $C$ on the crank slides within the window slot. Point $W$ on the window slides within the windowsill slot. Point $O$ in the figure has $x$ and $y$ coordinates (0, 0). The window is 0.62 m long and 0.04 m wide. The $y$ distance between point $O$ and point $W$ is 0.08 m. Point $E$ is located at coordinates (0.43, 0.11) with respect to point $O$. The crank is attached to the windowsill via a pin joint at point $O$. A 10-tooth gear is built into the base of the crank and is in mesh with a 5-tooth gear on the hand crank.

Part I: The window mechanism is to be driven using a motor to rotate the hand crank. Create a geared pair constraint using Working Model 2D between the crank base and the hand crank. (Note that it is not necessary to draw the gear teeth.) Use a sinusoidal function of angular position as an input for the motor. The mechanism simulation should begin in the initial position and end in the final position, as illustrated in the figure. Create a graph showing the angular position of the window with respect to the windowsill. What is the minimum length for the window slot, in order for the window mechanism to travel through the full range of motion?

Part II: For the written report, simulate the effects of a strong wind force on the window, while the window is opening. This can be done by applying a horizontal ($x$ direction) force at point $D$ on the window. The amplitude of the wind force should be a function of the angular position of the window. Program the wind force amplitude such that it is equal to 100 N when the window is in the final position, and 0 N when the window is in the initial position. Create a motor to rotate the hand crank and use a torque to drive the motor. What is the minimum torque required to open the window all the way? Prepare a graph of the torque versus window angular position in your report. Assume that only one-third of this torque could be supplied at the input. What changes could be made to the window mechanism so that it would still be able to open fully, against the wind? Is this an accurate simulation of wind, for the purpose of determining input torque for the mechanism? Why or why not?

Note: Do not include the wind simulation of Part II in your electronic submission. Submit only the mechanism as described in Part I.

## A.18  CONVERTIBLE AUTOMOBILE TOP

Create the convertible automobile top mechanism as shown in Figure A.18. The mechanism in the figure is in the fully extended position. The top panel and the rear panel, along with all other links, must be able to fold back and fit within a predefined space. This predefined space is shown in the diagram as a dashed rectangle. Point $O$ has $x$ and $y$ coordinates (0, 0). Point $O$ defines the top left of the rectangle, while point $E$ defines the lower right. Point $E$ is located at coordinates (1.36, −0.44). The top panel is 0.84 m long, measured from point $T$ to point $B$. The rear panel is 0.78 m long, measured from point $A$ to point $O$. Point $T$ on the top panel is located at coordinates (−1.275, 0.55) when the mechanism is in the fully extended position. Also, the top panel must be horizontal in the fully extended position.

The mechanism must be driven by a piston actuator, as shown in the figure. Use a sinusoidal function of length as an input for the actuator. Use the actuator to simulate the mechanism motion from the fully extended position to the fully collapsed position. When the mechanism is fully collapsed, all components, including the piston actuator, must fit within the predefined space described previously. Further, no part of the mechanism must pass through the sides or bottom of the predefined space.

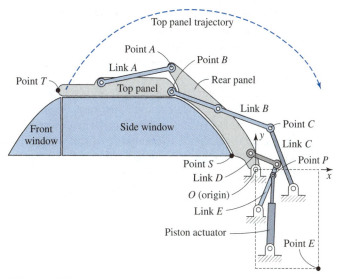

**Figure A.18** Convertible automobile top.

Create a graph showing the angular position of the top panel with respect to time. Using the **Tracking** feature of Working Model 2D, create a track of point $T$ as the mechanism collapses. What is the purpose of link $D$? What effect does the length of link $C$ have on the mechanism?

## A.19 HORIZONTAL-PLATFORM MECHANISM

Create the horizontal-platform mechanism as shown in Figure A.19. The purpose of this mechanism is to provide horizontal motion of the table top, within a specified tolerance. The table top must be able to travel at least 850 mm along the $x$ axis, during which the vertical translation of the table must be less than 4.0 mm. The angular deviation of the

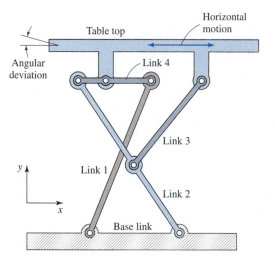

**Figure A.19** Horizontal-platform mechanism.

table top must be less than 1.0° from the horizontal, throughout the entire motion of the mechanism. In the starting position, as shown in the diagram, all the links and the table top must fit within a space that is 850 mm wide ($x$ direction) by 1050 mm tall ($y$ direction). None of the links may pass through the top of the table top during the motion of the mechanism.

This mechanism must be driven by a motor actuator with a constant velocity. The motor should be located at a joint between two of the links. Which two links should it be placed between? Explain your reasons. Create a graph that displays the table top angle with respect to time. Create two graphs to display (1) the table top vertical travel with respect time and (2) the table top horizontal travel with respect to time. What is the relationship between link 1 and link 3? What kind of application might this mechanism be used for?

The simulation must be able to complete one full cycle.

## A.20  WATER PUMP MECHANISM

Create the water pump mechanism as shown in Figure A.20. The objective of this mechanism is to rotate the handle (link 7) through an angular range of motion. The handle is connected to the pump mechanism to the left, which is illustrated by dashed lines. The objects that are illustrated by dashed lines should not be included in the simulation. In the starting position, $\theta$ (representing the angle of link 7) must be at 0°. At maximum rotation, $\theta$ must be at least 51°. Links 1 through 5 are contained in the rectangular mechanism

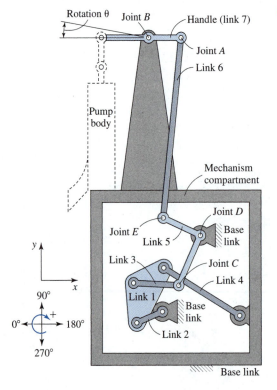

**Figure A.20**  Water pump mechanism.

compartment. The inside dimensions of the compartment are 260 mm wide ($x$ direction) by 265 mm tall ($y$ direction). During operation of the mechanism, links 1 through 5 are to remain within compartment space. No link (except link 6) may penetrate the walls of the compartment. With the handle in the horizontal position, the overall height of the entire mechanism is to be 750 mm or less. The length of the handle is 200 mm, and the length between joint $A$ and joint $B$ is 85 mm.

This mechanism must be driven by a motor actuator with a constant velocity. The motor should be located at a joint between two of the links. Which two links should it be placed between? Is the direction of rotation of the motor important? Explain your answer. Create a graph that shows the angular position of the handle with respect to the angular position of link 2. What function does link 5 serve? What is the effect of changing the distance between joint $D$ and joint $E$ on link 5?

The simulation submission must be able to complete one full cycle.

## A.21 STAIR-CLIMBING MECHANISM

Create the stair-climbing mechanism as shown in Figure A.21. The objective of this mechanism is to climb the staircase and finish with the entire mechanism standing on the upper platform. Create the staircase as shown in the figure. Each step is 254 mm long, and each riser is 178 mm tall. A single leg mechanism (SLM) is illustrated in the upper area of the diagram. The foot (link 2) has a triangular shape, with two sides 180 mm long, and one side 320 mm long. The complete climbing mechanism consists of four of the SLMs joined

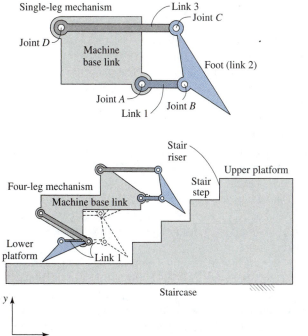

**Figure A.21** Stair-climbing mechanism.

together via their respective machine base links, as shown in the lower diagram. To understand the configuration of the complete climbing mechanism, think of the climbing mechanism configured as an automobile, where each wheel is replaced by an SLM. Since Working Model 2D software is two-dimensional, the two right SLMs are drawn as solid lines, while the two left SLMs are drawn as dashed lines. Note that the rotation of link 1 on the SLMs shown in dashed lines, is 180° from that of link 1 on the SLMs shown in solid lines. The entire mechanism at any given position should not be longer than 1150 mm and not taller than 750 mm. None of the links may pass through the staircase during the motion of the mechanism.

Each SLM must be driven by a motor actuator with a constant velocity. The speed of all four actuators must be the same. The motor of each SLM should be located at a joint between two of the links. Which two links should it be placed between? Is the direction of rotation of the motor important? Explain your answer. You will need a counterweight somewhere on the machine base link of the complete mechanism, to keep it from tipping backwards during climbing. Explain the rationale for placement and weight of this counterweight. Create one graph showing the rotational velocity of all four motors with respect to time.

The simulation must complete two full cycles (which consists of climbing two steps on the stairs).

## A.22  LOADING MACHINE SYSTEM

Create the loading machine system as illustrated in Figure A.22. The objective of this system is to lift the four parcels illustrated and drop them into the bin. The system consists of two separate mechanisms. The first is the bin loader mechanism (BLM) as illustrated at the top of the diagram. The second is the conveyor mechanism (CM) illustrated at the bottom of the diagram. The BLM consists of a fork, two links, and the bin. The BLM must lift parcels, which slide onto the fork, up and over into the bin. The bin is 1.6 m tall and 2.3 m long. The fork is 1.5 m long. The parcels are permitted to pass through the top of the bin but must not pass through the sides or the bottom of the bin. Also, no part of the fork link may penetrate into the bin.

The CM must carry the parcels onto the fork of the BLM. The CM consists of a straight-line Geneva mechanism. The CM will create intermittent $x$ translation of the conveyor sidewalls. This intermittent motion occurs when the peg driver, which is mounted on the crank wheel, engages the slot follower located on the follower link. The conveyor sidewalls are attached to the top of the follower link, and the parcels are on top of the conveyor sidewalls. The parcels are spaced 2.0 m apart (center-to-center) and are 400 mm tall and 500 mm long. The fork of the BLM is permitted to pass through the conveyor sidewalls during parcel loading, but not through the follower link. The follower link can translate only along the $x$ axis.

The BLM must be actuated by a motor using a sine function to control the rotation of link 2. The motor should be located at joint $A$ between the bin and link 2. The CM is to be actuated by a motor with a constant velocity, located at the center of the crank wheel. When the peg driver engages the slot follower, the follower link will be advanced by a specific amount along the $x$ axis. Create a graph showing the follower link velocity with respect to time. Explain the design strategy needed to create the peg driver and the follower link

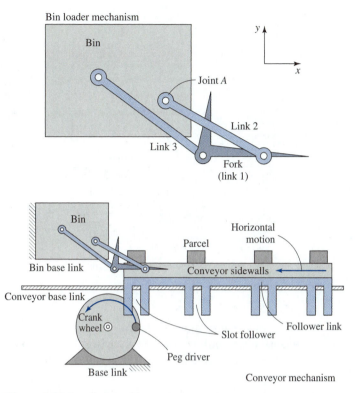

Figure A.22 Loading machine system.

system. Explain the relation between the parcel motion and the pickup by the fork. Why is the intermittent motion needed? How would it be possible to eliminate the intermittent motion of the conveyor?

The simulation must complete two full cycles (each consists of dropping two parcels into the bin).

## A.23 FLIPPER MECHANISM

Create the flipper mechanism as shown in Figure A.23. The purpose of this mechanism is to turn over a flat plate by passing it from the left pad to the right pad. The left pad and right pad rise simultaneously and meet on a line inclined from the vertical axis by angle $\theta$, at which point the flat plate is transferred. The flat plate is 15 mm thick and 240 mm long. In the starting position, as shown in the diagram, all links of the mechanism and the flat plate must fit within a space that is 880 mm wide ($x$ direction) by 200 mm tall ($y$ direction). The flat plate must start on the left pad. When the flip is complete, it is to rest on the right pad without protruding past the right edge of the right pad.

This mechanism must be driven by a motor with a constant velocity. The motor must be located at point $A$ and drive both crank $A$ and crank $B$. Create a graph to display the angle of the flat plate with respect to the angle of the left pad. What effect does the angle $\theta$

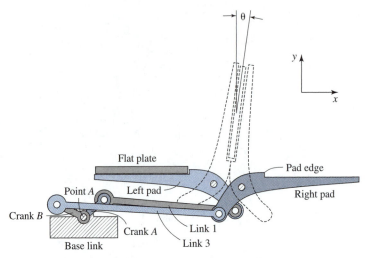

**Figure A.23** Flipper mechanism.

have on the operation of the mechanism? Explain the relevance of the location of the pad edge, for both pads. Replace link 1 with a rod element and create a graph showing the tension of the rod with respect to time. Describe possible applications of this mechanism.

The simulation must complete one full cycle.

## A.24 TRANSPORT MECHANISM

Create the transport mechanism as shown in Figure A.24. The objective of this mechanism is to push the parcel intermittently, along the surface of the fixed rail, from the start position to the end position. The parcel is a box 40 mm by 40 mm. The transport link has four posts that are used to push the parcel. The posts are spaced 120 mm apart (pitch distance), and are 40 mm tall. Note the dashed line that represents the approximate path of the transport link while the machine is in operation. With the exception of the posts, no links (including the transport link) are allowed to protrude above the fixed rail surface at any time during operation. The parcel must begin in the start position and after three push cycles must finish in the end position. There are two motion requirements for the transport link. Firstly, the angular deviation of the transport link must be less than 0.1° from the horizontal, throughout the motion cycle. Secondly, the horizontal path portion, indicated by the blue dashed line, must be at least 120 mm long, during which the transport link may not deviate more than 6.0 mm in the y direction. The entire mechanism, at any given position during its cycle, must fit within a space that is 740 mm in the x direction and 440 mm in the y direction.

This mechanism must be driven by a motor actuator with a constant velocity applied to the crank. Create a graph that shows the x position of the transport link with respect to its y position. Create a graph that shows the angular position of the transport link with

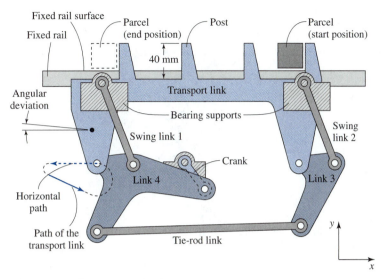

**Figure A.24** Transport mechanism.

respect to time. What function does the tie-rod link serve? Can you explain the role of the swing links, link 1 and link 2, on the mechanism, and their relation to each other?

The simulation must complete three cycles.

## A.25 TYPEWRITER MECHANISM

Create the typewriter mechanism as shown in Figure A.25. The objective of this mechanism is for the impact head to strike the drum by applying a force to the key. It is not necessary to include any of the objects drawn as dashed lines in the simulation. The type bar, which holds the impact head, must start in the horizontal position, as illustrated, and rotate 90° counterclockwise to strike the drum. The drum has a diameter of 160 mm. The impact head must be tangent to the drum during impact and must impact the drum at the midpoint on the right side. The center of the drum must be 157 mm above (positive *y* direction) point *A*, and 87 mm to the left (negative *x* direction) of point *A*. The entire mechanism (not including the drum) must fit within a space that is 380 mm in the *x* direction and 150 mm in the *y* direction, while it is in the starting position.

The mechanism is actuated by applying an intermittent vertical force on the key. Use a function in the **Active when** properties box for the applied force. A function in the form of $\sin(t) > n$, where n is a number between 0 and 1, may be useful. In order for your mechanism to work properly, you will need to add/adjust the following three elements: (1) Adjust the elastic coefficient between the impact head and the drum. Explain the significance of this coefficient and the effects on the mechanism. (2) Convert point *A* to a rotational damper and adjust the damping value as necessary, for the mechanism to operate properly. Explain the purpose of modelling point *A* as a rotational damper. Should all the rotational joints in the mechanism be modeled this way? (3) Use a spring between link 1 and ground (typewriter body) to have the mechanism return to the starting position when the key force

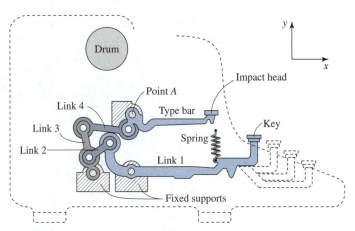

**Figure A.25**  Typewriter mechanism.

is not present. With regard to the key force, apply a force that you believe is reasonable during normal typing for the average person. Create a graph showing the contact force between the impact head and the drum as a function of time. What is the ratio between the peak contact force and the key input force? What does the ratio tell you about the mechanism?

The simulation must complete three typing cycles.

## A.26  GRAVITY GRIPPER MECHANISM

Create the gripper mechanism as illustrated in Figure A.26. The objective of the gripper is to lift the circular payload off the ground. There are no actuators to open or close the carrying arms. The gripper should operate as follows: (1) The gripper is lowered onto the payload, while centered above it. (2) As the gripping pads contact the payload, they will not rotate but will push the carrying arms outward. After the gripping pads pass over the maximum width of the payload, the carrying arms will fall inward. (3) As the gripper is raised, the gripping pads will rotate so that they are tangent to the payload. (4) The payload will remain in the gripper's grasp and be raised with the gripper. The gripper should be symmetrical, with the exception of the cross tie link, which connects the two lower links. Set the weight of the counterweights such that the carrying arms remain in a neutral position (as illustrated in the diagram) while at rest. While the gripper is at rest, the space between the gripping pads should be 160 mm. The payload has a diameter of 176 mm. In the starting position, the gripper is centered above the payload and is located at a height such that the pad pivot point is 220 mm above the ground. Use a rod for the cross tie link. When the gripper interacts with the payload, the motion of the carrying arms should be along an inclined line, as illustrated in the diagram.

Create a vertical slot, and rigidly join the lift link to the vertical slot. Use an actuator to connect the lift link to some fixed point above the lift link. Set **Actuator** to **length,** and enter a function of time in the form of [(starting length) + (desired travel)sin(t)] in the input box, to create the necessary up-and-down motion for the lift link, so it can grasp and lift the

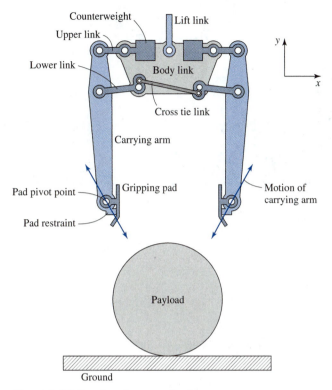

Counterweight

Lift link

Upper link

Lower link

Body link

Cross tie link

Carrying arm

Pad pivot point

Gripping pad

Motion of carrying arm

Pad restraint

Payload

Ground

**Figure A.26** Gravity gripper mechanism.

payload off the ground. Create a graph showing the tension in the cross tie link as a function of time. Adjust the weight of the payload to 10, 100, and 1000 kg. What is the effect on the simulation for each case? If the cross tie link has a cross-sectional area of 1.0 square inch and is made from piano wire–grade steel, what is the maximum payload that can be lifted? Will the gripper be able to pick up a 160 mm square object, or an inverted triangle with 160 mm sides? What are the requirements for the payload's shape, for this gripper to work?

The simulation must show the gripper securely grasping the circular payload and lifting it off the ground.

# APPENDIX
# B

# Commonly Employed Mechanisms and Machines

This appendix describes a diverse assortment of mechanisms and machines in common use.

## B.1 PARALLEL-MOTION MECHANISM

**Model B.1**
Parallel-Motion
Mechanism

It is possible to add links to a path-generating four-bar mechanism such that all points on one of the links will move in the same shape of path as the coupler point, and therefore the link always remains parallel to itself. This is called a *parallel-motion mechanism*. Figure B.1 is an illustration of such a mechanism. The four-bar mechanism, with a coupler point, consists of links 1 through 4. The additional links are numbered 5 through 8. Since points $CDFE$ and $O_4DFO_8$ form parallelograms, link 5 remains parallel to itself as the mechanism moves. Therefore, the orientation of a rigid body placed on link 5 stays constant.

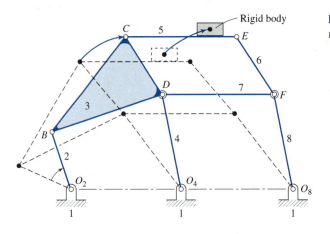

**Figure B.1** Parallel-motion mechanism [Model B.1].

424

## B.2 DWELL MECHANISM

*Dwell mechanisms* (also called *intermittent-motion mechanisms*) provide intervals of zero output motion while the input speed is continuous. Two such mechanisms are described below. Each employs a special shape of the coupler curve of a four-bar mechanism.

### B.2.1 Coupler Curve Dwell Mechanism (Straight Line)

> **Model B.2**
> Straight-Line
> Dwell
> Mechanism

Figure B.2 shows a four-bar mechanism for which a portion of the coupler curve is approximately a straight line. Link 5 is a slider attached to coupler point $C$. For the position shown, link 6 lies tangent to the "straight line" portion of the coupler curve. As link 2 starts rotating from the position shown, link 6 will momentarily remain stationary, and as the cycle continues, link 6 will have a pulsed rotation.

### B.2.2 Coupler Curve Dwell Mechanism (Circular Arc)

> **Model B.3**
> Circular-Arc
> Dwell
> Mechanism

In the four-bar mechanism shown in Figure B.3, the curve traced by coupler point $C$ is shown as a dashed line. Between points $F$ and $G$, this curve is approximately a circular arc, centered at $E$. Links 5 and 6 are added to the mechanism, as shown, sharing a moving pivot connection. The other end of link 5 is at $C$. As link 2 starts rotating from the position shown, link 6 remains stationary. Later in the cycle, link 6 will have a pulsed rotation.

### B.2.3 Geneva Mechanism

> **Model B.4**
> Geneva
> Mechanism

A *Geneva mechanism* shown in Figure B.4. If link 2 is driven at constant rotational speed, then the output motion of link 3 stops and starts in regular intervals [13]. Figures B.4(a) and B.4(b) show two positions of the mechanism. In Figure B.4(a), the *locking plate*, which is part of link 2, slides relative to a surface on link 3, the *slotted wheel*, and prevents any output motion. Figure B.4(b) shows another position for which the locking plate no longer

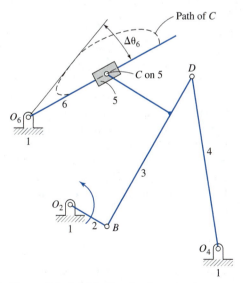

**Figure B.2** Straight-line dwell mechanism [Model B.2].

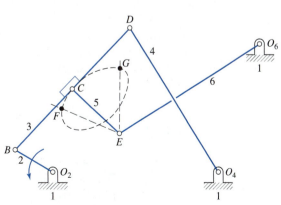

**Figure B.3** Circular-arc dwell mechanism [Model B.3].

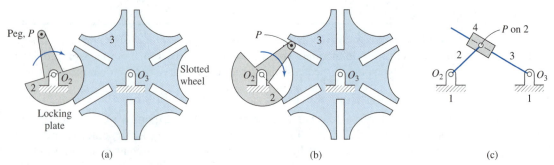

**Figure B.4** Geneva mechanism [Model B.4]: (a) wheel locked in position, (b) wheel driven by peg, (c) equivalent mechanism while peg is in slot.

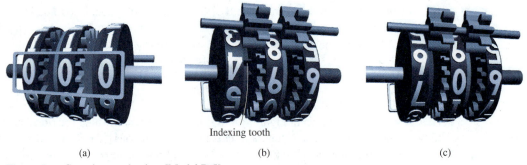

**Figure B.5** Counting mechanism [Model B.5].

holds link 3 stationary, and the *peg* on link 2 has entered one of the six equally spaced slots in link 3. The peg drives link 3 through one-sixth of a revolution (i.e., 60°) each time it has completed a cycle in a slot.

While the pin is in the slot, the skeleton diagram representation of this mechanism is shown in Figure B.4(c). In this skeleton diagram, links 2 and 3 are connected to the input and output of the mechanism, respectively, and link 4 is a slider.

Link 3 of the Geneva mechanism shown in Figure B.4(a) has six slots. A Geneva mechanism may also be designed to have either four or eight equally spaced slots in link 3. If there are four slots in the slotted wheel, then for each rotation of link 2, link 3 will rotate 90°.

A Geneva mechanism can be utilized in instances requiring an intermittent output motion to be produced from a uniform input motion. For example, it has been employed in movie cameras or projectors to periodically advance film. This mechanism would be an alternative to a four-bar mechanism with a coupler point, presented in Chapter 1, Section 1.2.2 (Figure 1.16).

### B.2.4 Counting Mechanism

Another example of an intermittent-motion mechanism is a *counting mechanism* shown in Figure B.5. The first dial has an indexing tooth (Figure B.5(b)), which, during each rotation, engages the adjacent wheel and advances it by one digit (Figure B.5(c)). This mechanism has application in odometers.

| Model B.5 |
| Counting |
| Mechanism |

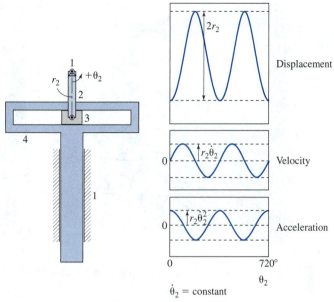

**Figure B.6** Scotch yoke mechanism [Model B.6].

## B.3 SCOTCH YOKE MECHANISM

**Model B.6**
Scotch Yoke
Mechanism

Figure B.6 depicts a *scotch yoke mechanism*. This mechanism transforms rotational motion into linear motion. The crank, link 2, is attached at each end using bearings, one end to link 1 and the other to link 3. As link 2 rotates, link 3 moves back and forth along a slide, link 4, also referred to as a *yoke*. Angular displacement of link 2, denoted as $\theta_2$, is measured with respect to a selected reference. By inspection, when $\theta_2 = 0$, link 4 reaches one of its extremum positions with respect to the base link. This is also referred to as a *limit position*. Link 4 has zero velocity when it is in a limit position. For this mechanism, if the angular motion of link 2 is constant, then the linear motion of link 4 is harmonic. The analysis of the limit positions of various mechanisms is presented in Chapter 2, Section 2.6.

**Model B.7**
Application of
Scotch Yoke
Mechanism

An application of a scotch yoke mechanism is in raising and lowering windows of an automobile. Figure B.7 shows such a mechanism. The window is attached to the yoke. Gears are added to this mechanism to reduce the amount of motion of the window compared to the regulator handle.

## B.4 ESCAPEMENT MECHANISM

An *escapement mechanism* is commonly employed in mechanical timepieces. It provides a timed release of output motion from stored energy. Energy may be supplied by using potential energy from lowering the height of a weight or the gradual unwinding of a spring.

**Model B.8**
Escapement
Mechanism

An illustration of an escapement mechanism is shown in Figure B.8. The escapement wheel operates against *pawls* that form part of the pendulum. For each rocking cycle of the pendulum, the wheel is allowed to advance one tooth on the wheel.

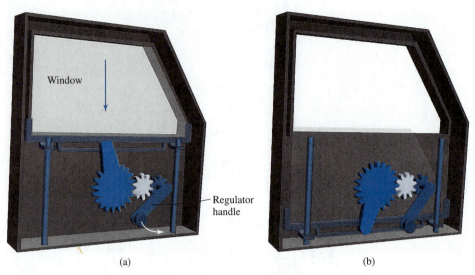

**Figure B.7** Application of a scotch yoke mechanism [Model B.7].

**Figure B.8** Escapement mechanism [Model B.8].

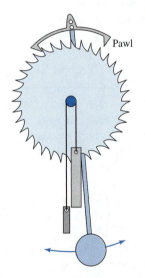

## B.5 IRIS MECHANISM

**Model B.9**
Iris Mechanism

An *iris mechanism* has application in controlling the amount of light allowed through camera lenses. Two configurations of a typical iris mechanism are shown in Figure B.9. One is in its fully open position (Figure B.9(a)), and the other has a reduced aperture (Figure B.9(b)).

An iris mechanism consists of a series of *vanes* equally spaced about the center of the opening. In Figures B.9(a) and B.9(b), a single vane has been isolated to reveal its motion. One end of each vane is pinned to the base link. The other end of each vane has a pin that slides in a slot of a concentric ring. Rotating the ring provides motion for all vanes.

**Video B.10**
Flow Regulator

Figure B.10 illustrates an iris mechanism that is employed to regulate the flow of a fluid.

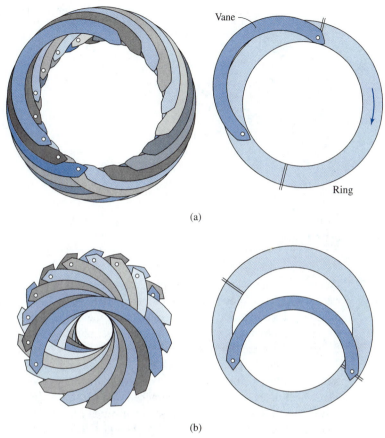

Vane

Ring

(a)

(b)

**Figure B.9** Iris mechanism [Model B.9].

**Figure B.10** Flow regulator [Video B.10].

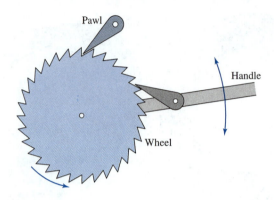

Pawl

Handle

Wheel

## B.6  ONE-WAY MECHANISM

### B.6.1  Ratchet Mechanism

**Model B.11**
Ratchet
Mechanism

Figure B.11 shows an illustration of a *ratchet mechanism*. For the arrangement shown, the ratchet wheel can only be rotated in the counterclockwise direction. In this case, the pawl engages with one tooth of the wheel. Moving the handle in the clockwise direction causes the pawl to slip over the ratchet teeth. Springs (not shown), located at the turning pairs of the pawls, keep the tips of the pawls in contact with the wheel.

**Model B.12**
Ratchet Wrench

A common application of this mechanism is in a *ratchet wrench*, where the driving torque can only be provided in one direction. Such a wrench is shown in Figure B.12(a). The wheel of the mechanism is attached to a socket, which would convey the torque to drive the bolt. Figure B.12(b) shows a close-up of the head of the wrench with the socket and bolt removed. The animation of **[Model B.12]** shows how the wrench can provide torque in the clockwise direction and undergo ratcheting action in the counterclockwise direction (Figure B.12(c)). By resetting the position of the pawl, torque and ratcheting are reversed (Figure B.12(d)).

### B.6.2  Sprag Clutch

**Model B.13**
Sprag Clutch

A *sprag clutch* is shown in Figure B.13. This mechanism includes inner and outer races. Surfaces between the races are circular. When the inner race is driven in the counterclockwise direction, *sprag elements* pivot to hold the inner and outer races together, and both races turn in unison. When the inner race is driven in the clockwise direction, the sprag elements move out of the way, and no motion is transmitted to the outer race.

## B.7  QUICK-RETURN MECHANISM

*Quick-return mechanisms* refer to those for which, if driven with a constant input speed, the time to complete the output motion in one direction differs from that of the opposite direction.

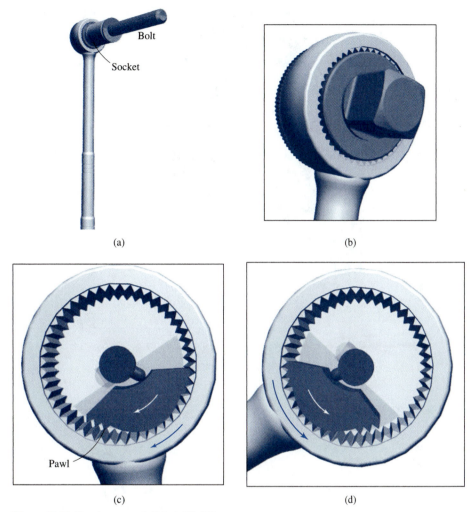

(a)

(b)

(c)

(d)

**Figure B.12**  Ratchet wrench [Model B.12].

There is a variety of quick-return mechanisms, one of which is shown in Figure B.14. Here, the mechanism is driven through rotation of link 2. We will assume that this link is driven at a constant rate. The output of this mechanism is link 6, a slider. Solid lines in Figure B.14 show an outline of the mechanism in one of its two limit positions. Dashed lines depict its other limit geometry. Both limit geometries occur when link 2 is perpendicular to link 4. At the instant when the mechanism is in a limit position, link 6 has zero velocity.

The two rotations of link 2 required to move the mechanism between the limit geometries are shown as $\Delta\theta_2$ and $2\pi - \Delta\theta_2$. Since these angles of rotation are not equal, the time required to move link 6 from one limit position to the other does not equal the time to return

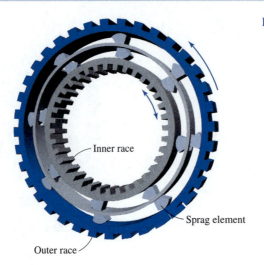

**Figure B.13** Sprag clutch [Model B.13].

Inner race

Sprag element

Outer race

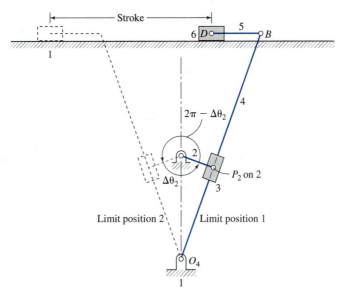

**Figure B.14** Quick-return mechanism.

the link back to its original position. For the illustration shown, if link 2 is turning in the counterclockwise direction, then link 6 will have a quicker motion to the right than to the left. A more detailed analysis of this and other mechanisms in their limit positions is presented in Chapter 2, Section 2.6.

**Model B.15**
Quick-Return
Mechanism

Figure B.15 shows an application of a quick-return mechanism used in a machine tool, providing a slower cutting motion than the return motion.

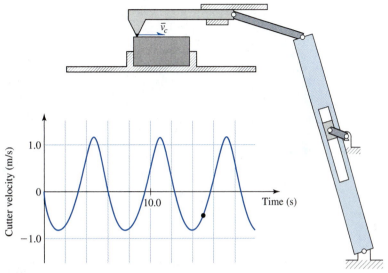

**Figure B.15** Quick-return mechanism [Model B.15].

# B.8 ENGINE

## B.8.1 Piston Engine

*Piston engines* incorporate *pistons*, which move inside of *cylinders* within a *cylinder block*. Figure B.16 illustrates a cross-sectional view of one form of such an engine. Four positions during a cycle of operation are shown. In Figure B.16(a), a mixture of air and fuel is drawn into the cylinder through opening the *intake valve*. The mixture is then compressed while the valves are closed, and the piston is driven upwards (Figure B.16(b)). Next, the fuel is ignited, causing a force downward on the piston and applying torque to the crankshaft in the direction of rotation (Figure B.16(c)). The final portion of each cycle is to evacuate the burned mixture through the *exhaust valve* (Figure B.16(d)).

> **Model B.16**
> Single-Cylinder Piston Engine Cross Section

A *sleeve engine*, illustrated in Figure B.17, is another form of a piston engine. This engine employs a concentric cylindrical *sleeve* located between the piston and the stationary cylinder wall. Holes in the sleeve allow passage of the mixture into and out of the cylinder. Motion of the sleeve is driven through a mechanism connected to the crankshaft. Holes in the sleeve replace the valves used for the piston engine illustrated in Figure B.16. These engines were used in vintage aircraft.

> **Model B.17**
> Sleeve Engine

Automotive engines generally have four, six, or eight cylinders. Cylinders may be arranged in a row (in-line), in two rows and set at an angle with respect to one another (V-type), or in two rows opposing each other (flat). Figure B.18 shows some typical arrangements of cylinders. Figure B.19 depicts a V-2 engine. Note that this illustration does not include the valve springs, shown in Figure 1.1. Such springs would be required for proper operation of this machine.

> **Model B.19**
> V-2 Engine

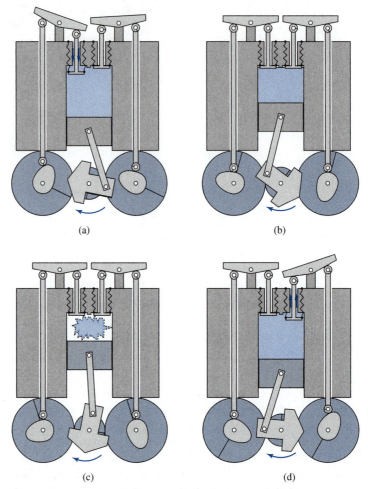

**Figure B.16** Positions of a single-cylinder engine throughout a cycle of motion [Model B.16]: (a) intake, (b) compression, (c) power, (d) exhaust.

**Model B.20A**
Radial Engine

**Model B.20B**
Radial Engine
with Propeller

**Video B.21**
Airplane Engine

**Model B.22**
Wankel Engine

A *radial engine* is another arrangement of cylinders for a piston engine. One is illustrated in Figure B.20(a). All pistons are constrained to move radially from a common point. Figure B.20(b) illustrates the same radial engine as shown in Figure B.20(a), except that a propeller has been added to the output shaft. Figure B.21 shows a radial engine incorporated in an airplane.

### B.8.2 Wankel Engine

A *Wankel engine* is illustrated in Figure B.22. At each of the three corners of the *rotor* is a seal that slides against the surface of the stationary housing. Part of the rotor

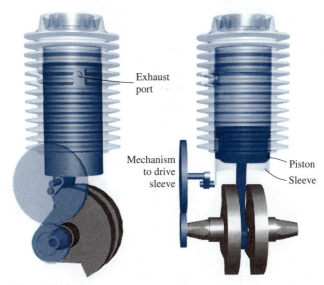

**Figure B.17** Sleeve engine [Model B.17].

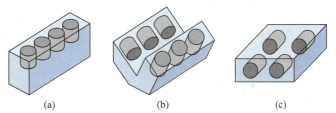

(a)           (b)           (c)

**Figure B.18** Typical configurations of a multiple-cylinder engine: (a) in-line 4, (b) V-6, (c) flat 4.

includes an internal gear that meshes with a fixed external gear. Figure B.22 shows four images of the positions of the rotor during a cycle consisting of intake, compression, power (i.e., combustion of the air-fuel mixture), and exhaust of the spent fuel. At intake, Figure B.22(a), the intake port is uncovered by the rotor. A mixture of air and fuel is drawn into the increasing volume between the rotor and the internal wall of the housing. The mixture is then closed off from the intake port and is compressed (Figure B.22(b)). Approaching maximum compression, the mixture is ignited by a spark plug (Figure B.22(c)). The combustion of the mixture provides a force on the side of the rotor, and the volume of the cavity is increased. Finally, the exhaust port is uncovered, while the burned mixture is expelled (Figure B.22(d)). The cycle is then repeated on an adjacent side of the rotor. There are three spark plug ignitions for each revolution of the rotor.

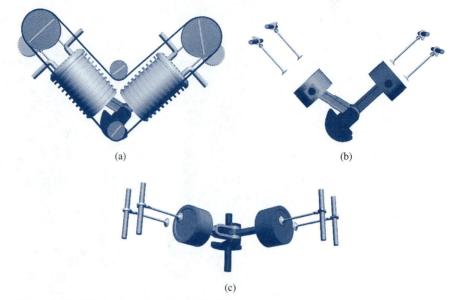

(a)                                        (b)

(c)

**Figure B.19**  V-2 piston engine [Model B.19].

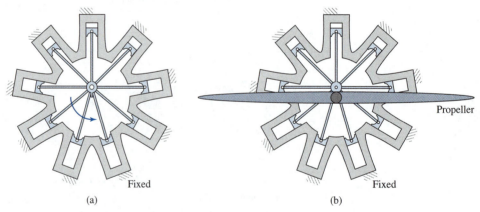

(a)                                        (b)

**Figure B.20**  (a) Radial engine [Model B.20A]. (b) Radial engine with propeller [Model B.20B].

**Figure B.21**  Airplane engine [Video B.21].

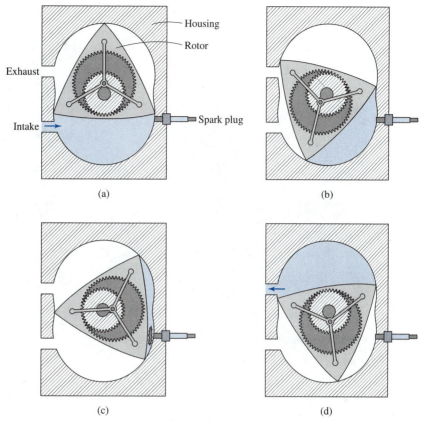

**Figure B.22** Wankel engine [Model B.22]: (a) intake, (b) compression, (c) power, (d) exhaust.

## B.9  MECHANISMS INCORPORATING A SWASH PLATE

### B.9.1  Compressor

Model B.24
Air-Conditioning
Compressor

An *air-conditioning compressor* is illustrated in Figures B.23 and B.24. As the input shaft rotates, the *swash plate* (follower) moves, causing the five pistons to sequentially undergo linear reciprocating motion within the cylinder block. *Flapper valves* allow for the expulsion of the compressed refrigerant.

### B.9.2  Main Rotor Drive of a Helicopter

Model B.25
Main Rotor of a
Helicopter

Figure B.25(a) illustrates the *main rotor blades* and *tail rotor blades* of a helicopter, and Figure B.25(b) shows a close-up of the main rotor blades. The cross section of each blade has the shape of an airfoil, as illustrated in Figure B.26. Spinning the main rotor blades about the axis of the *driveshaft* produces the lift forces required to fly the helicopter. The pilot can regulate the amount of lift force by adjusting the *angles of attack* of the blades. Figure B.25(b) illustrates the components used to change the angles of attack. Included in

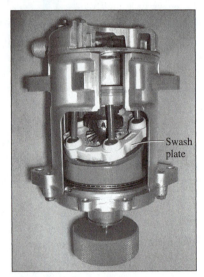

**Figure B.23** Air-conditioning compressor.

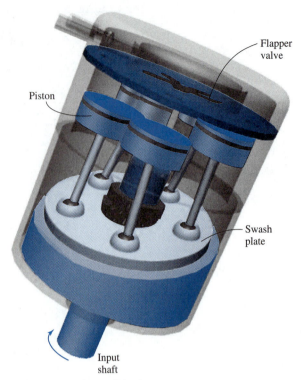

**Figure B.24** Air-conditioning compressor [Model B.24].

these components are two swash plates: the *rotating swash plate* and the *oscillating swash plate*. *Hydraulic actuators* can move the oscillating swash plate vertically or tilt it from the horizontal in any desired direction. The *scissor links* constrain the rotating swash plate to turn with the driveshaft and blades. The *bearing* allows the rotating swash plate to revolve with respect to the oscillating swash plate. Each *pitch control rod* is pin-jointed to the rotating swash plate and to a blade.

Motions of the oscillating swash plate can be controlled in two basic manners: *collective control* and *cyclic control*. In collective control (Figure B.25(c)), the oscillating swash plate is driven vertically by the hydraulic actuators, and in turn the pitch control rods alter the angles of attack (Figure B.26) of all blades simultaneously by the same amount. The lift forces generated by all blades are equal, driving the helicopter in a vertical direction. In cyclic control (Figure B.25(d)), the motions of the hydraulic actuators are not equal, which results in a tilting of the oscillating swash plate. Here, the angle of attack of every blade varies periodically during each revolution of the driveshaft. This results in unequal amounts of lift force on opposite sides of the driveshaft, and thrusting of the helicopter in a horizontal direction. The pilot employs cyclic control to either fly the helicopter in a desired horizontal direction or hover even when there is a lateral wind blowing.

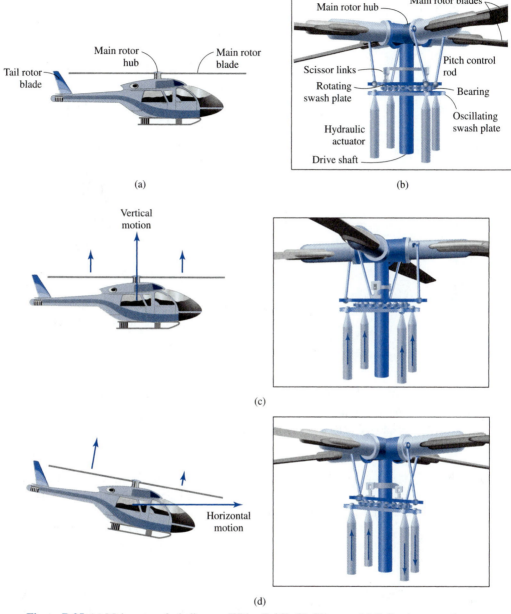

**Figure B.25** (a) Main rotor of a helicopter [Video B.25]. (b) Close-up. (c) Collective control. (d) Cyclic control.

**Figure B.26** Angle of attack of a helicopter blade.

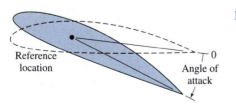

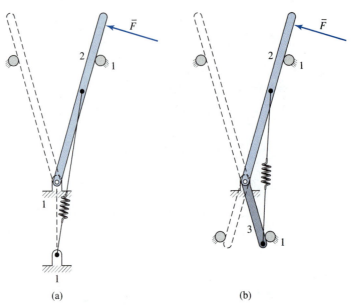

**Figure B.27**  Toggle mechanisms: (a) simple toggle [Model B27.A],
(b) double toggle [Model B27.B].

## B.10  TOGGLE MECHANISM

Two types of *toggle mechanisms* are described below. These mechanisms are commonly employed in mechanical electric switches.

### B.10.1  Simple Toggle

| |
|---|
| **Model B.27A** |
| Simple Toggle |
| Mechanism |

Figure B.27(a) illustrates a *simple toggle mechanism*. The lever, link 2, is held in position against one of two stops by a spring. If sufficient force is applied to the lever, the spring snaps the lever from one stop to the other.

A disadvantage of this mechanism is that it is possible to set the lever in a balanced vertical configuration between the two stops.

### B.10.2  Double Toggle

| |
|---|
| **Model B.27B** |
| Double Toggle |
| Mechanism |

Figure B.27(b) shows a *double toggle mechanism*. The arrangement of its links reduces the possibility of balancing the lever between the stops as with a simple toggle mechanism. Here, the short lever arm, link 3, can only be in equilibrium when it is in contact with either of its two stops. Because the movement of link 3 lags behind that of link 2, link 2 must be moved beyond the vertical position or link 3 will snap through from one stop to the other.

## B.11  SHAFT COUPLING

There are several mechanisms that allow transmission of rotational motion between two shafts that are not collinear. Some common types are described below.

**Figure B.28** Hooke's coupling [Model B.28].

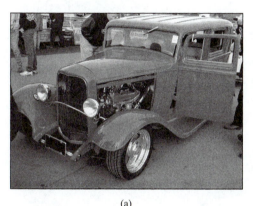

(a)

(b)

**Figure B.29** Steering linkage.

## B.11.1 Hooke's Coupling

An illustration of a *Hooke's coupling*, also referred to as a *universal joint*, is given in Figure B.28. This coupling can transmit motion between two shafts whose centerlines intersect but need not be parallel.

Figure B.29 shows an application of a Hooke's coupling in a customized automobile. The axis of the steering wheel column is misaligned with the shaft to the steering gearbox. A Hooke's coupling is employed to transmit motion between the shafts.

For the Hooke's coupling shown in Figure B.30(a), the input and output are connected to links 2 and 4, respectively. Their rotational speeds are designated as $\dot{\theta}_2$ and $\dot{\theta}_4$. If the angle between the centerlines of the shafts is $\beta$, then starting from the configuration illustrated, the *speed ratio*, $\dot{\theta}_4/\dot{\theta}_2$, between links 4 and 2 is illustrated in Figure B.30(b). The speed ratio is plotted for $\beta = 0$, $\beta = 20°$, and $\beta = 40°$. When $\beta = 0$, the shafts are collinear, and the speed ratio is unity for all values of $\theta_2$. However, when $\beta \neq 0$, and $\dot{\theta}_2$ is constant, $\dot{\theta}_4$ will fluctuate. The amount of variation of the speed ratio increases as $\beta$ enlarges.

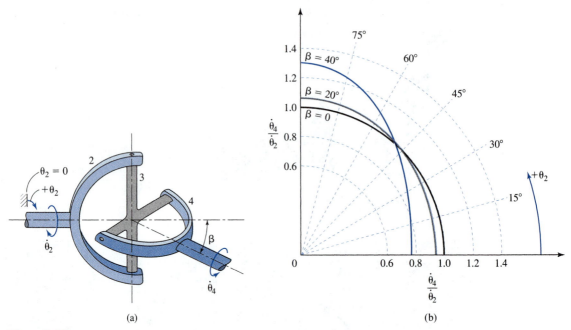

(a)

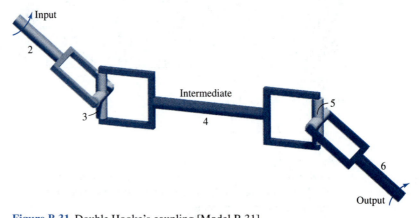

(b)

**Figure B.30** Hooke's coupling.

**Figure B.31** Double Hooke's coupling [Model B.31].

A single Hooke's coupling cannot maintain a constant speed ratio between the input and output, unless the shafts are collinear. However, by properly combining two Hooke's couplings, the variable speed ratio of one coupling may be counteracted by that of the other coupling. Figure B.31 shows such an arrangement. There will be a constant speed ratio between links 2 and 6 if the angles between the shafts of both Hooke's couplings are equal.

**Model B.31**
Double Hooke's
Coupling

### B.11.2 Oldham Coupling

An *Oldham coupling* can transmit rotational motion through shafts that are parallel but not necessarily collinear. Figure B.32(a) illustrates an exploded view of an Oldham coupling,

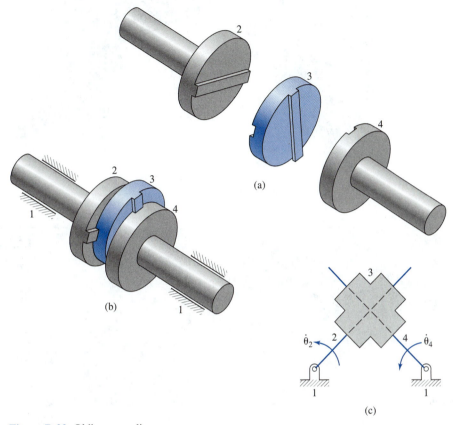

**Figure B.32** Oldham coupling.

and Figure B.32(b) shows the assembled form. A skeleton diagram of this coupling is given in Figure B.32(c). Another illustration of this coupling is shown in Figure B.33. The input and output of this device are links 2 and 4, respectively. Both links can transmit motion through an intermediate link 3, by means of a tongue and groove. Each tongue-and-groove pair allows relative sliding between the links. Since link 3 ensures that the angular difference between the tongue-and-groove pairs remains constant as the links rotate, links 2 and 4 always have the same rotational speed.

Figure B.34 shows an example application of an Oldham coupling employed in an automobile radio. A close-up of the coupling is provided in Figure B.34(b). Centerlines of the control knob shaft and tuner shaft are shown. The coupling is used to increase the space between the control knobs to accommodate a tape cassette, while at the same time permitting the tuner components to be moved inboard, thus leaving clearance with the casing.

### B.11.3 Offset Drive

Figure B.35 illustrates an *offset drive*. It can transmit rotational motion through shafts that are parallel. It consists of three discs of equal diameter. Bearings are equally spaced on the three discs. Three links, connected through bearings, remain parallel as the mechanism moves.

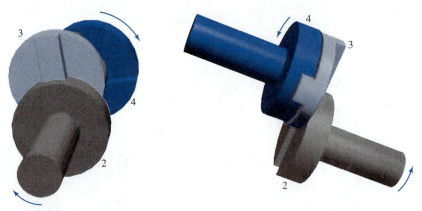

**Figure B.33** Oldham coupling [Model B.33].

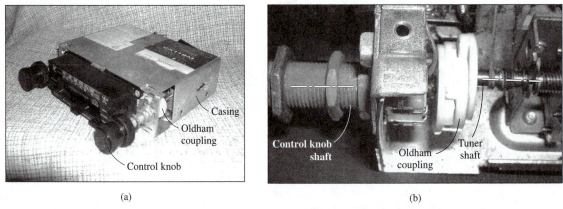

(a)                                                              (b)

**Figure B.34** Application of an Oldham coupling: (a) automobile radio, (b) close-up of Oldham coupling.

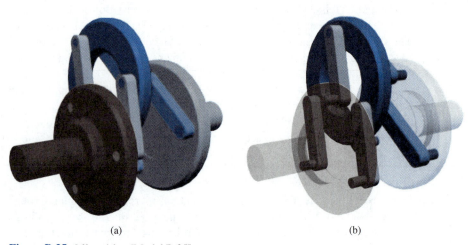

(a)                                                              (b)

**Figure B.35** Offset drive [Model B.35].

### B.11.4 Constant-Velocity Coupling

Model B.36
Constant-
Velocity
Coupling

A *constant-velocity coupling* can transmit rotational motion through shafts that intersect but are not necessarily parallel. An illustration of a constant-velocity coupling is shown in Figure B.36(a). As the animation from **[Model B.36]** progresses, a portion of the mechanism gradually disappears, revealing the inner components, as shown in Figure B.36(b). A cross section of the coupling is shown in Figure B.37(a). The input and output shafts are connected to the inner and outer journals of the coupling. The angle between the shafts is β. There are six grooves in each journal. Each of the six balls is inserted between two grooves, one on each journal. All grooves are at an oblique angle with respect to the centerline axis of a shaft. Two grooves for one ball are in opposite oblique directions. As illustrated in Figure B.37(b), each ball remains in the plane of symmetry between the two shafts, regardless of the value of β. The perpendicular distance $a$ to the center of the ball is the same for both shafts. Under operating conditions, forces are transmitted from one set of grooves, through each of the balls, and onto the other set of grooves. Since forces and motions are transmitted through the balls, the input and output shafts rotate at the same rotational speed.

This type of joint is relatively compact and is commonly used in front wheel drive vehicles. Smooth transmission of motion is achieved despite changes in alignment of the shafts.

(a)                                      (b)

**Figure B.36** Constant-velocity coupling [Model B.36].

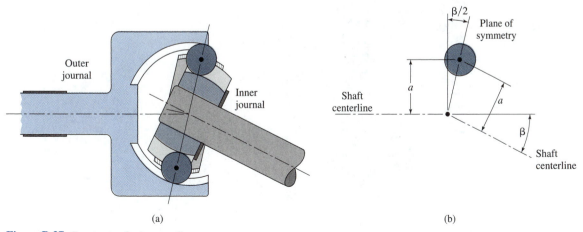

(a)                                      (b)

**Figure B.37** Constant-velocity coupling.

The animation provided through **[Model B.36]** shows the case where the axes of rotation of the shafts remain stationary. However, the angle between the centerlines of the shafts may change as the shafts rotate, but the input and output rotational speeds would remain equal.

## B.12  ZERO-MAX DRIVE

Figure B.38(a) shows a *zero-max drive*, which can convert constant input speed to a selected value of output speed. The drive consists of a combination of several similar mechanisms. Motion is transmitted from the input shaft through all of the mechanisms to a *one-way clutch* on the output shaft. Each linkage provides motion to the output shaft in sequence to ensure the output motion is continuous. Rotational inertia of the load permits the one-way clutches to overrun the driving links and smooth out pulsations imparted by each of the mechanisms.

Figure B.38(b) shows a skeleton form of one of the mechanisms. The location of base pivot $O_4$ is adjustable. If it is moved along the track, then link 6 oscillates through a smaller angle for each input rotation, and the speed ratio between the input and output is increased. When the base pivot is moved to $O_4'$, the output shaft remains stationary.

## B.13  SYNCHRONIZER

A *synchronizer* is used to rigidly couple or uncouple a gear and the shaft on which it is mounted, and it is commonly employed in manual transmissions (see Chapter 6, Section 6.2) of automobiles. When a gear and shaft are rigidly coupled, they act as one solid component capable of transmitting power through a meshing gear. When uncoupled, the gear can only spin freely relative to the shaft, and no power may be transmitted. The coupling process is carried out without movement of the gear along the axis of the shaft. Therefore, using a synchronizer, gears are always in mesh, and an operator can change speed ratios without clashing the gear teeth.

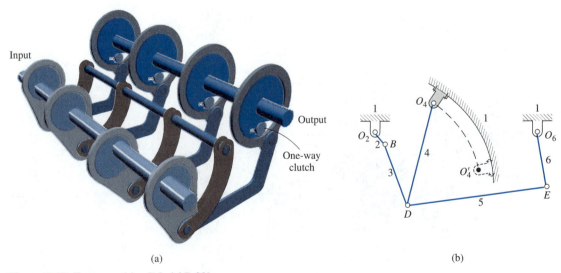

(a)

(b)

**Figure B.38**  Zero-max drive [Model B.38].

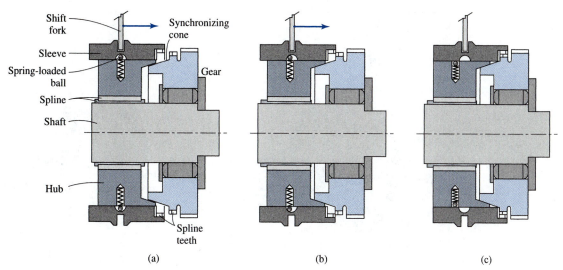

Figure B.39 Cross section of a synchronizer.

Figures 6.9 and 6.10 show a manual transmission incorporating synchronizers. Gear 3 remains in mesh with gear 7 for all speed ratios. However, only when the transmission is in "second gear" (Figure 6.10(d)), in which one of the synchronizers rigidly couples gear 7 to the output shaft, can power be transmitted through gears 3 and 7. For all other speed ratios of the transmission, gear 7 is uncoupled from the output shaft, and alternative gears are used to transmit power.

Figure B.39 shows cross-sectional views of a synchronizer in three distinct configurations that occur during the coupling process. To commence coupling, the operator must first cut off torque being delivered to the transmission, by disengaging the clutch (Figure 6.9). Figure B. 39(a) illustrates the uncoupled configuration in which the gear is free to spin on its shaft. Components of the synchronizer include the *hub*, which is splined to the shaft, and the *sleeve*, which is splined to the hub. By pushing the *shift fork* in the direction shown, the hub and sleeve move toward the gear. *Spring-loaded balls* located in detents in the sleeve deter relative movement between the hub and sleeve. Figure B.39(b) shows the condition of initial contact between *synchronizing cones* on the gear and hub. As the cones are pressed together, frictional forces cause the gear, hub, and sleeve to rotate at the same speed. Further movement of the shift fork forces the sleeve to slide, relative to the hub, toward the gear, and the internal spline teeth on the sleeve slide over the external spline teeth on the gear. Since the rotational speeds of the sleeve and gear are the same, this is accomplished without clashing of the spline teeth. When the sleeve moves with respect to the hub, the balls move down against their springs. Now the gear is locked to the shaft through the sleeve, and the coupling is complete, as shown in Figure B. 39(c).

Another illustration of a synchronizer is shown in Figure B.40, in the uncoupled and rigidly coupled configurations. In the animation provided through **[Model B.40]**, when the synchronizer is uncoupled, the meshing gears stop rotating. However, in this instance, they are actually free to rotate at any speed.

**Model B.40**
Synchronizer

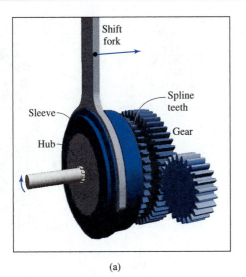

(a)                                                                  (b)

**Figure B.40**  Synchronizer [Model B.40]: (a) uncoupled, (b) rigidly coupled.

## B.14  CLUTCH

A *clutch* may be used to connect and disconnect rotational motions between two collinear shafts. Figure B.41 shows a cross section of a clutch, known as a *plate clutch*, used in a vehicle equipped with a manual transmission. The engaged and disengaged configurations are illustrated. To disengage the clutch, the pedal is depressed as shown in Figure B.41(a). Upon releasing the pedal, the diaphragm spring forces the friction pad against the flywheel, thus providing a direct connection between the engine and transmission (not shown), as illustrated in Figure B.41(b). Another illustration of a plate clutch is shown in Figure B.42.

> **Model B.42**
> Clutch
> Mechanism

## B.15  POSITIVE-DISPLACEMENT PUMP

> **Model B.43A**
> Positive-
> Displacement
> Pump
> Incorporating
> Crescent

Figure B.43 shows two versions of a *positive-displacement pump*. In both cases, fluid is drawn into totally enclosed pockets and is expelled from the pump as the pockets collapse. For the pump illustrated in Figure B.43(a), fluid is entrapped in the spaces between gear teeth and on either side of the *crescent*. For the pump illustrated in Figure B.43(b), fluid is constrained between vanes that are allowed to move radially with respect to the *rotor*. All vanes maintain contact with the inside surface of the housing. Another form of a positive-displacement pump employing gears is presented in Chapter 5, Section 5.6.

> **Model B.43B**
> Positive-
> Displacement
> Pump
> Incorporating
> Vanes

## B.16  LOCK

### B.16.1  Key Lock

> **Model B.44**
> Key Lock

A common form of *key lock*, known as a *pin tumbler lock* [14], is illustrated in Figure B.44. It includes a *cam*, *plug*, and *housing*. The cam and plug are rigidly connected. The housing

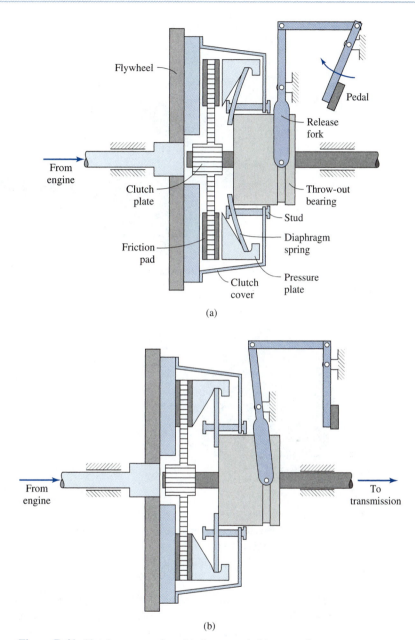

**Figure B.41** Clutch cross section: (a) disengaged, (b) engaged.

remains stationary. The cam provides locking action by engaging another component (not shown). The lock can be opened if the plug and cam are rotated. As the animation from **[Model B.44]** proceeds, the housing becomes translucent, and *pins* of varying lengths are revealed (Figure B.44(b)). They are constrained to slide in *pin chambers* within the housing and plug. While in a locked configuration, the pins are arranged in pairs, placed end to

**Figure B.42**  Clutch mechanism [Model B.42].

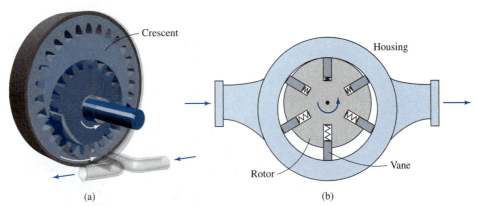

**Figure B.43**  Positive-displacement pumps: (a) incorporating crescent [Model B.43A], (b) incorporating vanes [Model B.43B].

end, and are forced downward by compressed *cylinder springs* (shown in Figure B.44(b), but not part of the animation). The *keyway* has small tabs that prevent the pins from moving fully downward, across the entire keyway.

The plug cannot rotate unless all dividing surfaces between pairs of pins are in alignment with the cylindrical face of the plug. As the tip of the key is inserted into the keyway, the pins are moved in their axial directions. Even when the key is partially inserted (Figure B.44(b)), the pins prevent the plug from rotating. Only when the key is fully inserted (Figure B.44(c)), where the shoulder of the key is pushed against the end of the plug, are all dividing surfaces even with the face of the plug. The head of the key can then be rotated by hand, along with the plug and cam (Figure B.44(d)).

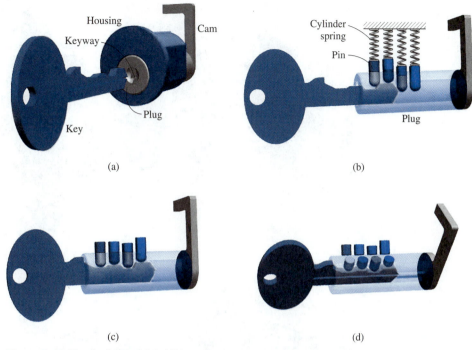

Figure B.44  Key lock [Model B.44].

## B.16.2 Combination Lock

**Model B.45**
Combination
Lock

A *combination lock* is shown in Figure B.45(a). It includes a *dial* that may be rotated manually relative to a stationary concentric *ring*. Numbers and serrations are inscribed on the periphery of the dial. They are used for aligning numbers of the combination with a reference mark located on the ring. The design shown requires that three numbers be input in proper sequence in order for the lock to be opened. The lock also includes four *wheels*. The dial is rigidly connected to wheel 1 through the *spindle*. Wheels 2, 3, and 4 are free to rotate with respect to the spindle. Motion can be transmitted from one wheel to the next through *extension arms*, which engage tabs located on the back sides of the wheels (Figure B.45(b)). Each wheel has a small section removed from its periphery, referred to as a *gate*. The *side bar* is rigidly connected to the *fence*, which in turn is pin-jointed to the *bolt*. It is the bolt that provides locking action by protruding into the recess of an adjacent component (not shown). The lock may be opened by retracting the bolt.

If the dial is rotated at least three complete revolutions in one direction, then all extension arms contact a tab on an adjacent wheel, and all wheels move in unison. The first number of the combination is input by continuing to turn the dial until it is aligned with the *stationary mark* on the ring. This brings the gate on wheel 4 to line up under the side bar (Figure B.45(b)). Then, the dial is turned in the opposite direction until the extension arms on wheels 1 and 2 contact tabs on an adjacent wheel. The dial then continues to be turned until the second number of the combination is aligned with the stationary mark. This brings the gate on wheel 3 to line up under the side bar. The dial is then rotated in the opposite direction until the third number of the combination is aligned with the stationary mark, and

**Figure B.45** Combination lock [Model B.45].

the gate on wheel 2 is aligned under the side bar. Next, the gate on wheel 1 is aligned under the tab on the fence. The side bar and fence then move downward under action of the compressed spring (shown in Figure B.45(c), but not part of the animation). Finally, providing a small turn of the dial, the bolt is retracted (Figure B.45(c)).

## B.17  CHUCK

### B.17.1  Drill Chuck

Model B.47
Drill Chuck

A powered hand drill is shown in Figure B.46(a). A close-up of the associated *drill chuck* is shown in Figure B.46(b). Figure B.47(a) shows another drill chuck in assembled form,

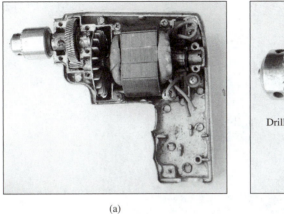

(a)                                                                              (b)

**Figure B.46** Powered hand drill.

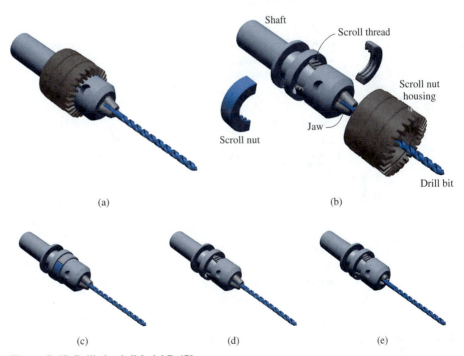

(a)                                                                              (b)

(c)                                    (d)                                    (e)

**Figure B.47** Drill chuck [Model B.47].

and Figure B.47(b) shows an exploded view of the same unit. When assembled, the *scroll nut housing* is press-fit over the *scroll nut*, and these components move as a rigid body. Internal threads on the two halves of the scroll nut mesh with threads on each of the *jaws*. When the scroll housing is rotated, meshing takes place between the threads, and the three jaws are forced to move with respect to the centerline of the shaft and drill bit.

**Figure B.48**  Manual lathe chuck.

Figure B.47(c) shows the drill chuck without the scroll nut housing. In Figures B.47(d) and B.47(e), the scroll nut is also removed, and the jaws are shown in closed and open positions with respect to a drill bit.

### B.17.2  Manual Lathe Chuck

**Model B.49**
Manual Lathe Chuck

A *manual lathe chuck* is shown in Figure B.48. Another illustration is provided in Figure B.49. Tightening is accomplished by rotating the wrench while in one of the three sockets. Each socket is attached to a *pinion gear*, which in turn rotates the *scroll*. The scroll is machined with three starts of spiral grooves. These grooves mesh with teeth on the three jaws. As the scroll rotates, all jaws move radially to and from the center of the scroll.

### B.17.3  Power Lathe Chuck

**Model B.50**
Power Lathe Chuck

A *power lathe chuck* is shown in Figure B.50. This unit is activated from a hydraulic cylinder (not shown), which drives a central piston axially along the centerline of the chuck. Motion is transferred from the piston to the three jaws through *wedge blocks*. Angled serrations on the piston, wedge blocks, and jaws allow the three jaws to move radially with respect to the centerline of the chuck.

## B.18  COMPLIANT MECHANISM

Most mechanisms presented so far in this appendix consist of a number of links connected together by pin joints and sliders. However, *compliant mechanisms* are often made of a single piece of material. They are designed to undergo elastic deflections when subjected to applied force. Elastic deflections provide the required motions. When the forces are removed, the mechanism generally returns to its original configuration.

Figure B.51 shows an extension spring, an example of a compliant mechanism. It is made of a single piece of wire wound into a coil, and with a loop at each end. By applying

(a)                                                          (b)

Socket
and
pinion

Workpiece

Wrench

Scroll                                    Jaw

(c)

**Figure B.49**  Manual lathe chuck [Model B.49].

a force at each end, one end translates with respect to the other, and when the force is removed, the spring retracts to its original length.

Figures B.52 and B.53 show additional examples of compliant mechanisms. Figure B.52 shows a snap-lock connector used on a telephone extension cord. When the connector is inserted into the mating slot of a telephone receptacle, the clip locks the connector in place. To remove the connector, the clip must be depressed. Figure B.53 shows a CD (compact disc) jewel case. A typical case incorporates two compliant mechanisms. The first is the lid release mechanism (Figure B.53(a)). In order to open the case, a person must use one hand to depress the lid release mechanism on either side of the case, while the other hand is used to lift the lid. The second compliant mechanism, illustrated in Figure B.53(b), is the CD holder mechanism. This consists of a set compliant radial clips that deflect inward to clear the hole of a CD, and then expand outward to hold onto the CD. To release the CD, the radial clips must be depressed by a finger, while a hand lifts on the CD.

Figure B.54 illustrates a camera lens cover made of two compliant elements. The first element is the cap cover that protects the lens (Figure B.54(a)), and the second element is

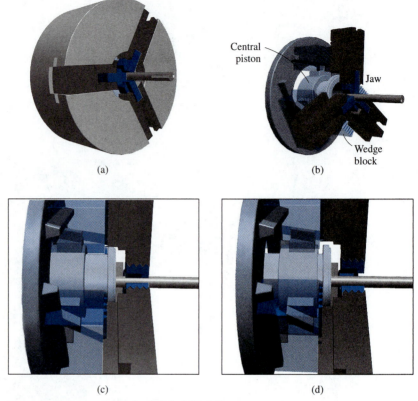

**Figure B.50** Power lathe chuck [Model B.50].

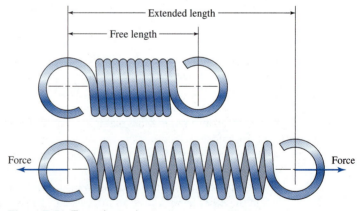

**Figure B.51** Extension spring.

the thread catcher that locks against the threads surrounding the lens. By applying force on the outer edges of the thread catcher, as shown in Figure B.54(b), the thread catcher slides within the guide tracks of the cap cover.

Some compliant mechanisms do not return to their original position after removal of the applied forces. Figure B.55(a) shows a shampoo bottle cap that is made of a single

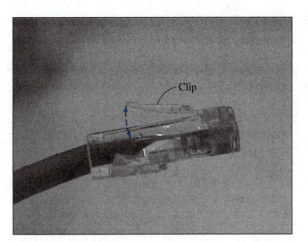

**Figure B.52** Snap-lock connector.

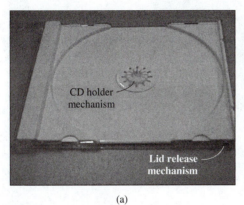

(a)

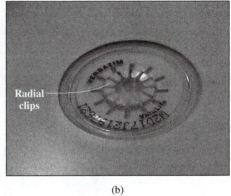

(b)

**Figure B.53** CD jewel case.

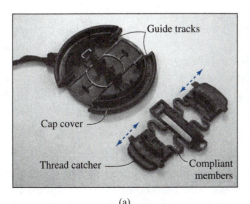

(a)

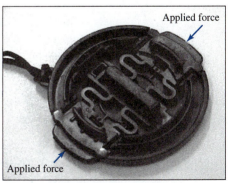

(b)

**Figure B.54** Camera lens cover: (a) disassembled, (b) assembled.

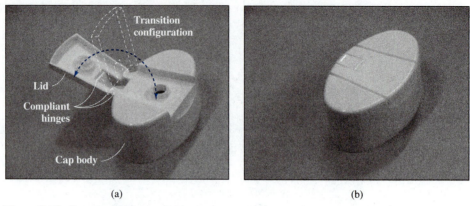

(a)                                         (b)

**Figure B.55**  Shampoo bottle cap: (a) open, (b) closed.

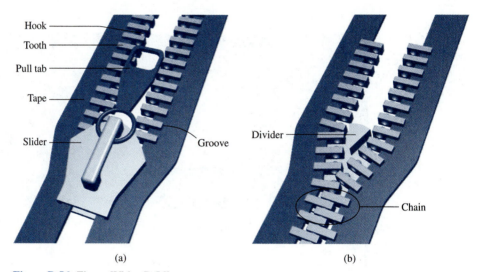

(a)                                         (b)

**Figure B.56**  Zipper [Video B.56].

piece of material. It incorporates a number of thin sections that can easily bend and act as compliant hinges. In order to close the lid, it must be deflected past the transition configuration shown in Figure B.55(b). If the applied force is removed before the lid reaches the transition configuration, it will return to its open position. If the applied force is removed after the lid passes the transition configuration, the lid will continue on its own to move toward the closed position. Similarly, to open the lid, a force must be applied beyond the transition configuration. Otherwise, it will return toward the closed position.

**Video B.56**
Zipper

A *zipper*, illustrated in Figure B56(a), is another example of a commonly employed compliant mechanism. Each side of a zipper has a row of *teeth* attached to a flexible *tape*. The tape is in turn connected to the fabric (not shown in figure). The *pull tab* is used to grasp and move the *slider*. Depending on the direction of motion of the slider, the teeth are either joined together or divided. During zipping up, each tooth in one row has a *hook* that interlocks with a *groove* in a tooth located in the other row. After zipping up, the two rows form a *chain*. Figure B56(b) shows a view with part of the slider removed. The *divider* is used either to separate the teeth while unzipping or guide the two rows together during zipping up.

# B.19 MICROELECTROMECHANICAL SYSTEM (MEMS)

A *microelectromechanical system (MEMS)* is a combination of miniature mechanical components fabricated on a silicon chip. Components such as gears and compliant mechanisms may be included. Figure B.57(a) illustrates a MEMS gear train produced at Sandia National Laboratories [15], and Figure B.57(b) is a picture of microgrippers developed at the University of Toronto [16]. A MEMS is often manufactured using the same technology and materials as those employed in building an integrated circuit composed of miniature transistors, resistors, and capacitors.

A MEMS is used in an accelerometer chip in deployment of an air bag. The MEMS chip detects the extremely high deceleration of a vehicle during a collision and provides a signal to deploy the air bag. Figure B.58 illustrates a simplified layout of a MEMS accelerometer. Figure B.58(a) shows the configuration when there is no acceleration. An inertial mass is suspended by four flexible beams, which are also fixed to the chip at anchor points. When deceleration occurs, as shown in Figure B.58(b), there is relative motion between the inertial mass and the rest of the chip. If the deceleration is great enough, then changes in gap $\delta x$ are sensed by the capacitive sensor, and a signal is sent to the microelectronics located on the chip (not shown in the figure) to deploy the air bag. The

60 μm

(a)

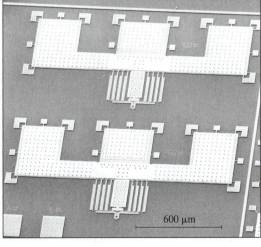

600 μm

(b)

**Figure B.57** Examples of MEMS: (a) gear train (courtesy of Sandia National Laboratories, Albuquerque, NM), (b) microgrippers.

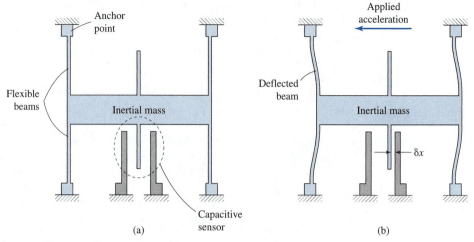

**Figure B.58**  MEMS accelerometer.

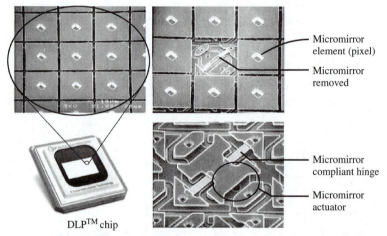

**Figure B.59**  DLP chip illustrating array of micromirrors (courtesy of Texas Instruments, Dallas, TX).

micromechanical components complement the microelectronics to produce a microsensor that is not possible to construct using microelectronics alone.

Another example of a MEMS is the digital micromirror device produced by Texas Instruments, referred to as the DLP™, which is used in many video projection systems. A picture of a DLP chip is shown in Figure B.59. This MEMS consists of a matrix of micromirrors. There is a micromirror for each pixel of a projected display. Under each micromirror is a microactuator that is used to tilt its micromirror back and forth. A light source is continuously directed at the chip. Using a microactuator, light reflected from a micromirror may be alternately directed to and away from the projected image. This effectively allows each pixel of the display to be switched on and off.

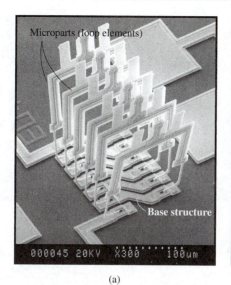

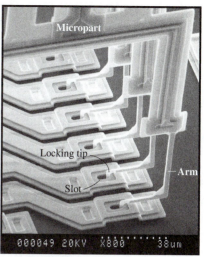

(a)                                                        (b)

**Figure B.60** (a) Microcoil. (b) Close-up.

The fabrication of a MEMS consists of successive deposition, doping, and etching of thin films of material. This yields micromechanical components that have low aspect ratios (i.e., height versus planar length), usually between 1:10 and 1:1,000. Recent research into microassembly at the University of Toronto has shown the potential of fabricating a far more complex MEMS. It is now possible to construct three-dimensional microstructures [17]. Figure B.60(a) shows a three-dimensional microcoil developed at the University of Toronto that can be used to convert electromagnetic radiation, such as radio waves, into electrical signals. Figure B.60(b) shows a close-up view of the microjoints that are formed by compliant locking tips on the ends of the microparts.

The microcoil is constructed from a set of microparts that start out planar on the surface of a chip, as shown in Figure B.61(a). A compliant microgripper, solder-bonded to the end-effector of a robotic manipulator, as shown in Figure B.61(b), is used to grasp the microparts from the chip. Figure B.61(c) shows the microgripper lining up with a micropart, and Figure B.61(d) shows the microgripper grasping the micropart.

## B.20 THREADED CONNECTION

**Model B.62**
Screw and Nut

**Model B.63**
Recirculating
Ball Screw

Two *threaded connections* are shown in Figures B.62 and B.63. Figure B.62 shows a traditional screw and thread. This device can convert rotary motion of either the screw or the nut into relatively slow linear motion of the mating member along the axis of rotation. For each rotation of the nut with respect to the screw, there is relative linear motion along the screw equal to the pitch of the thread.

Figure B.63 illustrates a *recirculating ball screw*. The kinematics of a recirculating ball screw are identical to those of a traditional screw. Therefore, a distinction is not required when performing a kinematic analysis. However, a recirculating ball screw has significantly

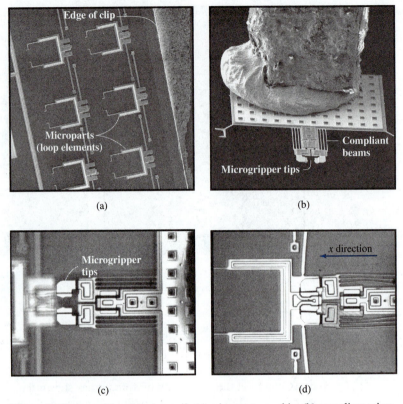

(a)

(b)

(c)

(d)

**Figure B.61** Assembly of a microcoil: (a) microparts on chip, (b) compliant microgripper on manipulator, (c) microgripper aligned to micropart, (d) microgripper grasping micropart.

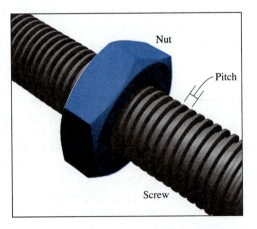

**Figure B.62** Screw and nut [Model B.62].

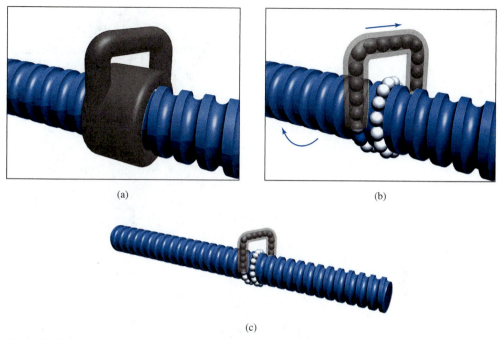

(a)

(b)

(c)

**Figure B.63** Recirculating ball screw [Model B.63].

less friction compared to a traditional screw and thread. Instead of the sliding contact between the screw and the nut, there is rolling contact of balls in grooves along the screw.

## B.21 BRAKING SYSTEM

### B.21.1 Disc Brake

**Model B.64**
Disc Brake

Figure B.64 illustrates a *disc brake*, which includes *brake pads* located on each side of the *rotor*. The *floating caliper* is mounted on rubber bushings (not shown), allowing small side-to-side movements as indicated in Figure B.64(a). When the brake is applied, the piston is extended, and it pushes one brake pad against a side of the rotor. Pressure in the piston cylinder also causes the floating caliper to shift to the left (Figure B.64(b)). This movement brings the other brake pad, mounted on the floating caliper, into contact on the opposite side of the rotor. As a result, the brake pads pinch the rotor tightly. Friction between the brake pads and rotor will reduce the rate of rotation. After the brake is released (Figure B.64(c)), there is no spring to pull the brake pads away from the rotor. The brake pads either stay in very light contact with the rotor or negligible wobble in the rotor will push the brake pads a small distance away, clear of the rotor. Most automobile disc brakes contain vents between the sides of the rotor. They pump air radially through the rotor to dissipate frictional heat energy.

Older designs of disc brakes incorporated two pistons and a rigidly fixed caliper. This design has been largely eliminated because the floating caliper design is less expensive and more reliable.

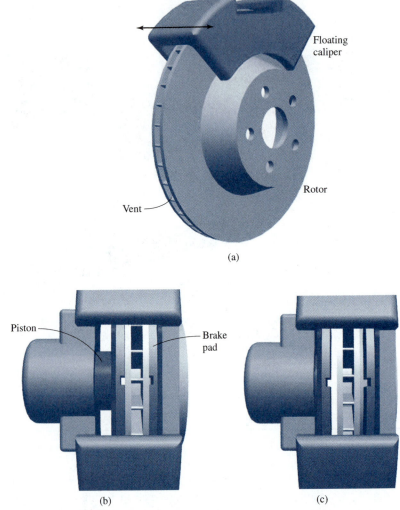

**Figure B.64**  (a) Disc brake [Model B.64]. (b) Brake applied. (c) Brake released.

### B.21.2  Drum Brake

Model B.65
Drum Brake

Figure B.65 illustrates a drum brake. It has two *brake shoes* and a *piston*. When the brake is activated, the piston pushes the brake shoes against the *drum* to provide the braking action. As the brake shoes contact the drum, there is a wedging action that increases the friction force (Figure B.65(b)). After the brakes are released, the shoes are pulled away from the drum by springs. Other springs are used to hold the brake shoes in place.

For drum brakes to function correctly, excessive travel of the pistons must be avoided. However, as the brake shoes wear, a wider gap will form between the shoe and the drum, requiring adjustment to be made. This is why most drum brakes incorporate an *adjuster*. Each time the vehicle stops by braking, after travelling in reverse, the shoes are pushed

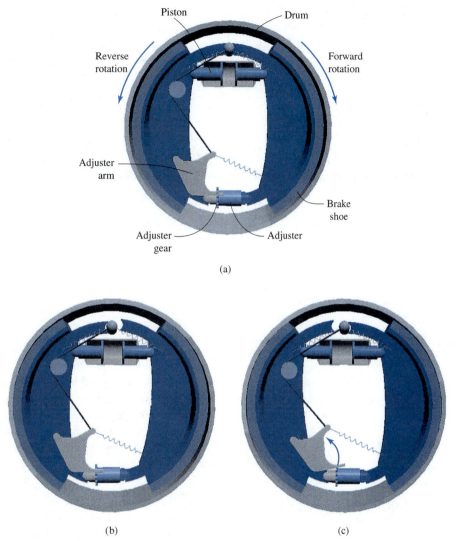

**Figure B.65** Drum brake [Model B.65].

against the drum, and the adjuster lever is pulled (Figure B.65(c)). Whenever the gap has enlarged sufficiently by wear, the adjuster lever is permitted to have enough rocking motion to advance the adjuster gear by one tooth. This in turn causes movement in the threaded connection of the adjuster, causing it to slightly lengthen and thereby reduce the gap.

# C

# Scalars and Vectors

## C.1 INTRODUCTION

A *scalar* is a quantity that has magnitude but no direction associated with it. Mass, energy, and temperature are examples of scalars. Relationships between scalars may be dealt with using algebraic expressions. A *vector*, however, has both direction and magnitude. Examples of vectors are force, displacement, velocity, and acceleration.

Vector quantities can be represented graphically by straight lines with arrowheads indicating directions, such as vectors $\bar{b}$ and $\bar{c}$ shown in Figure C.1. The length of each line is proportional to the magnitude of the vector.

## C.2 COMPONENTS OF A VECTOR

A vector may be described by specifying its components. The components are aligned along directions of *unit vectors* defined below.

Two sets of unit vectors are employed in this textbook. They are described in the following subsections.

### C.2.1 Cartesian Components

Consider vector $\bar{b}$, shown in Figure C.2. It may be described by its *Cartesian components* $b_x$, $b_y$, and $b_z$ so that

$$\bar{b} = b_x\bar{i} + b_y\bar{j} + b_z\bar{k}$$

(C.2-1)

where $\bar{i}$, $\bar{j}$, and $\bar{k}$ are unit vectors in the $x$, $y$, and $z$ directions, respectively.

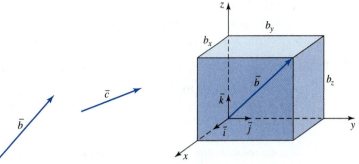

**Figure C.1** Representation of vectors.

**Figure C.2** Cartesian components of a vector.

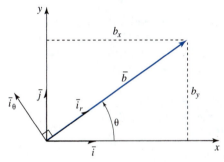

**Figure C.3**

The *magnitude* of the vector may be expressed as

$$b = |\bar{b}| = \left(b_x^2 + b_y^2 + b_z^2\right)^{1/2}$$ (C.2-2)

When $b_z = 0$, as shown in Figure C.3, the direction of the vector may be described by angle $\theta$ with respect to the reference shown. The components of this vector are

$$b_x = b\cos\theta; \qquad b_y = b\sin\theta$$ (C.2-3)

and therefore

$$\theta = \tan^{-1}\left(\frac{b_y}{b_x}\right), \qquad -\frac{\pi}{2} \leq \theta \leq \frac{\pi}{2}$$ (C.2-4)

## C.2.2 Radial-Transverse Component

Vector $\bar{b}$, illustrated in Figure C.3, may be described as

$$\bar{b} = b\bar{i}_r$$ (C.2-5)

where $\bar{i}_r$ is the *unit radial vector*. Also associated with this vector is the *unit transverse vector*, $\bar{i}_\theta$, which is perpendicular to $\bar{i}_r$.

## C.3  VECTOR OPERATIONS

### C.3.1  Addition and Subtraction

*Addition* of vectors involves adding the $x$, $y$, and $z$ components of the vectors. For instance, if

$$\bar{d} = \bar{b} + \bar{c} \qquad \text{(C.3-1)}$$

then the result expressed in terms of the components is

$$\boxed{\bar{d} = (b_x + c_x)\bar{i} + (b_y + c_y)\bar{j} + (b_z + c_z)\bar{k}} \qquad \text{(C.3-2)}$$

Graphically, this addition is accomplished by positioning vectors head to tail. Two vectors are shown in Figure C.4(a), and their addition is illustrated in Figure C.4(b).

*Subtraction* of vectors involves finding the differences of the $x$, $y$, and $z$ components individually. If

$$\bar{d} = \bar{b} - \bar{c} \qquad \text{(C.3-3)}$$

then the result, expressed in terms of the components, is

$$\bar{d} = (b_x - c_x)\bar{i} + (b_y - c_y)\bar{j} + (b_z - c_z)\bar{k} \qquad \text{(C.3-4)}$$

Subtraction of vectors may be regarded as the addition of the negative value of one vector to the positive value of another, that is,

$$\bar{d} = \bar{b} - \bar{c} = \bar{b} + (-\bar{c}) \qquad \text{(C.3-5)}$$

Graphically, the difference between two vectors is illustrated in Figure C.4(c).

### C.3.2  Dot Product

If we consider two vectors $\bar{b}$ and $\bar{c}$, then their *dot product* or *scalar product* is defined as

$$\boxed{\bar{b} \cdot \bar{c} = |\bar{b}||\bar{c}| \cos \alpha = b_x c_x + b_y c_y + b_z c_z} \qquad \text{(C.3-6)}$$

where $\alpha$ is the angle between the two vectors as shown in Figure C.5. The dot product of two vectors is a scalar quantity.

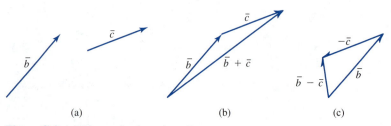

(a)          (b)          (c)

**Figure C.4** Addition and subtraction of vectors.

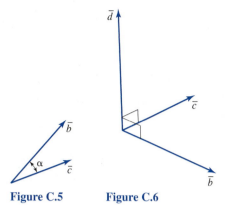

**Figure C.5**     **Figure C.6**

### C.3.3  Cross Product

If we consider two vectors $\overline{b}$ and $\overline{c}$, then their *cross product* is defined as

$$
\overline{d} = \overline{b} \times \overline{c} = \begin{vmatrix} \overline{i} & \overline{j} & \overline{k} \\ b_x & b_y & b_z \\ c_x & c_y & c_z \end{vmatrix}
$$
$$
= (b_y c_z - b_z c_y)\overline{i} + (b_z c_x - b_x c_z)\overline{j} + (b_x c_y - b_y c_x)\overline{k}
$$

(C.3-7)

The cross product of two vectors is a vector quantity. The direction of the resultant vector is perpendicular to the two vectors that are employed in the cross product (Figure C.6). For the special case that $b_z = 0$ and $c_z = 0$, then vector $\overline{d}$ is in the direction of vector $\overline{k}$.

## C.4  COMPLEX NUMBERS AND VECTORS IN THE COMPLEX PLANE

The *rectangular form* of a complex number is

$$
R = x + iy
$$

(C.4-1)

where
   $x$ = the *real component* of the complex number
   $y$ = the *imaginary component* of the complex number

and

$$
i = (-1)^{1/2}
$$

(C.4-2)

The *polar form* of a complex number is

$$
R = re^{i\theta} = r(\cos\theta + i\sin\theta)
$$

(C.4-3)

where
   $r$ = the *magnitude* of the complex number
   $\theta$ = the *argument* of the complex number

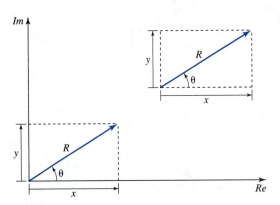

**Figure C.7**  Graphical representation of a complex number.

A complex number may be represented graphically by a vector in the *complex plane* as shown in Figure C.7, where quantities $x$, $y$, and $\theta$ are illustrated. For instance, $\theta$ is the angle of the vector relative to the horizontal, measured from its tail in the counterclockwise direction.

Using Equations (C.4-1) and (C.4-3), the following expressions relate the rectangular and polar forms of a complex number:

$$x = r \cos \theta; \qquad y = r \sin \theta$$

$$r = (x^2 + y^2)^{1/2}; \qquad \theta = \tan^{-1}\left(\frac{y}{x}\right), \qquad -\frac{\pi}{2} \leq \theta \leq \frac{\pi}{2} \tag{C.4-4}$$

Alternatively, if the real and imaginary components of the complex number are kept separate, then it is possible to determine the argument over the extended range as

$$\theta = \tan 2^{-1}(x, y), \qquad -\pi \leq \theta \leq \pi \tag{C.4-5}$$

# Mechanics: Statics and Dynamics

## D.1 INTRODUCTION

*Mechanics* is a science that predicts the conditions of either rest or motion of bodies under the action of forces and moments. As indicated in Figure D.1, mechanics may be divided into the topics of *statics*, which deals with bodies at rest, and *dynamics*, which is concerned with bodies in motion. Dynamics may be further subdivided into kinematics and kinetics. *Kinematics* is the description and analysis of the motion of objects without consideration of the applied forces that cause them. *Kinetics* is the study of the effects of forces on the motions of objects.

A *particle* implies that the dimensions of an object are so small that all mass may be assumed to be concentrated at a single point. For a *body*, however, having finite size, different points within the body can have different velocities. It may be considered as an assembly of particles.

A body will always deform to some extent when acted on by a force. However, a body is considered to be *rigid* when these deformations are small enough to be neglected, and all particles are assumed to remain at fixed distances in relation to one another. In nearly all

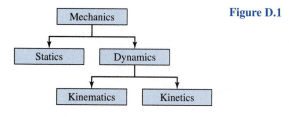

**Figure D.1**

instances in this textbook, bodies are assumed to be rigid. The valve spring shown in Figure 1.1 is an example of a body that cannot be considered rigid.

This appendix covers the kinematics and kinetics of particles and rigid bodies. The material provides background needed in this textbook dealing with the analysis and design of mechanisms. Analyses will be restricted to planar motion.

## D.2  KINEMATICS OF A POINT USING CARTESIAN COORDINATES

### D.2.1  Position

Consider point $B$, in Figure D.2, which is constrained to undergo planar motion. The location of the point is depicted by means of a *position vector*, relative to fixed point $O$.

Employing Cartesian axes incorporating unit vectors $\bar{i}$ and $\bar{j}$, in the $x$ and $y$ directions respectively, the position of point $B$ with respect to point $O$ is

$$\bar{r}_{BO} = (r_{BO})_x\bar{i} + (r_{BO})_y\bar{j}$$ (D.2-1)

### D.2.2  Velocity

The *velocity* of point $B$ is found by taking the time derivative of Equation (D.2-1). If the coordinate system and unit vectors are fixed, then the expression for velocity is

$$\bar{v}_{BO} = \frac{d}{dt}(\bar{r}_{BO}) = \frac{d}{dt}[(r_{BO})_x\bar{i} + (r_{BO})_y\bar{j}] = (\dot{r}_{BO})_x\bar{i} + (\dot{r}_{BO})_y\bar{j}$$ (D.2-2)

where a dot represents differentiation with respect to time.

In Equation (D.2-2), $\bar{v}_{BO}$ represents the velocity of $B$ relative to point $O$. However, since point $O$ is fixed, $\bar{v}_{BO}$ is also the absolute velocity of $B$. The motion of point $B$ is the same with respect to any fixed point, and therefore we may express this velocity as

$$\bar{v}_{BO} = \bar{v}_B$$ (D.2-3)

**Figure D.2**

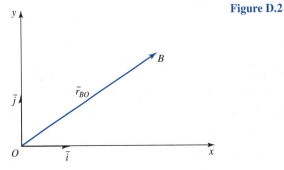

### D.2.3 Acceleration

The *acceleration* of point $B$ is found by taking the time derivative of Equation (D.2-2). The result is

$$\bar{a}_{BO} = \bar{a}_B = \frac{d}{dt}(\bar{v}_{BO}) = \frac{d}{dt}[(v_{BO})_x\bar{i} + (v_{BO})_y\bar{j}] = (\ddot{r}_{BO})_x\bar{i} + (\ddot{r}_{BO})_y\bar{j} \qquad (D.2\text{-}4)$$

## D.3  KINEMATICS OF A POINT USING RADIAL-TRANSVERSE COORDINATES

### D.3.1 Position

Consider point $B$, in Figure D.3, which is constrained to undergo planar motion. The location of the point is depicted by means of a position vector, relative to fixed point $O$. Employing the radial-transverse vectors, it is expressed as

$$\bar{r}_{BO} = r_{BO}\bar{i}_r \qquad (D.3\text{-}1)$$

where $\bar{i}_r$ is the *unit radial vector*. Also shown in Figure D.3 is $\bar{i}_\theta$, the *unit transverse vector*, which points 90° counterclockwise from the direction of $\bar{i}_r$.

### D.3.2 Velocity

Taking the time derivative of Equation (D.3-1), we find the velocity of point $B$ to be

$$\bar{v}_{BO} = \bar{v}_B = \frac{d}{dt}(\bar{r}_{BO}) = \frac{d}{dt}(r_{BO}\bar{i}_r) = \dot{r}_{BO}\bar{i}_r + r_{BO}\frac{d}{dt}(\bar{i}_r) \qquad (D.3\text{-}2)$$

Equation (D.3-2) includes a term involving the time derivative of $\bar{i}_r$. This is because, for the radial-transverse coordinate system, the radial direction can be time dependent.

To determine an expression for the time derivative of $\bar{i}_r$, we consider a small change of the position of point $B$ as it moves along its path. After a short time interval, it has moved to $B'$ as shown in Figure D.4(a).

Figure D.4(b) illustrates two sets of unit vectors corresponding to the position vectors for points $B$ and $B'$. From this figure

$$\Delta\bar{i}_r = \bar{i}_r' - \bar{i}_r; \qquad \Delta\bar{i}_\theta = \bar{i}_\theta' - \bar{i}_\theta \qquad (D.3\text{-}3)$$

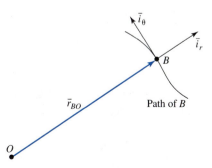

**Figure D.3** Position of a point using radial-transverse coordinates.

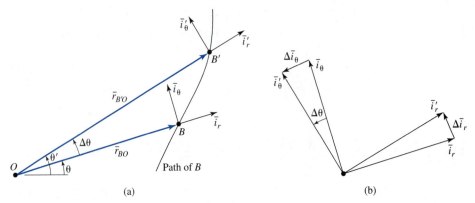

**Figure D.4** Two closely spaced positions of a point: (a) radial-transverse coordinate systems, (b) superimposed coordinate systems.

By definition

$$\frac{d}{d\theta}(\bar{i}_r) = \lim_{\Delta\theta \to 0} \frac{\Delta\bar{i}_r}{\Delta\theta} \tag{D.3-4}$$

The derivative given in Equation (D.3-4) is a vector quantity and must have both direction and magnitude. With regard to the magnitude, we note that since $\bar{i}_r$ has unit length, then for small values of $\Delta\theta$

$$|\Delta\bar{i}_r| \approx 1 \times \Delta\theta = \Delta\theta \tag{D.3-5}$$

and therefore

$$\left|\frac{d}{d\theta}(\bar{i}_r)\right| \approx \frac{\Delta\theta}{\Delta\theta} = 1 \tag{D.3-6}$$

Furthermore, by inspection of Figure D.4(b), the direction of the vector quantity in Equation (D.3-4) is perpendicular to the radial vector. Thus, it is in the transverse direction.

Combining the information related to the direction and magnitude,

$$\frac{d}{d\theta}(\bar{i}_r) = \bar{i}_\theta \tag{D.3-7}$$

Applying the chain rule of differentiation, in which we employ Equation (D.3-7),

$$\frac{d}{dt}(\bar{i}_r) = \frac{d}{d\theta}(\bar{i}_r)\dot{\theta} = \dot{\theta}\bar{i}_\theta \tag{D.3-8}$$

where parameter $\dot{\theta}$ is the *angular velocity* of the position vector.

Similarly, the time derivative of the unit transverse vector is

$$\frac{d}{dt}(\bar{i}_\theta) = -\dot{\theta}\bar{i}_r \tag{D.3-9}$$

By substituting Equation (D.3-8) in Equation (D.3-2), we obtain the velocity expressed in terms of its radial and transverse components as follows:

$$\bar{v}_B = \dot{r}_{BO}\bar{i}_r + r_{BO}\dot{\theta}\bar{i}_\theta \tag{D.3-10}$$

### D.3.3 Acceleration

Returning to Equation (D.3-10), and differentiating this with respect to time, we find the acceleration of point $B$ to be

$$\bar{a}_{BO} = \bar{a}_B = \frac{d}{dt}(\bar{v}_B) = \frac{d}{dt}(\dot{r}_{BO}\bar{i}_r + r_{BO}\dot{\theta}\bar{i}_\theta)$$

$$= \ddot{r}_{BO}\bar{i}_r + \dot{r}_{BO}\frac{d}{dt}(\bar{i}_r) + \dot{r}_{BO}\dot{\theta}\bar{i}_\theta + r_{BO}\ddot{\theta}\bar{i}_\theta + r_{BO}\dot{\theta}\frac{d}{dt}(\bar{i}_\theta) \tag{D.3-11}$$

Substituting Equations (D.3-8) and (D.3-9), and rearranging,

$$\bar{a}_B = (\ddot{r}_{BO} - r_{BO}\dot{\theta}^2)\bar{i}_r + (r_{BO}\ddot{\theta} + 2\dot{r}_{BO}\dot{\theta})\bar{i}_\theta \tag{D.3-12}$$

where $\ddot{\theta}$ is the *angular acceleration* of the position vector.

## D.4 DISPLACEMENT MATRICES

Figure D.5 shows a point $Q$ on a rigid body that undergoes planar motion from position 1 to position $n$ ($n = 2, 3, 4, \ldots$). The coordinates of the point before and after this motion may be related by

$$\begin{Bmatrix} x_n \\ y_n \\ 1 \end{Bmatrix} = [D_{1n}] \begin{Bmatrix} x_1 \\ y_1 \\ 1 \end{Bmatrix} \tag{D.4-1}$$

where

$[D_{1n}] = 3 \times 3$ *displacement matrix* [18]
$x_1, y_1 =$ coordinates of the point in position 1
$x_n, y_n =$ coordinates of the point in position $n$

The position vector of the point after the displacement is

$$\bar{r}_n = x_n\bar{i} + y_n\bar{j} \tag{D.4-2}$$

Successive displacements may be described using displacement matrices. For instance, if a point moves from position 1 to position 2, and then from position 2 to position 3, then using displacement matrices,

$$\begin{Bmatrix} x_2 \\ y_2 \\ 1 \end{Bmatrix} = [D_{12}] \begin{Bmatrix} x_1 \\ y_1 \\ 1 \end{Bmatrix}; \qquad \begin{Bmatrix} x_3 \\ y_3 \\ 1 \end{Bmatrix} = [D_{23}] \begin{Bmatrix} x_2 \\ y_2 \\ 1 \end{Bmatrix} \tag{D.4-3}$$

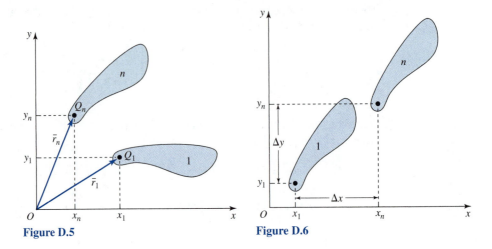

**Figure D.5**                    **Figure D.6**

Combining the above equations,

$$\left\{\begin{array}{c} x_3 \\ y_3 \\ 1 \end{array}\right\} = [D_{23}][D_{12}]\left\{\begin{array}{c} x_1 \\ y_1 \\ 1 \end{array}\right\} \tag{D.4-4}$$

Listed below are forms of the displacement matrix corresponding to typical motions.

### D.4.1 Translation

Consider the body shown in Figure D.6, which undergoes pure translation (i.e., no rotation). The change in the coordinates of any point on the body is

$$x_n = x_1 + \Delta x; \qquad y_n = y_1 + \Delta y \tag{D.4-5}$$

The corresponding displacement matrix is

$$[D_{1n}] = [D_{1n}]_T = \begin{bmatrix} 1 & 0 & \Delta x \\ 0 & 1 & \Delta y \\ 0 & 0 & 1 \end{bmatrix} \tag{D.4-6}$$

### D.4.2 Rotation About a Fixed Origin

Consider the body shown in Figure D.7, which undergoes rotation about a fixed origin. In this instance, it is convenient to employ the polar form of a complex number in order to determine the coordinates of a point before and after the motion has taken place. We

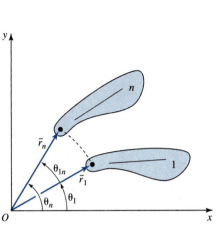

Figure D.7

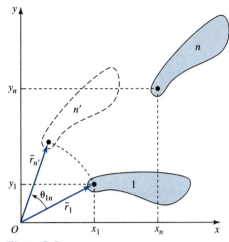

Figure D.8

recognize that

$$|R_1| = |R_n| = r \tag{D.4-7}$$

$$
\begin{aligned}
R_n &= re^{i\theta_n} = re^{i(\theta_1+\theta_{1n})} = re^{i\theta_1}e^{i\theta_{1n}} \\
&= r(\cos\theta_1 + i\sin\theta_1)(\cos\theta_{1n} + i\sin\theta_{1n}) \\
&= (x_1\cos\theta_{1n} - y_1\sin\theta_{1n}) + i(x_1\sin\theta_{1n} + y_1\cos\theta_{1n}) \\
&= x_n + iy_n
\end{aligned}
\tag{D.4-8}
$$

where

$$x_1 = r\cos\theta_1; \qquad y_1 = r\sin\theta_1 \tag{D.4-9}$$

Using Equation (D.4-8), the displacement matrix is

$$
[D_{1n}] = [D_{1n}]_R =
\begin{bmatrix}
C_{1n} & -S_{1n} & 0 \\
S_{1n} & C_{1n} & 0 \\
0 & 0 & 1
\end{bmatrix}
\tag{D.4-10}
$$

where

$$C_{1n} = \cos\theta_{1n}; \qquad S_{1n} = \sin\theta_{1n} \tag{D.4-11}$$

## D.4.3 Rotation About a Fixed Origin Followed by a Translation

Consider the body shown in Figure D.8, which undergoes rotation about a fixed origin (from position 1 to position $n'$), followed by a translation (from position $n'$ to position $n$).

In this instance, we employ the above results to obtain

$$[D_{1n'}]_R = \begin{bmatrix} C_{1n} & -S_{1n} & 0 \\ S_{1n} & C_{1n} & 0 \\ 0 & 0 & 1 \end{bmatrix}; \quad [D_{n'n}]_T = \begin{bmatrix} 1 & 0 & (x_n - x_{n'}) \\ 0 & 1 & (y_n - y_{n'}) \\ 0 & 0 & 1 \end{bmatrix} \quad \text{(D.4-12)}$$

Combining the above expressions, we obtain

$$[D_{1n}] = [D_{n'n}]_T [D_{1n'}]_R = \begin{bmatrix} C_{1n} & -S_{1n} & (x_n - x_{n'}) \\ S_{1n} & C_{1n} & (y_n - y_{n'}) \\ 0 & 0 & 1 \end{bmatrix} \quad \text{(D.4-13)}$$

However

$$\begin{Bmatrix} x_{n'} \\ y_{n'} \\ 1 \end{Bmatrix} = [D_{1n'}]_R \begin{Bmatrix} x_1 \\ y_1 \\ 1 \end{Bmatrix} = \begin{Bmatrix} x_1 C_{1n} - y_1 S_{1n} \\ x_1 S_{1n} + y_1 C_{1n} \\ 1 \end{Bmatrix} \quad \text{(D.4-14)}$$

Combining Equations (D.4-13) and (D.4-14),

$$[D_{1n}] = \begin{bmatrix} C_{1n} & -S_{1n} & A_{13n} \\ S_{1n} & C_{1n} & A_{23n} \\ 0 & 0 & 1 \end{bmatrix} \quad \text{(D.4-15)}$$

where

$$A_{13n} = x_n - x_1 C_{1n} + y_1 S_{1n}; \quad A_{23n} = y_n - x_1 S_{1n} - y_1 C_{1n} \quad \text{(D.4-16)}$$

## D.5  MOMENT OF A FORCE

An important concept that arises in many problems is the *moment of a force* about some specified point or axis. Referring to Figure D.9, the moment $\overline{M}_O$ of force $\overline{F}$ about point $O$ is defined as

$$\overline{M}_O = \overline{r} \times \overline{F} \quad \text{(D.5-1)}$$

## D.6  COUPLE

Figure D.10 illustrates a body subjected to two forces that are equal in magnitude and opposite in direction. The perpendicular distance between the two lines of action is $d$. The moment produced by these forces is referred to as a *couple*. The couple about an axis normal to its plane and passing through point $O$ in the plane is

$$M = F(a + d) - Fa \quad \text{or} \quad M = Fd \quad \text{(D.6-1)}$$

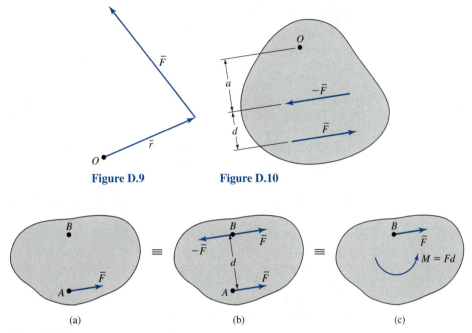

Figure D.9                    Figure D.10

(a)                         (b)                         (c)

Figure D.11

The second expression in Equation (D.6-1) is independent of dimension $a$, which locates the forces with respect to point $O$. It follows that the moment of a couple has the same value for all points.

Figure D.11(a) illustrates a body subjected to a force $\bar{F}$, which passes through point $A$. This system is equivalent to that shown in Figure D.11(b), in which forces $\bar{F}$ and $-\bar{F}$ are added through point $B$. It is recognized that force $\bar{F}$ through $A$ and force $-\bar{F}$ through $B$ provide zero net force on the body and a counterclockwise couple of $M = Fd$. The system therefore is in turn equivalent to that shown in Figure D.11(c), where the given force $\bar{F}$ acting at point $A$ is replaced by an equal force acting at point $B$ and a couple. Using similar reasoning, it is also possible to replace a force and a couple acting on a body with a force that has an alternate point of application.

## D.7 NEWTON'S LAWS

Sir Isaac Newton formulated three fundamental laws of mechanics, which were first published in 1687. They provide the basis of studying kinetics.

For purposes of this textbook, *Newton's laws* may be expressed as follows:

1. Every particle remains at rest, or continues to move in a straight line with constant velocity, if there is no resultant force on it.
2. If a resultant force acts on a particle, then the particle will accelerate in the direction of the force, at a rate proportional to that force.
3. For every action between two adjoining bodies, there are actions on each body that are equal, opposite, and simultaneous.

The first law may be regarded as a special case of the second.

The second law is expressed mathematically as

$$\overline{F} = m\overline{a} = m(\ddot{x}\overline{i} + \ddot{y}\overline{j})$$  (D.7-1)

where $\overline{F}$ is the resultant force on the particle, $m$ is the mass, and $\overline{a}$ is the acceleration. Vectors $\overline{F}$ and $\overline{a}$ have the same direction. Common sets of units for Equation (D.7-1) are presented in Chapter 2.

Equation (D.7-1) is applied to the analysis of a particle. In the next section, this equation will be employed in the development of the equations of motion for a rigid body.

## D.8  KINETICS OF A RIGID BODY UNDERGOING PLANAR MOTION

### D.8.1  Linear Motion

Figure D.12(a) illustrates a rigid body subjected to an external force. We consider the body to be an assembly of particles. A typical particle of mass $m_k$ ($k = 1, 2, \ldots$) is shown in Figure D.12(b).

Associated with the body is its *mass center* having coordinates $(x_G, y_G)$ with respect to a fixed coordinate system. They are defined as

$$x_G = \frac{\sum_k m_k x_k}{\sum_k m_k} = \frac{\sum_k m_k x_k}{m}; \qquad y_G = \frac{\sum_k m_k y_k}{m}$$  (D.8-1)

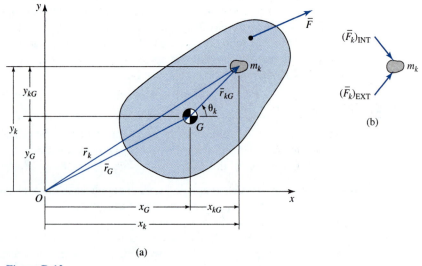

(a)

(b)

**Figure D.12**

where the *mass* of the body is

$$m = \sum_k m_k \tag{D.8-2}$$

The position vector to the center of mass is

$$\bar{r}_G = x_G \bar{i} + y_G \bar{j} \tag{D.8-3}$$

Forces on a particle in a body, as shown in Figure D.12(b), may be distinguished by those which are external and those which are internal. External forces are those which are applied to the particle from outside of the body. Examples include gravity and reaction forces from adjacent bodies. Internal forces are those which arise from adjacent particles in the body. Using Equation (D.7-1), the net result is

$$\bar{F}_k = (\bar{F}_k)_{\text{EXT}} + (\bar{F}_k)_{\text{INT}} = m_k \bar{a}_k \tag{D.8-4}$$

Now we sum equations for all of the particle masses in the system:

$$\sum_k \bar{F}_k = \sum_k (\bar{F}_k)_{\text{EXT}} + \sum_k (\bar{F}_k)_{\text{INT}} = \sum_k m_k \bar{a}_k \tag{D.8-5}$$

However, based on Newton's third law of mechanics, it is necessary that

$$\sum_k (\bar{F}_k)_{\text{INT}} = \bar{0} \tag{D.8-6}$$

and therefore Equation (D.8-5) reduces to

$$\sum_k (\bar{F}_k)_{\text{EXT}} = \sum_k m_k \bar{a}_k \tag{D.8-7}$$

Dropping the subscript EXT from Equation (D.8-7), and breaking it into its scalar components, gives

$$\sum_k (F_k)_x = \sum_k m_k \ddot{x}_k; \qquad \sum_k (F_k)_y = \sum_k m_k \ddot{y}_k \tag{D.8-8}$$

Rearranging Equations (D.8-1), and differentiating twice with respect to time,

$$m\ddot{x}_G = \sum_k m_k \ddot{x}_k; \qquad m\ddot{y}_G = \sum_k m_k \ddot{y}_k \tag{D.8-9}$$

Combining Equations (D.8-8) and (D.8-9),

$$\sum_k (F_k)_x = m\ddot{x}_G; \qquad \sum_k (F_k)_y = m\ddot{y}_G \tag{D.8-10}$$

Combining scalar Equations (D.8-10) into vector form,

$$\boxed{\bar{F} = \sum_k (F_k)_x \bar{i} + \sum_k (F_k)_y \bar{j} = m\bar{a}_G} \tag{D.8-11}$$

where

$$\boxed{\bar{a}_G = \ddot{x}_G \bar{i} + \ddot{y}_G \bar{j}} \tag{D.8-12}$$

Equation (D.8-11) indicates that the net external force on a body equals the mass of the body multiplied by the acceleration of the center of mass. This result is independent of the location at which the force is applied.

## D.8.2 Angular Motion

A body of finite size can also have angular motions. For planar motion, we measure moments about a point. For a particle of mass $m_k$ ($k = 1, 2, \ldots$), we can break the moments into internal and external components, that is,

$$(\overline{M}_O)_k = (\overline{M}_O)_{k,\text{EXT}} + (\overline{M}_O)_{k,\text{INT}} \tag{D.8-13}$$

Using Equation (D.5-1), the moment is

$$(\overline{M}_O)_k = \overline{r}_k \times \overline{F}_k \tag{D.8-14}$$

where

$$\overline{r}_k = x_k \overline{i} + y_k \overline{j} \tag{D.8-15}$$

$$\overline{F}_k = (F_k)_x \overline{i} + (F_k)_y \overline{j} \tag{D.8-16}$$

Substituting Equation (D.7-1) for the $k$th mass in Equation (D.8-16),

$$\overline{F}_k = m_k \ddot{x}_k \overline{i} + m_k \ddot{y}_k \overline{j} \tag{D.8-17}$$

Combining Equations (D.8-14), (D.8-15), and (D.8-17),

$$(\overline{M}_O)_k = \overline{r}_k \times \overline{F}_k = \begin{vmatrix} \overline{i} & \overline{j} & \overline{k} \\ x_k & y_k & 0 \\ m_k \ddot{x}_k & m_k \ddot{y}_k & 0 \end{vmatrix} = m_k(x_k \ddot{y}_k - y_k \ddot{x}_k)\overline{k} \tag{D.8-18}$$

Summing the moments for all particle masses that make up the rigid body,

$$\overline{M}_O = \sum_k (\overline{M}_O)_{k,\text{EXT}} + \sum_k (\overline{M}_O)_{k,\text{INT}} = \sum_k m_k(x_k \ddot{y}_k - y_k \ddot{x}_k)\overline{k} \tag{D.8-19}$$

However, based on Newton's third law,

$$\sum_k (\overline{M}_O)_{k,\text{INT}} = \overline{0} \tag{D.8-20}$$

Also, since the moment is in the direction of $\overline{k}$, we can consider Equation (D.8-19) as a scalar equation, that is,

$$M_O = \sum_k m_k(x_k \ddot{y}_k - y_k \ddot{x}_k) \tag{D.8-21}$$

From Figure D.12(a), we may express

$$\begin{aligned} x_k &= x_G + x_{kG}; & \ddot{x}_k &= \ddot{x}_G + \ddot{x}_{kG} \\ y_k &= y_G + y_{kG}; & \ddot{y}_k &= \ddot{y}_G + \ddot{y}_{kG} \end{aligned} \tag{D.8-22}$$

Substituting Equations (D.8-22) in Equation (D.8-21),

$$M_O = \sum_k m_k x_{kG} \ddot{y}_{kG} + x_G \sum_k m_k \ddot{y}_{kG} + \ddot{y}_G \sum_k m_k x_{kG} + x_G \ddot{y}_G \sum_k m_k$$
$$- \sum_k m_k y_{kG} \ddot{x}_{kG} - y_G \sum_k m_k \ddot{x}_{kG} - \ddot{x}_G \sum_k m_k y_{kG} - y_G \ddot{x}_G \sum_k m_k \tag{D.8-23}$$

We recognize that

$$\sum_k m_k x_{kG} = 0; \qquad \sum_k m_k \ddot{x}_{kG} = 0$$
$$\sum_k m_k y_{kG} = 0; \qquad \sum_k m_k \ddot{y}_{kG} = 0 \tag{D.8-24}$$

and

$$x_{kG} = r_{kG} \cos \theta_k; \qquad y_{kG} = r_{kG} \sin \theta_k \tag{D.8-25}$$

Substituting Equations (D.8-24) and (D.8-25) in Equation (D.8-23), and simplifying,

$$M_O = I_G \ddot{\theta} + m(x_G \ddot{y}_G - \ddot{x}_G y_G) \tag{D.8-26}$$

where

$$I_G = \sum_k m_k r_{kG}^2 \tag{D.8-27}$$

is the *polar mass moment of inertia* of the body with respect to its mass center. Typical values of the polar mass moment of inertia are provided in the following section.

## D.9  POLAR MASS MOMENT OF INERTIA

If each particle mass is infinitesimal, Equation (D.8-27) may be expressed as

$$I_G = \int_m r_G^2 \, dm \tag{D.9-1}$$

where $r_G$ is the distance of mass element $dm$ from the mass center.

The polar mass moment of inertia about the mass center may be expressed in terms of its *radius of gyration*, $k$, using the following relation

$$I_G = mk^2 \tag{D.9-2}$$

where the mass of the body is

$$m = \int_m dm \tag{D.9-3}$$

The following three examples show applications of Equations (D.9-1) and (D.9-2) for geometries often encountered in mechanisms.

## EXAMPLE D.1  Polar Mass Moment of Inertia of a Slender Rod

For the uniform slender rod of length $L$ shown in Figure D.13(a), determine the polar mass moment of inertia with respect to its mass center, and its radius of gyration.

### SOLUTION

The center of mass of the rod is located at its midpoint. We thus conveniently set the origin of a coordinate system at the center of mass. Furthermore, we consider an element of mass of

$$dm = \rho A \, dx \qquad\qquad (D.9\text{-}4)$$

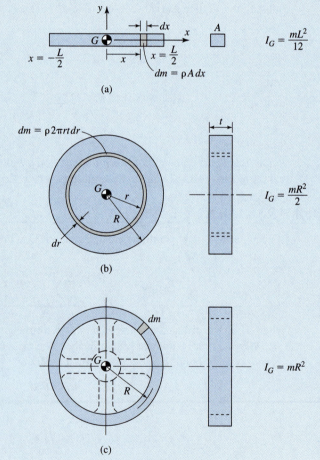

**Figure D.13**  (a) Slender rod. (b) Solid disc. (c) Rim disc.

where $\rho$ is the density, and $A$ is the cross-sectional area. Substituting Equation (D.9-4) in Equation (D.9-1), and using the appropriate limits of integration,

$$I_G = \int_m r_G^2 \, dm$$

$$= \int_{x=-L/2}^{x=L/2} x^2 \rho A \, dx = \rho A \frac{x^3}{3} \bigg|_{x=-L/2}^{x=L/2} = \rho A \frac{L^3}{12} = \frac{mL^2}{12}$$

(D.9-5)

where

$$m = \rho A L$$

(D.9-6)

Using Equations (D.9-2) and (D.9-5), the radius of gyration is

$$k = \left(\frac{I_G}{m}\right)^{1/2} = \left(\frac{mL^2/12}{m}\right)^{1/2} = \frac{L}{\sqrt{12}}$$

(D.9-7)

## EXAMPLE D.2  Polar Mass Moment of Inertia of a Solid Disc

For a solid disc of radius $R$ and thickness $t$ shown in Figure D.13(b), determine the polar mass moment of inertia with respect to its mass center.

### SOLUTION

The center of mass of the disc is located at the center of its cross section. We consider an element of mass of

$$dm = 2\pi r t \rho \, dr$$

(D.9-8)

Substituting Equation (D.9-8) in Equation (D.9-1), and using the appropriate limits of integration,

$$I_G = \int_m r_G^2 \, dm = \int_0^R r^2 2\pi r t \rho \, dr = 2\pi t \rho \frac{r^4}{4} \bigg|_{r=0}^{r=R} = \frac{mR^2}{2}$$

(D.9-9)

where

$$m = \rho \pi R^2 t$$

(D.9-10)

## EXAMPLE D.3  Polar Mass Moment of Inertia of a Rim Disc

For a rim disc of radius $R$ shown in Figure D.13(c), assume the mass is uniformly distributed at the periphery of the rim. Neglect the effects of the spokes, which are shown as dashed lines in the figure. Determine the polar mass moment of inertia with respect to its mass center.

SOLUTION

The center of mass of the disc is located at the center of the circular rim. The polar mass moment of inertia is

$$I_G = \int_m r_G^2 \, dm = R^2 \int dm = mR^2 \qquad \text{(D.9-11)}$$

# D.10  WORK, ENERGY, AND POWER

### D.10.1  Kinetic Energy

The *kinetic energy* due to the motion of a particle is

$$T_k = \tfrac{1}{2} m_k v_k^2 \qquad \text{(D.10-1)}$$

Considering a rigid body to be a composition of a series of particles, it can be shown that the kinetic energy of the body is

$$T = \sum_k T_k = \tfrac{1}{2} m v_G^2 + \tfrac{1}{2} I_G \dot{\theta}^2 \qquad \text{(D.10-2)}$$

where $v_G$ is the speed of the mass center.

### D.10.2  Work Done by a Force

When applying Newton's second law of motion, the forces and accelerations are those at an instant and thus give instantaneous solutions; however, in many cases it is necessary to determine the change in motion over a finite time interval. The *work-energy* method is one way of doing this, where the interval is one of displacement, and the change of motion is defined in terms of velocity.

Consider a force $\overline{F}$ applied to a particle or rigid body, as shown in Figure D.14, and suppose that the point of application of $\overline{F}$ moves a small distance $\delta \overline{s}$, during which the force is assumed to remain constant. The work done by $\overline{F}$ during this displacement is

$$\delta W = \overline{F} \cdot \delta \overline{s} = F \delta s \cos \phi \qquad \text{(D.10-3)}$$

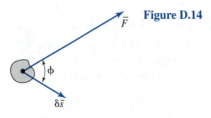

**Figure D.14**

where $\phi$ is the angle between $\overline{F}$ and $\delta\overline{s}$. By summing all such small displacements over a finite distance, the work is

$$W = \int_s F \cos \phi \, ds \qquad \text{(D.10-4)}$$

and is measured in Newton-meters (N-m) or joules (J). For angular motions, the infinitesimal amount of work done may be expressed as

$$dW = M \, d\theta \qquad \text{(D.10-5)}$$

### D.10.3 Power

*Power* is the time rate of change of doing work. For a body undergoing linear translation

$$\frac{dW}{dt} = Fv \cos \phi \qquad \text{(D.10-6)}$$

where $v$ is the speed of the body.

For rotating systems, using Equation (D.10-3),

$$\frac{dW}{dt} = M\frac{d\theta}{dt} = M\omega \qquad \text{(D.10-7)}$$

where $\omega$ is the rotational speed of the body.

For the SI system, the unit of power is

$$1 \text{ watt} = 1 \text{ N-m/sec} \qquad \text{(D.10-8)}$$

# Index of CD-ROM

This appendix contains a listing of all the files for demonstration included on the CD-ROM that accompanies this textbook. Also indicated is the type of software on which each file is based.

Many files are based on Working Model 2D and Visual Nastran and have been stored in VRML format. Files employing Working Model 2D require either a copy of the software or the file player version of the software. Files stored in VRML format can be shown using Flux Player.

Also included on the CD-ROM are video clips of real mechanical systems in operation. This software requires QuickTime Movie Player.

## CHAPTER 1 Introduction

| File Name | Working Model 2D Player | Flux Player (VRML) | QuickTime Movie Player |
|---|:---:|:---:|:---:|
| Model 1.1: Single-Cylinder Piston Engine | | • | |
| Model 1.4: Engine Assembly | | • | |
| Model 1.5: Slider Crank Mechanism | • | | |
| Model 1.7: Slider Crank Mechanism with Offset | • | | |
| Model 1.8: Four-Bar Mechanism | • | | |
| Video 1.10: Equivalent Four-Bar Mechanisms | | | • |
| Model 1.12: Washing Machine Mechanism | • | | |
| Model 1.14A: Four-Bar Mechanism with Coupler Point | • | | |
| Model 1.14B: Four-Bar Mechanism with Coupler Point | • | | |
| Model 1.15A: Chebyshev Straight-Line Mechanism | • | | |

| | |
|---|---|
| Model 1.15B: Robert Straight-Line Mechanism | • |
| Model 1.15C: Watt Straight-Line Mechanism | • |
| Model 1.16: Film Transport Mechanism | • |
| Model 1.21: Toothed Gears | • |
| Model 1.22: Internal Gear | • |
| Model 1.23: Rack and Pinion | • |
| Model 1.24: Rack and Pinion Steering | • |
| Model 1.25: Disc Cam Mechanism | • |
| Model 1.26: Disc Cams | • |
| Model 1.27A: Wedge Cam Mechanism | • |
| Model 1.27B: Cylindrical Cam Mechanism | • |
| Model 1.27C: End Cam Mechanism | • |
| Model 1.28: Fishing Reel | • |
| Video 1.29: Prosthetic Hand | • |
| Model 1.38: Front-End Loader Mechanism | • |
| Model 1.39: Slider Crank Mechanism and Its Three Inversions | • |
| Model 1.41: Four-Bar Mechanism and Its Three Inversions | • |
| Model 1.45: Parallelogram Four-Bar Mechanism | • |
| Model 1.47: Variable Base Link Four-Bar Mechanism | • |
| Model 1.48: Variable-Offset Slider Crank Mechanism | • |
| Model 1.51: Cognates of a Four-Bar Mechanism | • |
| Model 1.54: Cognate of a Slider Crank Mechanism | • |

## CHAPTER 3  Graphical Kinematic Analysis of Planar Mechanisms

| File Name | Working Model 2D Player | Flux Player (VRML) | QuickTime Movie Player |
|---|---|---|---|
| Video 3.6: Relative Velocity | | | • |

## CHAPTER 4  Analytical Kinematic Analysis of Planar Mechanisms

- Mathcad program: fourbarkin, kinematic analysis of a four-bar mechanism
- Mathcad program: slidercrankkin, kinematic analysis of a slider crank mechanism
- Mathcad program: fourbarcpkin, kinematic analysis of a four-bar mechanism with a coupler point
- Mathcad program: slidercrankcpkin, kinematic analysis of a slider crank mechanism with a coupler point

## CHAPTER 5  Gears

| File Name | Working Model 2D Player | Flux Player (VRML) | QuickTime Movie Player |
|---|---|---|---|
| Model 5.2: Continuously Variable Traction Drive | | • | |
| Model 5.4: Continuously Variable Belt Drive | | • | |

*(Continued)*

| File Name | Working Model 2D Player | Flux Player (VRML) | QuickTime Movie Player |
|---|---|---|---|
| Model 5.5A: Straight Spur Gears | | • | |
| Model 5.5B: Helical Spur Gears | | • | |
| Model 5.7: Miter Gears | | • | |
| Model 5.8A: Narrow Herringbone Gears | | • | |
| Model 5.8B: Wide Herringbone Gears | | • | |
| Model 5.9A: Plain Bevel Gears | | • | |
| Model 5.9B: Spiral Bevel Gears | | • | |
| Video 5.10: Hypoid Gear Set | | | • |
| Model 5.11A: Worm and Wheel Gears, Right-Hand Worm | | • | |
| Model 5.11B: Worm and Wheel Gears, Left-Hand Worm | | • | |
| Model 5.12A: Double-Start Right-Hand Worm | | • | |
| Model 5.12B: Double-Start Left-Hand Worm | | • | |
| Model 5.15: Involute Generation | • | | |
| Model 5.18: External-External Involute Gears Meshing | • | | |
| Model 5.19: External-Internal Involute Gears Meshing | • | | |
| Model 5.20: Rack and Pinion | • | | |
| Model 5.22: Cycloid Generation | • | | |
| Video 5.23: Gerotor Pump | | | • |
| Model 5.25: Gerotor Pump | • | | |
| Model 5.26: Supercharger | • | | |
| Model 5.27: External-External Cycloidal Gears Meshing | • | | |
| Model 5.32: Antibacklash Gear | | • | |
| Video 5.43: Form Milling | | | • |
| Model 5.45: Straight Spur Gear Hobbing | | • | |
| Video 5.48: Straight Spur Gear Hobbing | | | • |
| Model 5.50: Helical Spur Gear Hobbing | | • | |
| Video 5.51: Helical Spur Gear Hobbing | | | • |
| Model 5.53A: Shaping of a Straight Spur Gear | | • | |
| Model 5.53B: Shaping of an Internal Gear | | • | |
| Video 5.54A: Shaping of a Straight Spur Gear | | | • |
| Video 5.54B: Shaping of an Internal Gear | | | • |
| Video 5.56: Planing of a Plain Bevel Gear | | | • |
| Model 5.58: Rotary Broaching | | • | |
| Model 5.67: Undercut Gear | • | | |

## CHAPTER 6  Gear Trains

| File Name | Working Model 2D Player | Flux Player (VRML) | QuickTime Movie Player |
|---|---|---|---|
| Model 6.4A: Simple Gear Train | | • | |
| Model 6.6A: Reverted Gear Train | | • | |
| Video 6.7: Winch Gear Train | | | • |

## CHAPTER 7 Cams

| File Name | Working Model 2D Player | Flux Player (VRML) | QuickTime Movie Player |
|---|:---:|:---:|:---:|
| Video 7.2: Moving-Headstock Milling Machine | | | • |
| Video 7.3A: Single Knife-Edge Follower, Disc Cam Mechanism | | | • |
| Video 7.3B: Multiple Knife-Edge Followers, Disc Cam Mechanism | | | • |
| Video 7.3C: Roller Follower, Disc Cam Mechanism | | | • |
| Video 7.4: Cylindrical Cam Mechanism | | | • |
| Video 7.5: Face Cam Mechanism | | | • |
| Model 7.6: Disc Cam Mechanisms | • | | |
| Model 7.25: Positive-Motion Cam Mechanism | • | | |
| Model 7.26: Constant-Breadth Cam Mechanism | • | | |
| Model 7.28: Movie Projector Mechanism | | • | |
| Video 7.53A1: Manufacture of a Disc Cam—Scribing Radial Lines | | | • |
| Video 7.53A2: Manufacture of a Disc Cam—Scribing Circular Arcs | | | • |
| Video 7.53B: Manufacture of a Disc Cam—Scribing Rises and Falls | | | • |
| Video 7.53C: Manufacture of a Disc Cam—Rough Cut | | | • |
| Video 7.53D: Manufacture of a Disc Cam—Fine Cut | | | • |

• Cam Design, kinematic analysis of disc cam mechanisms

## CHAPTER 8 Graphical Force Analysis of Planar Mechanisms

| File Name | Working Model 2D Player | Flux Player (VRML) | QuickTime Movie Player |
|---|:---:|:---:|:---:|
| Model 8.10: Front-End Loader | • | | |
| Model 8.16: Dynamic Force Analysis | • | | |

## CHAPTER 9 Analytical Force Analysis and Balancing of Planar Mechanisms

• Mathcad program: fourbarforce, dynamic analysis of a four-bar mechanism
• Mathcad program: slidercrankforce, dynamic analysis of a slider crank mechanism
• Mathcad program: fourbarbalance, balancing of a four-bar mechanism
• Mathcad program: slidercrankbalance, balancing of a slider crank mechanism

## CHAPTER 10  Flywheels

| File Name | Working Model 2D Player | Flux Player (VRML) | QuickTime Movie Player |
|---|:---:|:---:|:---:|
| Model 10.4: Engine Flywheel | • | | |

## CHAPTER 11  Synthesis of Mechanisms

| File Name | Working Model 2D Player | Flux Player (VRML) | QuickTime Movie Player |
|---|:---:|:---:|:---:|
| Model 11.2: Corkscrew Mechanisms | • | | |
| Model 11.21A: Headlight Cover, Undesirable Design | | • | |
| Model 11.21B: Headlight Cover, Desirable Design | | • | |

- Mathcad program: fourbarfuncsyn, function synthesis of a four-bar mechanism
- Mathcad program: slidercrankfuncsyn, function synthesis of a slider crank mechanism
- Mathcad program: fourbarrbg, rigid-body guidance synthesis of a four-bar mechanism

## APPENDIX B  Commonly Employed Mechanisms and Machines

| File Name | Working Model 2D Player | Flux Player (VRML) | QuickTime Movie Player |
|---|:---:|:---:|:---:|
| Model B.1: Parallel-Motion Mechanism | • | | |
| Model B.2: Straight-Line Dwell Mechanism | • | | |
| Model B.3: Circular-Arc Dwell Mechanism | • | | |
| Model B.4: Geneva Mechanism | • | | |
| Model B.5: Counting Mechanism | | • | |
| Model B.6: Scotch Yoke Mechanism | • | | |
| Model B.7: Application of Scotch Yoke Mechanism | | • | |
| Model B.8: Escapement Mechanism | • | | |
| Model B.9: Iris Mechanism | • | | |
| Video B.10: Flow Regulator | | | • |
| Model B.11: Ratchet Mechanism | • | | |
| Model B.12: Ratchet Wrench | | • | |
| Model B.13: Sprag Clutch | | • | |
| Model B.15: Quick-Return Mechanism | • | | |
| Model B.16: Single-Cylinder Piston Engine Cross Section | • | | |
| Model B.17: Sleeve Engine | | • | |
| Model B.19: V-2 Engine | | • | |
| Model B.20A: Radial Engine | • | | |
| Model B.20B: Radial Engine with Propeller | • | | |
| Video B.21: Airplane Engine | | | • |

## PROBLEM WORKSHEETS

# Trigonometric Identities

$$\cos\left(u \pm \tfrac{\pi}{2}\right) = \mp \sin u; \qquad \sin\left(u \pm \tfrac{\pi}{2}\right) = \pm \cos u$$

$$\sin(u + \pi) = -\sin u; \qquad \sin(-u) = -\sin u$$

$$\cos(u + \pi) = -\cos u; \qquad \cos(-u) = \cos u$$

$$\tan\left(u + \tfrac{\pi}{2}\right) = -\frac{1}{\tan u}; \qquad \tan(-u) = -\tan u$$

$$\sin^2 u + \cos^2 u = 1$$

$$\tan^2 u + 1 = \sec^2 u$$

$$\cot^2 u + 1 = \csc^2 u$$

$$\sin^2 u = \tfrac{1}{2}(1 - \cos 2u)$$

$$\cos^2 u = \tfrac{1}{2}(1 + \cos 2u)$$

$$\sin(u \pm v) = \sin u \cos v \pm \cos u \sin v$$

$$\cos(u \pm v) = \cos u \cos v \mp \sin u \sin v$$

$$\sin^{-1} u = \cos^{-1}(1 - u^2)^{1/2}$$

$$\tan 2^{-1}(x, y) = \tan^{-1}\left(\frac{y}{x}\right), \qquad -\pi \le \tan 2^{-1}(x, y) \le \pi$$

$$\frac{d}{dx}\sin u = \cos u \frac{du}{dx}; \qquad \frac{d}{dx}\cos u = -\sin u \frac{du}{dx}$$

$$\frac{d}{dx}\sin^{-1} u = \frac{1}{\sqrt{1 - u^2}}\frac{du}{dx}, \qquad -\tfrac{\pi}{2} \le \sin^{-1} u \le \tfrac{\pi}{2}$$

$$\frac{d}{dx}\cos^{-1} u = -\frac{1}{\sqrt{1 - u^2}}\frac{du}{dx}, \qquad 0 \le \cos^{-1} u \le \pi$$

$$\frac{d}{dx}\tan^{-1}u = \frac{1}{1+u^2}\frac{du}{dx}, \qquad -\frac{\pi}{2} \le \tan^{-1}u \le \frac{\pi}{2}$$

$$\frac{d}{dx}\cot^{-1}u = -\frac{1}{1+u^2}\frac{du}{dx}, \qquad -\frac{\pi}{2} \le \cot^{-1}u \le \frac{\pi}{2}$$

## Cosine Law

$$a^2 = b^2 + c^2 - 2bc\cos\alpha$$

## Sine Law

$$\frac{a}{\sin\alpha} = \frac{b}{\sin\beta} = \frac{c}{\sin\gamma}$$

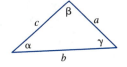

# Answers to Selected Problems

## Chapter 1
P1.1(a): $m = 1$
P1.3(a): $m = 1$
P1.6(b): links may form a mechanism; change point
P1.10(a): $r_3 \geq 3.5$

## Chapter 2
P2.1(a): 50 rpm @ $\theta_2 = 20°$ and $200°$
P2.3(a): $\theta_2 = 53.1°$ and $306.9°$
P2.4(a): 7.58 rad/sec
P2.4(b): 7.58 rad/sec

## Chapter 3
P3.1: 60 in/sec; 39.4 in/sec; 13.1 in/sec
P3.3: 6.28 cm/sec
P3.15(c)(ii): 27.1 rad/sec$^2$ CW; 13.9 rad/sec$^2$ CCW

## Chapter 4
P4.1: 1.53 in/sec $\searrow$ 70°; 0.212 rad/sec CCW
P4.2: 42.30 rad/sec$^2$ CCW
P4.4: 102.3 cm/sec $\nwarrow$ 119.4°
P4.5(c): 179 cm/sec$^2$ $\searrow$ 18.1°

## Chapter 5
P5.1: 5.97 in
P5.3: 66.3 rpm
P5.4(b): 30
P5.14(b): 1.48

## Chapter 6

P6.1(b): 56.25 rpm CCW

P6.10: 56 rpm (in the same direction as shaft A)

P6.11(b): 3

P6.31(a): 34.9 rpm CCW

## Chapter 7

P7.10: 2.19 cm; 1.19 cm; 0.060 cm

P7.12: 3.39 cm; 0.090 cm

## Chapter 8

P8.1: 1,800 N  ↑; 1,500 N  ↘ 77°

P8.2(b): 129 N  ↓

P8.13(a): 9,300 in-lb CW

P8.14: 0.32 N-m CW

## Chapter 9

P9.4: 622 gr @ $-175.8°$; 728 gr @ 170°

P9.6(a): 3.6 gr

## Chapter 10

P10.2: 0.1053; 490 kg

P10.3: 307.5 rpm; 292.5 rpm

P10.4(b): 0.0488

P10.9(c): 174.2 kg

## Chapter 11

P11.1: $r_2 = 0.786$ cm; $r_3 = 1.36$ cm; $r_4 = 1.38$ cm

P11.2: $r_1 = 1.70$ cm; $r_2 = 2.33$ cm; $r_4 = 4.05$ cm

# References

1. Dechev, N., Cleghorn, W. L., and Naumann, S., "Multiple Finger, Passive Grasp Prosthetic Hand," *Mechanism and Machine Theory,* Vol. 36, No. 10, pp. 1157–1173, 2001.

2. Uicker, J. J., Pennock, G. R., and Shigley, J. E., *Theory of Machines and Mechanisms,* Third Edition, Oxford University Press, 2003.

3. Mabie, H. H., and Reinholtz, C. F., *Mechanisms and Dynamics of Machinery,* Fourth Edition, Wiley, 1987.

4. Beer, F. P., and Johnson, E. R., *Vector Mechanics for Engineers,* McGraw-Hill, 1988.

5. Cleghorn, W. L., and Podhorodeski, R. P., "Disc Cam Design Using a Microcomputer," *International Journal of Mechanical Engineering Education,* Vol. 16, No. 4, pp. 235–250, 1988.

6. Lévai, Z., "Theory of Epicycle Gears and Epicycle Change-Speed Gearing," Budapest Technical University of Building, Civil Transport Engineering, Professorate for Motor Vehicles, Doctorate dissertation, No. 29, 1966.

7. Cleghorn, W. L., and Tyc, G., "Kinematic Analysis of Multiple Stage Planetary Gear Trains Using a Microcomputer," *International Journal of Mechanical Engineering Education,* Vol. 15, No. 1, pp. 57–69, 1987.

8. Chocholek, S. E., "The Development of a Differential for the Improvement of Traction Control," In *Traction Control and Anti-wheel-spin Systems for Road Vehicles,* ImechE, Automobile Division, pp. 75–82, 1988.

9. Kraus, H., and Babbidge, H. D., "Symptoms of Withdrawal: An Assessment of Mechanical Corkscrews," *Chartered Mechanical Engineer,* Vol. 24, No. 12, pp. 70–75, 1977.

10. Hartenberg, R. S., and Denavit, J., *Kinematic Synthesis of Mechanisms,* McGraw-Hill, 1964.

11. Hrones, J. A., and Nelson, G. L., *Analysis of the Four Bar Linkage,* Wiley, 1951.

12. Erdman, A. G., and Sandor, G. N., *Mechanism Design: Analysis and Synthesis,* Prentice-Hall, 1997.

13. Sclater, N., and Chironis, N. P., *Mechanisms and Mechanical Devices Sourcebook,* McGraw-Hill, 2001.

14. Atth, M., *All About Locks and Locksmithing,* Hawthorn, 1972.

15. Sandia National Laboratories, www.sandia.gov.

16. Dechev, N., Cleghorn, W. L., and Mills, J. K., "Microassembly of 3D Microstructures Using a Compliant, Passive Microgripper," *Journal of Microelectromechanical Systems,* Vol. 13, No. 2, pp. 176–189, 2004.

17. Dechev, N., Cleghorn, W. L., and Mills, J. K., "Microassembly of Microelectromechanical Components into 3D MEMS," *Canadian Journal of Electrical Computing Engineering,* Vol. 27, No. 1, pp. 7–15, 2002.

18.  Suh, C. H., and Radcliffe, C. W., "Synthesis of Plane Linkages with Use of the Displacement Matrix," *ASME J. Engineering for Industry,* Vol. 84, No. 2, pp. 206–214, 1967.

19.  Berkof, R. S., and Loewen, G. G., "A New Method for Completely Force Balancing Simple Linkages," *ASME J. Engineering for Industry,* Vol. 91, No. 1, pp. 21–26, 1969.

# Index

# CONVERSION FACTORS FOR TORQUE OR MOMENT OF FORCE

| multiply number of → <br><br> to <br> obtain ↓     by ↘ | Dyne-centimeters (dyne-cm) | Gram-force–centimeters (grf-cm) | Kilogram-force–meters (kgf-m) | Pound-feet (lb-ft) | Newton-meters (N-m) |
|---|---|---|---|---|---|
| Dyne-centimeters (dyne-cm) | 1 | 980.7 | $9.807(10^7)$ | $1.356(10^7)$ | $1(10^7)$ |
| Gram-force–centimeters (grf-cm) | $1.020(10^{-3})$ | 1 | $1(10^5)$ | $1.383(10^4)$ | $1.020(10^4)$ |
| Kilogram-force–meters (kgf-m) | $1.020(10^{-8})$ | $1(10^{-5})$ | 1 | 0.1383 | 0.1020 |
| Pound-feet (lb-ft) | $7.376(10^{-8})$ | $7.233(10^{-5})$ | 7.233 | 1 | 0.7376 |
| Newton-meters (N-m) | $1(10^{-7})$ | $9.807(10^{-5})$ | 9.807 | 1.356 | 1 |

# CONVERSION FACTORS FOR MASS MOMENT OF INERTIA

| multiply number of → <br><br> to <br> obtain ↓     by ↘ | Gram–centimeters squared (gr-cm²) | Kilogram–meters squared (kg-m²) | Pound mass–inches squared (lbm-in²) | Pound mass–feet squared (lbm-ft²) | (lb-in-sec²) |
|---|---|---|---|---|---|
| Gram–centimeters squared (gr-cm²) | 1 | $1(10^7)$ | $2.927(10^3)$ | $4.214(10^5)$ | $1.130(10^6)$ |
| Kilogram–meters squared (kg-m²) | $1(10^{-7})$ | 1 | $2.927(10^{-4})$ | $4.214(10^{-2})$ | 0.1130 |
| Pound mass–inches squared (lbm-in²) | $3.417(10^{-4})$ | $3.417(10^3)$ | 1 | 144 | 386.0 |
| Pound mass–feet squared (lbm-ft²) | $2.373(10^{-6})$ | 23.73 | $6.944(10^{-3})$ | 1 | 2.681 |
| (lb-in-sec²) | $8.853(10^{-7})$ | 8.853 | $2.591(10^{-3})$ | 0.3730 | 1 |